Incremental Delta-Sigma ADCs

Maurits Ortmanns · Paul Kaesser ·
Johannes Wagner

Incremental Delta-Sigma ADCs

Theory, Architectures and Design

Springer

Maurits Ortmanns
University of Ulm
Ulm, Germany

Paul Kaesser
University of Ulm
Ulm, Germany

Johannes Wagner
University of Ulm
Ulm, Germany

ISBN 978-3-032-09467-4 ISBN 978-3-032-09468-1 (eBook)
https://doi.org/10.1007/978-3-032-09468-1

This Springer imprint is published by the registered company Springer Nature Switzerland AG
The registered company address is: Gewerbestrasse 11, 6330 Cham, Switzerland

To Maja, Moritz and Susanne

To Rita and Dietmar

To Rebecca

To the curious minds who ask why, and to the mentors who taught us to listen.

Preface

This book is the outcome of several years of academic research conducted at the Institute of Microelectronics at the University of Ulm. What began as a series of focused investigations gradually developed into a comprehensive body of work that we now have the pleasure of presenting in this volume. The inspiration for writing this book arose from the growing need to consolidate and share the insights gained during our collaborative research efforts, as well as from the realization that our findings could serve as a valuable resource for both academic researchers and industry professionals working in this field.

A significant part of the material presented in this book stems from the Ph.D. theses work of Johannes and Paul, whose in-depth investigations and rigorous approaches provided a strong foundation for many of the chapters. In addition, the research contributions of Patrick Vogelmann, Mohamed Mokhtar, and Omar Ismail—each of whom worked on distinct but complementary aspects of the overarching topic—also enriched the content and scope of this book. We are sincerely grateful for the commitment, creativity, and perseverance they brought to their work during their time as doctoral candidates.

We gratefully acknowledge the financial support provided by the German Research Foundation (DFG), which enabled much of the research documented here. Their funding within the projects OR245/10 and OR245/13 and their trust in our scientific direction were instrumental in bringing this project to life.

Our heartfelt thanks also go to our colleagues at the Institute of Microelectronics, especially Dr. John Kauffman and Dr. Joachim Becker, whose feedback, discussions, and technical support were invaluable throughout this journey. We would also like to express our deep gratitude to our families and friends, whose patience, encouragement, and unwavering support sustained us through the long and often challenging process of completing both the research and this manuscript.

We hope that readers will find this book both informative and thought-provoking, and that it will inspire further exploration and innovation in the exciting field of microelectronics.

Ulm, Germany
July 2025

Maurits Ortmanns
Paul Kaesser
Johannes Wagner

Contents

Acronyms

AAF	Anti-aliasing filter
ADC	Analog-to-digital converter
AISAD	AI assisted MATLAB tool for the design of delta sigma modulators
ANN	Artificial neural network
BIBO	Bounded input, bounded output
CIC	Cascade-of-integrators comb
CIFB	Cascade-of-integrators with feedback compensation
CIFF	Cascade-of-integrators with feedforward compensation
CMOS	Complementary metal-oxide-semiconductor
CoI	Chain of integrators
CRFB	Cascade-of-resonators with feedback compensation
CRFF	Cascade-of-resonators with feedforward compensation
CT	Continuous-time
DAC	Digital-to-analog converter
DC	Direct current
DCF	Digital cancellation filter
DEM	Dynamic element matching
DNL	Differential nonlinearity
DR	Dynamic range
DS	Delta-Sigma
DSM	Delta-Sigma modulator
DT	Discrete-time
DTFT	Discrete-time Fourier transform
DWA	Data-weighted averaging
EC	Extended counting
ELD	Excess loop delay
ENOB	Effective number of bits
ER	Extended range
FB/FF	Feedback/feedforward
FIR	Finite impulse response
FoM	Figure of merit

GBW	Gain-bandwidth product
GPU	Graphics processing unit
IBN	Inband noise
IC	Integrated circuit
I-DS	Incremental Delta-Sigma
I-DSM	Incremental Delta-Sigma modulator
IIT	Impulse-invariant transform
INL	Integral nonlinearity
ISG	Inter-stage gain
LDP	Low-distortion path
LE	Linear-exponential
LP	Lowpass
LSB	Least significant bit
LTI	Linear time-invariant
LTV	Linear time-variant
LUT	Look-up table
MASH	Multistage noise shaping
MINLP	Mixed integer nonlinear program
MLD	Mixed-logical dynamical
MSA	Maximum stable amplitude
NPF	Noise penalty factor
NRZ	Non-return-to-zero
NTF	Noise transfer function
OBG	Out-of-band gain
OOB	Out-of-band
OpAmp	Operational amplifier
OSR	Oversampling ratio
OTA	Operational transconductance amplifier
PDF	Probability density function
PSD	Power spectral density
QN	Quantization noise
QTZ	Quantizer
RP	Recuperation phase
RZ	Return-to-zero
S/H	Sample and hold
SAR	Successive approximation register
SC	Switched-capacitor
SFDR	Spurious-free dynamic range
SIMSIDES	Simulink-based Sigma-Delta simulator
SINAD	Signal-to-interference ratio including noise and distortion
SMASH	Sturdy MASH
SNDR	Signal-to-noise-and-distortion ratio
SNR	Signal-to-noise ratio
SoA	State of the art
SOR	Single-opamp resonator

SQNR	Signal-to-quantization-noise ratio
SR	Slew-rate
STF	Signal transfer function
THD	Total harmonic distortion
VBW	Variable bit width
VCO	Voltage controlled oscillator

Symbols

A_0	DC gain
$A(q)$	Noise power amplification factor
B	Quantizer bit width
β_{npf}	Thermal noise penalty factor
$\hat{\beta}_{\text{npf}}$	Thermal noise penalty factor approximation
$\beta_{\text{npf,ct}}$	Thermal noise penalty factor for CT architectures
$Ш_{T_s}(t)$	Impulse comb in time domain
$Ш_{1/T_s}$	Impulse comb in frequency domain
Δ	Quantizer step size
D_l	Sinc decimation factor
$D[n]$	Digital Nyquist-rate sample
$e[k]$	Quantization error
$e_{\text{resid}}[k]$	Residuum error after last integrator
f_{b}	Bandwidth
$F_{\text{c}}(z)$	Compensation FIR filter in feedback path
$F(z)$	FIR filter in feedback path
f_{N}	Nyquist rate
f_{s}	Sampling frequency
FS	Full-scale signal (peak-to-peak)
f_{sig}	(Input) signal frequency
$H_{\text{brick}}(f)$	Ideal brickwall filter
$H_{\text{ct}}(s)$	CT loop filter
$H_{\text{dac}}(s)$	Laplace transform of a general DAC pulse
$H_{\text{dec}}(f)$	Decimation filter
$H_{\text{dt}}(z)$	DT loop filter
$H_{\text{dt, eq}}(z)$	Equivalent DT filter of CT loop filter
$h_{\text{fir}}[k]$	FIR equivalent of an I-DS ADC
$h_{\text{fir, ntf}}[k]$	Impulse response of the overall NTF
$h_{\text{fir, stf}}[k]$	Impulse response of the overall STF
$h_{\text{free}}[k]$	Overall impulse response of a I-DS ADC
$H_{\text{LP}}(z)$	Transfer function of the digital lowpass filter

$h_{\mathrm{rec}}[k]$	Impulse response of the reconstruction filter
$H_{\mathrm{rec}}(z)$	Transfer function of the reconstruction filter
$h_{\mathrm{dsm}}[k]$	Impulse response of the free-running DSM
IBN_{inc}	Integrated inband noise of an I-DS ADC
IBN_{OSR}	Integrated inband noise of oversampled spectrum
IBN_{shaped}	Integrated inband noise of shaped spectrum
k	DT variable
k_{E}	Feedback factor in exponential phase
L	Modulator order
L_{d}	Order of the digital reconstruction filter
M	Number of samples
M_{EP}	Number of samples in exponential phase
M_{LP}	Number of samples in linear phase
M_{on}	Number of active samples before the recuperation phase
M_{reset}	Number of reset samples
M_{RP}	Number of samples in recuperation phase
n	Index variable, $n = 1 \ldots \infty$
N_{FIR}	Number of taps of a FIR DAC
$NTF_{\mathrm{dsm}}(f)$	Noise transfer function of a freely running DSM
$NTF_{\mathrm{free}}(z)$	Noise transfer function of an I-DSM without reset
$NTF_{\mathrm{ct,\,free}}(z)$	Noise transfer function of a CT I-DSM without reset
$NTF_{\mathrm{inc}}(f)$	Overall noise transfer function of an I-DS ADC
$NTF_{\mathrm{ct,\,inc}}(f)$	Overall noise transfer function of a CT I-DS ADC
$NTF_{\mathrm{dt,\,inc}}(z)$	Overall noise transfer function of a DT I-DS ADC
$NTF_{\mathrm{ct,\,dsm}}(f)$	Noise transfer function of a freely running CT DSM
$NTF_{\mathrm{dt,\,dsm}}(z)$	Noise transfer function of a freely running DT DSM
OBG_{sc}	Out-of-band gain of a scaled modulator's free-running NTF
OBG_{unsc}	Out-of-band gain of an unscaled modulator's free-running NTF
ω_{I}	CT integrator corner frequency
$P_{\mathrm{N,\,free},l}$	Noise at the output of free-running DS from the l-th integrator input
$P_{\mathrm{N,\,inc},l}$	Noise at the output of incremental DS from the l-th integrator input
$P_{\mathrm{N},l}$	Noise from the l-th integrator input
P_{noise}	Quantization noise power
P_{sig}	Input signal power
$P_{\mathrm{sig,\,out}}$	Input signal power at the output of the ADC
q	Quantizer gain
e	Quantization noise in time domain
$\mathrm{rect}_M\,[k]$	DT rectangular window function
$\mathrm{rect}_T\,(t)$	CT rectangular window function
$r(t)$	CT feedback signal
σ_{e}^2	Total quantization noise power
$S_{\mathrm{e}}(f)$	Power spectral density of the quantization noise
$SQNR_{\mathrm{OSR}}$	Signal-to-quantization-noise ratio of oversampling ADC
SNR_{p}	Peak signal-to-noise ratio
m	Number of stages in a multi-stage noise shaping architecture

$STF_{\mathrm{ct,free}}(f)$	Signal transfer function of a CT I-DSM without reset
$STF_{\mathrm{ct,inc}}(f)$	Overall signal transfer function of a CT I-DS ADC
$STF_{\mathrm{dt,inc}}(z)$	Overall signal transfer function of a DT I-DS ADC
$STF_{\mathrm{ct,rp}}(f)$	Overall signal transfer function of a CT I-DS ADC with recuperation phase
$STF_{\mathrm{dt,rp}}(z)$	Overall signal transfer function of a DT I-DS ADC with recuperation phase
$STF_{\mathrm{ct,dsm}}(f)$	Signal transfer function of a freely running CT DSM
$STF_{\mathrm{dt,dsm}}(z)$	Signal transfer function of a freely running DT DSM
$S_{\mathrm{u_s}}(f)$	Power spectral density of the sampled input signal
$S_{\mathrm{u}}(f)$	Power spectral density of the input signal
τ_{ELD}	Excess loop delay
$TF_{\mathrm{free}}(z)$	Overall transfer function of an I-DS ADC without reset
$TF_{\mathrm{dt,inc}}(f)$	Overall transfer function of an I-DS ADC
$TF_{\mathrm{dsm},l}(f)$	Inner transfer function of a DSM
$TF_{l}(f)$	Inner transfer function of an I-DS ADC
$TF_{\mathrm{dt,dsm}}(z)$	Transfer function of a free-running DSM
T_{on}	Active time before the recuperation phase
T_{s}	Sampling period
$u_{\mathrm{s}}(t)$	CT representation of sampled signal
$\hat{u}_{\mathrm{const}}$	Constant input signal
$u[k]$	DT DSM input signal
$u_n[k]$	DT I-DSM input signal of *n*th sample
$u_n(t)$	CT I-DSM input signal of *n*th sample
$u(t)$	Continuous-time input signal
$v[k]$	Quantizer output
$v_n[k]$	Quantizer output of *n*th sample (I-DSM)
$w_{\mathrm{lin}}[k]$	Filter weights of linear reconstruction filter
$X_{\mathrm{rect},M}(f)$	Frequency response of DT rectangular window function
$X_{\mathrm{rect},T}(f)$	Frequency response of CT rectangular window function
$X'_{\mathrm{rect},T}(f)$	Frequency response of CT rectangular window function shifted by half of its width
$y[k]$	DT quantizer input signal
$y_n[k]$	DT quantizer input signal of *n*th sample (I-DSM)
$y_n(t)$	CT quantizer input signal of *n*th sample (I-DSM)

Chapter 1
Introduction

1.1 History of Incremental Delta-Sigma ADCs

Inose et al. laid the foundation for the Delta-Sigma (DS) analog-to-digital converter (ADC) with their paper "A Telemetering System by Code Modulation—Δ-Σ Modulation" in 1962 [1]. It was the first time that the name and principle of DS modulation was proposed as an extension of the then existing Δ modulation for telemetering systems. Although the later breakthrough of DS modulation was largely based on switched-capacitor (SC) implementations in integrated circuits (ICs), this first discrete system consisted of a continuous-time integrator with current mode input together with a Schmitt trigger as quantizer. This setup could achieve a signal-to-noise ratio (SNR) of up to 35 dB in a signal bandwidth of about 5 kHz. Since then, DS ADCs have received a lot of attention and have become one of the fundamental ADC architectures [2–5]. Initially, DS ADCs were widely used in applications requiring high relative accuracy, such as low-bandwidth audio processing [6], their continuous-time (CT) implementation allowed ever-higher bandwidths, mainly driven by RF CMOS and wireless transceivers. Today's DS ADCs achieve several hundred MHz of bandwidth [7, 8].

It took more than a decade after Inose's fundamental paper until the appearance of a DS ADC in literature which could be called an incremental ADC [9–11]: In his work, van de Plassche et al. presented a first-order current input DS modulator that passively integrated the input current on a capacitor that was simultaneously driven by the current mode feedback of the modulator. As a result, this node became very quiet, depending on the quality of the feedback signal, and resembled a virtual ground node. This allowed the modulator to multiplex between different channels because, to a first-order approximation, there was no memory in the analog part of the modulator from the previously converted input. The digital counter was run for a certain amount of time to build a high-resolution digital output and then explicitly reset before the

M. Ortmanns et al., *Incremental Delta-Sigma ADCs*,
https://doi.org/10.1007/978-3-032-09468-1_1

next conversion started. The author also presented a technique using an additional loop around the modulator, including a digital-to-analog converter (DAC). This loop was placed in a coarse conversion phase and removed a large portion of the input signal during the fine conversion. According to the author, this technique allowed an increase in resolution at the cost of a slight increase in conversion time.

About ten years later, Robert, Temes, Deval et al. published a paper entitled "A 16-bit Low-Voltage CMOS A/D Converter", which achieved an outstanding accuracy of 16 bits through circuit-level innovations involving the use of a SC integrator together with compensation techniques for error reduction [12]. They introduced an explicit reset of the capacitors, erasing the memory in the analog Delta-Sigma modulator (DSM) and thus allowing true Nyquist operation. A year later, the same group published "A Second-Order High-Resolution Incremental A/D Converter with Offset and Charge Injection Compensation" describing an improved ADC that outperformed their previous work in terms of bandwidth by nearly a factor of 100 [13], thus laying the foundation for the *incremental* DS ADC. Subsequently, many systems requiring high absolute accuracy—i.e. low offset and gain errors—such as measurement and instrumentation applications, and at the same time Nyquist-operation relied on this variation of the original DS ADC and were referred to as *incremental DS ADCs*.

Over the past decades, these converters have found applications in many fields. Their ability to operate as Nyquist converters makes them very suitable for image sensors, biomedical sensor arrays, industrial applications, and many more, since the incremental Delta-Sigma (I-DS) ADC can operate on multiplexed inputs, in time interleaved fashion, and does not require an initial settling period to produce its first valid output, making it suitable for one-shot conversion. This also led to a more recent trend to return to this type of ADC, and circuit and system innovations have been continuously contributed [14].

Although several reference books have been published for free-running DSM over the last three decades, including [2–5], there is no reference book for incremental converters that explains their basic operation, their unique architectural techniques and their differences from the free-running counterpart, and summarizes their design approach. The incremental operation has only received small attention in [5]. A complete book was dedicated to them in [15] covering their basic working principle, their time-domain description and analysis, and only two digital reconstruction filter variants were covered. It also only shortly revised the principle of two extended architectures and finally elaborated on an integrated circuit example. Very often, state of the art (SoA) references the PhD thesis of Markus [16]. This book tries to contribute a comprehensive and complete overview of incremental DS converters.

1.2 State of the Art of Published Incremental ADCs

When it comes to ADC surveys and comparisons, most research papers refer to the ADC survey collected by Boris Murmann [17]. His survey only takes into account papers published at ISSCC and VLSI conferences, thus leaving out many individual designs that contributed new architectural or circuit ideas to ADCs in general and to I-DS ADCs in particular. Adding publications from JSSC, ESSERC, CICC, and other conferences and journals gives a much larger picture but does not change the trend visible from the ADC survey in [17]. For the sake of completeness, it should be mentioned that a more extensive overview has been provided in [18] and updated in [19], as the data collection in [18] has been discontinued.

The most common scatter plots generated with the SoA ADC survey in [17] are shown in Figs. 1.1, 1.2 and 1.3. They show trends using the energy plot (ADC resolution versus energy per Nyquist sample), the aperture plot (ADC resolution versus Nyquist conversion rate), and the efficiency plot (ADC resolution versus figure of merit (FoM)). Murrmann [17] notes that low-power designs excel in the energy plot, while the aperture plot favors designs that achieve a good overall speed-resolution product.

Many FoMs have been contributed to the state of the art, of which only two have survived in research papers. The Walden FoM was introduced by [20], and the thermal-noise-limited FoM was originally proposed in [21] and later used in [3], when it received the name Schreier-FoM. While the Walden FoM was based on technology and thus $C_{min}V^2$-limited designs, where power increases only $\times 2$ per additional bit, the Schreier FoM is ideally suited for thermal noise limited designs, as it treats $\times 2$ in noise (i.e. dynamic range (DR)), a $\times 2$ in bandwidth and a $\times 2$ in power equally with 3 dB each, cf. Appendix A. Interestingly, Sundstrom et al. introduced yet another trend, inferring that matching limited designs would not increase power by $\times 4$ for one bit, but even by $\times 8$ [22]. In recent years, it has been increasingly

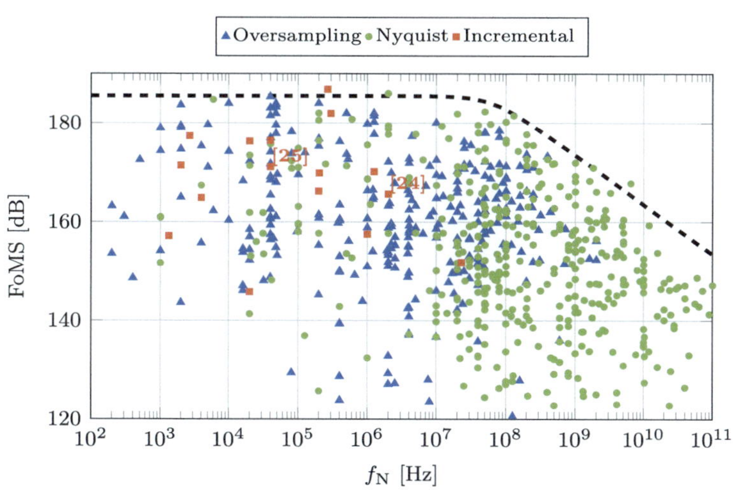

Fig. 1.1 Schreier FoM versus Nyquist conversion rate of published ADCs at ISSCC and VLSI from 1997 to 2025 based on [17] and two additional designs from [24, 25].

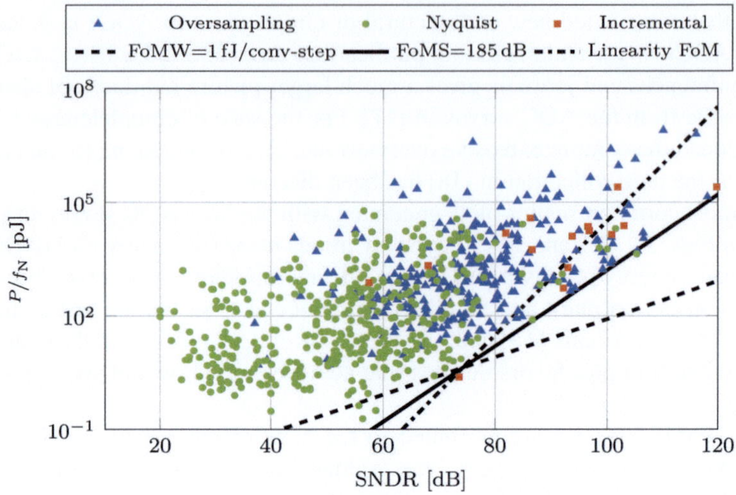

Fig. 1.2 Energy per Nyquist sample (P/f_N) versus resolution of published ADCs at ISSCC and VLSI from 1997 to 2025 based on [17]. Trend lines indicate the Walden FoM of 1 fJ/bit, the Schreier FoM of 185 dB, and the trend line indicating the $\times$ 8 power increase per additional bit for matching limited designs following [22].

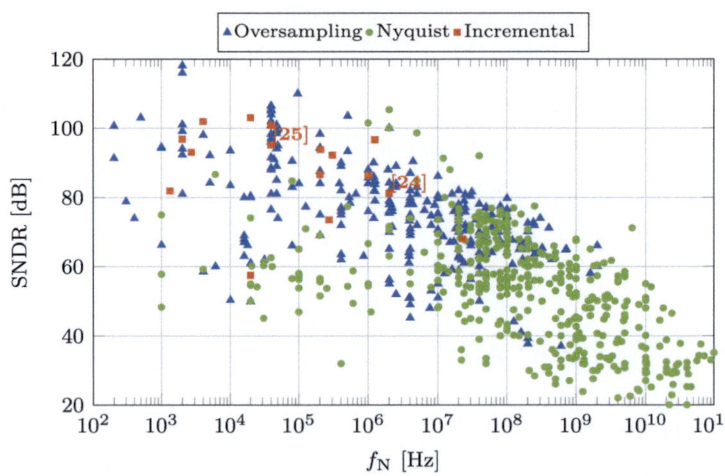

Fig. 1.3 Resolution versus Nyquist conversion rate of published ADCs at ISSCC and VLSI from 1997 to 2025 based on [17] and two additional designs from [24, 25].

recognized that while these charts provide a good indication of trends and achievable resolution and conversion rates, they must be used with caution. The comparison only works over a certain range of resolution and speed. Also, technology nodes are not included. Also, features like true Nyquist conversion are not necessarily included. This list of arguments does not even begin to cover on- or off-chip calibration, the use of MATLAB to get the best results, robustness to temperature, voltage and process variations or aging, and much more. The reader is strongly advised to always take a close look, especially when individual designs stand out from the crowd! In the following, we have sorted the data from [17] into oversampling ADCs, Nyquist ADCs, and incremental ADCs.

Schreier FoM (FoMS) over Nyquist Frequency Looking at Fig. 1.1, we can first observe that incremental ADCs mostly populate the low speed range. On the faster side, [23] sticks out with a 23 MS/s resetting ADC with a FoMS≈150 dB, and more recently, [24] showed a 2 MS/s incremental ADC with a more SoA FoMS≈180 dB. However, it is clear from the designs presented that the Nyquist capability of the incremental ADCs comes at a cost, and they do not necessarily achieve the best-in-class FoM.

Energy over Resolution A few more interesting facts can be derived by looking at the Fig. 1.2. First, we see that the Walden FoM no longer clearly indicates a design trend, and only for very low-resolution designs could it still claim to be useful. Second, we see that the high-resolution SoA is truly noise-limited, as it follows the thermal noise trend line (FoMS) very closely. Third, another well-known fact is that most Nyquist-rate ADCs are found in the lower resolution domain, while oversampling ADCs populate the high resolution domain. In between is where the incremental converters are found, i.e. they offer true Nyquist conversion for high resolution. In the same region, there are very few non-oversampling successive approximation register (SAR) ADCs along with the incremental ADCs. Another observation is that the majority of high-resolution Nyquist-rate ADCs is far away from the thermal noise trend line. Sundstrom et al. [22] derived a $\times$ 8 power per bit trend for matching limited high-linearity designs, which is included in Fig. 1.2 as linearity FoM, but it is seen that the actual trend line for most of the highest resolution and linearity Nyquist converters is even worse. The resulting "Nyquist gap", i.e. the region between most high-resolution Nyquist ADCs and the noise-limited low-energy oversampling ADCs, is the region where many of the published incremental ADCs are found.

Resolution over Nyquist As we discussed with Fig. 1.1, incremental converters are not among the fastest ADCs, but rather populate the very high resolution domain. If we look at the latest SoA in Fig. 1.3, they achieve the highest resolution among the Nyquist ADCs for low-speed designs. Interestingly, in the region above $f_{nyq} \approx 1MHz$, several very high resolution Nyquist-rate SAR ADCs have been proposed in recent years that surpass all SoA in resolution and FoM, but this has been achieved by using very mature CMOS nodes, high I/O supply voltages, and extremely complex digital linearization, while the highest linearity is achieved only at comparatively low input frequencies [26].

Consequently, it will be interesting to see how SoA evolves and whether incremental ADCs will claim superior performance in this domain in the future.

This book attempts to provide a fundamental knowledge base for researchers and designers to pursue this quest.

Chapter 2
Overview of Delta-Sigma ADCs

Since incremental Delta-Sigma (I-DS) analog-to-digital converter (ADC) can be seen as a subtype of the more commonly used free-running Delta-Sigma (DS) ADC, in this chapter we will first give a general overview of the basics of ADCs, quantization in time and amplitude, introduce oversampling and noise shaping as basic techniques used in all DS ADCs, and then discuss the operating behavior of free-running DS ADCs to provide the theoretical background for understanding the following chapters. The chapter will also cover typical performance measures of ADCs in general and DS ADCs in particular, as well as provide an overview of the vast topic of DS ADC synthesis and optimization.

2.1 Basics of Analog-to-Digital Conversation

All ADCs include the principle operation to convert an analog, time- and value-continuous signal to a digital, time- and value-discrete signal. This operation relies on sampling in the time domain and discretization of the continuous value range into a finite number of output levels, commonly referred to as quantization. The process of digitization is illustrated in Fig. 2.1. The encoding of the discrete amplitude levels into any digital representation is often implicitly assumed and not further mentioned, as encoding does not further alter the signal properties. On the contrary, both sampling and quantization largely alter the signal properties. They can be described independently and are explained in the following as they are the basis for all further considerations. In Appendix A related relevant performance measures and definitions are explained in detail.

M. Ortmanns et al., *Incremental Delta-Sigma ADCs*,
https://doi.org/10.1007/978-3-032-09468-1_2

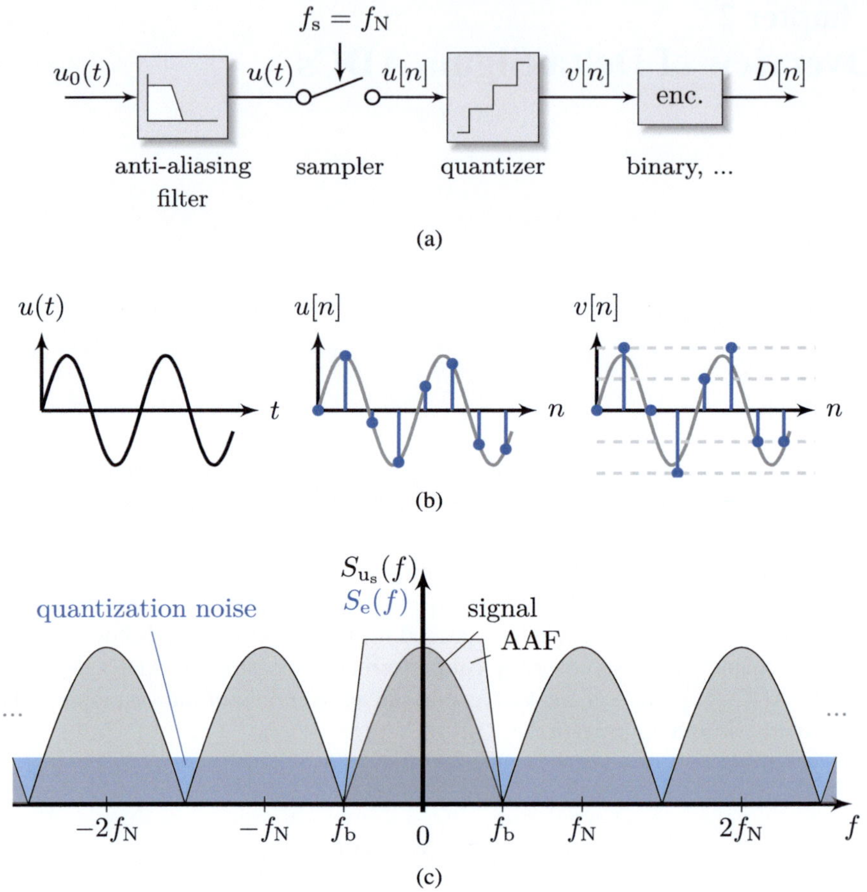

Fig. 2.1 (a) Block diagram of a Nyquist-rate ADC consisting of an anti-aliasing filter, a sampling operation, and a quantizer, (b) an illustration of exemplary signals at the individual nodes of the ADC, and (c) illustration of images of the spectrum of the band-limited input signal at multiples of the sampling frequency f_s disturbed by quantization noise (blue), cf. Sect. 2.1.2. ©2021 Wagner. Reprinted, with permission, from Wagner [27].

2.1.1 Sampling at Nyquist Rate

Nyquist sampling is shortly reviewed next, whereas a more detailed description can be found, e.g. in [28]. The process of periodically sampling a continuous-time (CT) signal $u(t)$ with a period T_s in order to obtain the corresponding discrete-time (DT) sequence $u[n]$ is described by

$$u[n] = u(t)|_{t=nT_s}, \qquad n \in \mathbb{Z}. \tag{2.1}$$

All elements of the sequence $u[n]$ are called samples, which are obtained with the sampling frequency $f_s = \frac{1}{T_s}$. In the time domain, this sampling operation can be also described by the multiplication of the CT signal with a time-domain impulse comb $\text{III}_{T_s}(t)$:

$$u_s(t) = u(t) \cdot \text{III}_{T_s}(t) = \sum_{n=-\infty}^{\infty} u(nT_s) \cdot \delta(t - nT_s). \tag{2.2}$$

The resulting CT signal $u_s(t)$ has only nonzero values at the sampling points, as it consists of weighted Dirac pulses and is equivalent to the DT sequence $u[n]$. In the frequency domain, this multiplication with an impulse comb in the time domain corresponds to the convolution of the Fourier transform of the CT signal $S_u(f)$ and the impulse comb $\text{III}_{1/T_s}(f)$ as mathematically expressed by

$$S_{u_s}(f) = S_u(f) * \frac{1}{T_s}\text{III}_{1/T_s}(f) = \frac{1}{T_s} \sum_{n=-\infty}^{\infty} S_u(f - \frac{n}{T_s}]. \tag{2.3}$$

The result of the convolution in (2.3) is visualized in Fig. 2.1c: the frequency-domain representation $S_u(f)$ is centered around zero and its images are at multiples of f_s. In order to not corrupt the signal due to aliasing of the images into the baseband, the Nyquist-Shannon sampling theorem

$$f_s \geqq f_N = 2f_b \tag{2.4}$$

has to be fulfilled: the sampling frequency f_s has to be chosen larger than twice the signal bandwidth f_b, called the Nyquist frequency f_N. It is the lowest sampling rate that allows an ideal reconstruction of the input signal and thus avoids aliasing [28]. To enforce the Nyquist sampling condition and to avoid aliasing, commonly all ADCs use a suitable upfront low-pass CT anti-aliasing filter (AAF) to obtain the bandlimited analog input signal $u(t)$. It is illustrated in the block diagram in Fig. 2.1a and by its frequency response in Fig. 2.1c in light gray. An ADC operating at a sampling frequency of $f_s = f_N$ is usually referred to as a Nyquist-rate converter.

It is important to note that by using a nonideal AAF with a finite slope, basically every ADC has to use a sampling frequency larger than the ideal Nyquist frequency in order to separate the repetitive DT spectra in Fig. 2.1c by some amount and thus allow for a finite slope of the AAF. Using such a choice of sampling frequency over the Nyquist frequency in an ADC does not yet lead to an oversampling converter, which will be explained in Sect. 2.2.

2.1.2 Quantization and White Noise Model

Widrow states in his early work on quantization in [29]:

> Quantization is an operation that takes place when a physical quantity is represented numerically. It is the assignment of an integral value to a physical quantity corresponding to the nearest number of units contained in it. Quantization is like sampling-in-amplitude.

During quantization, the possible range of values of a signal is reduced from a continuous, infinite number into a limited amount of quantized values. Those values are then encoded into digital values. The process of quantization and its modeling though is independent of the process of encoding, as indicated in Fig. 2.1. Illustrated in Fig. 2.2, the quantizer introduces a quantization error $e[n]$ to its analog input signal $y[n]$. Depending on the number of quantization levels and thus the step width Δ, the ideal transfer characteristic differs more or less from the quantized output.

Thereby, under the general assumption of equally sized quantizer steps, Δ becomes

$$\Delta = \frac{FS}{2^B - 1} \tag{2.5}$$

with B being the number of quantizer bits and FS being its full-scale input range. The analog value Δ is represented by the weight of one least significant bit (LSB) of the quantization process. As long as the input signal remains within the full-scale range of the quantizer, the quantization error $e[n]$ remains bounded within

$$e[n] \in [-\Delta/2, \Delta/2]. \tag{2.6}$$

Quantization is highly nonlinear, and understanding as well as modeling it as a quasi-linear system has been a research topic for a long time [29–31]. A detailed overview of quantization in general can be found in [32]. The quantization process with regard to DS ADCs is covered in more detail, e.g. in [2–5]. In the following, we only cover the topic to the extent necessary for this book.

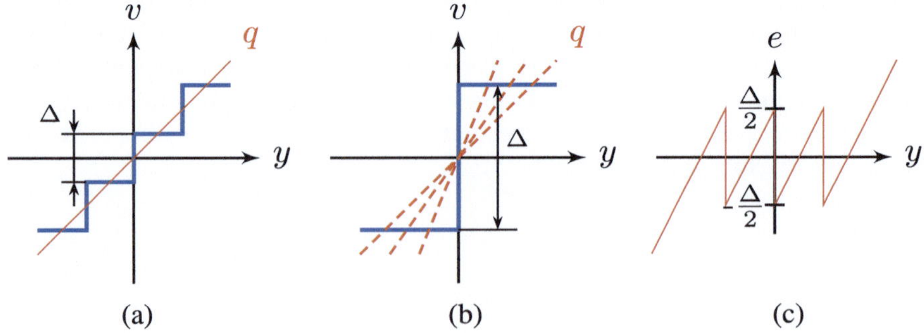

Fig. 2.2 Illustration of the quantizer gain for (a) a multi-bit and (b) a single-bit quantizer and (c) the quantization error.

2.1.2.1 Linear Quantizer Model

In modeling quantization, the quantizer (QTZ) can firstly be replaced by a summation of the analog input value $y[n]$ and the quantization error $e[n]$ as shown in Fig. 2.3. In case of Nyquist-rate ADCs, the Nyquist-rate digital output sample $D[n]$ equals (the digitally encoded) quantizer output $v[n]$ and the quantizer input signal $y[n]$ equals the sampled input signal of the ADC $u[n]$, cf. Fig. 2.1a.

In order to derive a quasi-linear model of quantization, a statistical model for the quantization error $e[n]$ is needed, and also the question needs to be answered which transfer characteristic, i.e. *effective quantizer gain* q, is suitable in Fig. 2.3. The linearized model is then described by

$$v[n] = q \cdot y[n] + e[n]. \tag{2.7}$$

When observing Fig. 2.2, the purpose of the additional quantizer gain q, defining the slope of the quantizer transfer curve, becomes obvious [33]. For a multi-bit quantizer, the slope and thus q is easily understood to be one, if the quantization is done with many levels and as long as the quantizer is not overloaded. However, for a single-bit quantizer, it is not obvious how a best-fit slope should imitate the sign-operation. Thus, the quasi-linear modeled quantizer gain q has to be statistically described. The average power of the quantization error can be written as

$$\sigma_\mathrm{e}^2 = \langle e, e\rangle \tag{2.8}$$
$$= \langle v - qy, v - qy\rangle \tag{2.9}$$
$$= \langle v, v\rangle - 2q\langle v, y\rangle + q^2\langle y, y\rangle. \tag{2.10}$$

By minimizing this equation, the best fitting slope of the linear model is obtained, which yields [3]:

$$q = \frac{\langle v, y\rangle}{\langle y, y\rangle} = \frac{\mathrm{E}[vy]}{\mathrm{E}[yy]} \tag{2.11}$$

with the inner products $\langle v, y\rangle$ and $\langle y, y\rangle$ defined as

$$\langle a, b\rangle = \lim_{N\to\infty}\left[\frac{1}{N}\sum_{n=0}^{N} a[n]b[n]\right]. \tag{2.12}$$

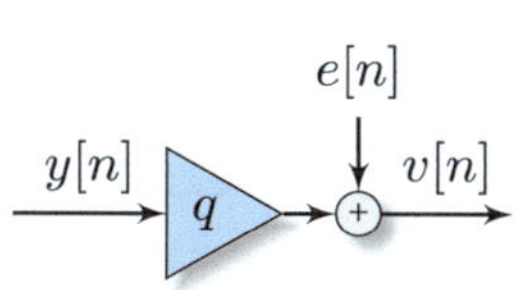

Fig. 2.3 Linear quantizer model.

Using (2.11), the quantizer gain is obtained by numerical simulations using sufficiently large sample size N, i.e. with long simulation times and by applying different input signals [3–5].

It should be noted that more sophisticated models were derived in the past, e.g. in [34], a model based on two parameters was suggested. There, the quantization error was described by one parameter for scaling a DC component and a second for scaling a random component. Although it yielded more accurate results, the higher complexity is omitted in most state of the art and the easier quasi-linear model is the de-facto standard.

2.1.2.2 Quantization Noise Modeled as White Noise

The modeling of the signal properties of the quantization error seems impossible, as it depends on the actual input signal at any time instant. For a DC input signal, $e[n]$ is DC as well. For a periodic input signal, $e[n]$ becomes periodic and thus tonal. Many authors have derived theories to model the quantization error [30, 31]. Under the assumptions of a sufficiently active input, which spans over a large enough range of the full scale, a non-overload condition (cf. (2.6)), a large number of quantization steps, and moreover of a quantization error uncorrelated to the input signal, the quantization error $e[n]$ can be modeled as white noise that is equally distributed over all frequencies with a uniform probability density function (PDF). The error e is then referred to as quantization noise [30]. With the boundary limitation of e as in (2.6), and uniform PDF, the probability density of e becomes $\frac{1}{\Delta}$. The resulting overall quantization noise power then yields:

$$\sigma_{\mathrm{e}}^2 = \mathrm{E}\left[e^2\right] = \int_{-\infty}^{+\infty} \mathrm{PDF}(e) \cdot e^2 \, \mathrm{d}e = \int_{-\frac{\Delta}{2}}^{+\frac{\Delta}{2}} \frac{1}{\Delta} \cdot e^2 \, \mathrm{d}e = \frac{\Delta^2}{12} \tag{2.13}$$

only relying on the QTZ step width Δ in (2.5). With the white noise model and the equal distribution in the sampled frequency range $[-f_{\mathrm{s}}/2, f_{\mathrm{s}}/2]$, the power spectral density (PSD) of the quantization noise becomes

$$S_{\mathrm{e}}(f) = \frac{\Delta^2}{12} \frac{1}{f_{\mathrm{s}}}. \tag{2.14}$$

This white quantization noise spectrum was already illustrated in Fig. 2.1c. With this approximation, a first performance measure of any ADC can be derived: the signal-to-quantization-noise ratio (SQNR). It is defined as the ratio of the input power, usually using a single tone sinusoidal input signal, and the quantization noise power expressed in decibel. Thereby, it is assumed that the sinusoidal input signal has a maximum peak-peak amplitude equal to the quantizer FS. Note that an input signal with $S = 0dBFS$ can be applied to a general quantizer as in Fig. 2.2 without overload, while other ADCs, such as Delta-Sigma modulators (DSMs), need a back-off to remain in a non-overload condition. By dividing the resulting signal power by the quantization noise $P_{\text{noise}} = \sigma_e^2$ from (2.13), this yields

$$SQNR_{\text{Nyquist}} = 10\,\text{dB} \cdot \log_{10}\left(\frac{P_{\text{sig}}}{P_{\text{noise}}}\right) = 6.02\,\text{dB} \cdot B + 1.76\,\text{dB}. \tag{2.15}$$

Thus, the achievable SQNR can be directly related to the quantizer bit width B and an additional quantizer bit is equivalent to increasing the SQNR by 6dB and vice versa.

2.2 Oversampling and Noise-shaping ADCs

2.2.1 Oversampling ADCs

From (2.14) the idea of oversampling can be directly deduced as a way to reduce the quantization noise PSD. In an oversampling ADC, as depicted in Fig. 2.4a, the sampling and thus quantization operate at $f_s \gg f_N$. Note that we refer to the DT instants at Nyquist rate as $[n]$, whereas the DT instants at the oversampled frequency as $[k]$. The oversampling ratio (OSR) is defined as

$$OSR = \frac{f_s}{f_N} = \frac{f_s}{2f_b}. \tag{2.16}$$

Due to oversampling, two advantages appear: first, the low-pass-filtered input signal is periodically repeated with the larger f_s. Thus, the resulting "gap" between the periodically repeated spectra of the band-limited input signal can be used to relax the requirements of the AAF, cf. Sect. 2.1.1. Second, oversampling with subsequent digital filtering can be used to reduce the quantization error in the frequency band of interest, i.e. the inband noise (IBN), since the total quantization noise power σ_e^2 is only dependent on the quantizer step width Δ in (2.13), independent of OSR. However, the quantization noise PSD is inversely proportional to the sampling frequency, cf. (2.14). Thus, by using a digital low-pass filter $H_{\text{LP}}(z)$ subsequent to the quantizer in Fig. 2.4a, the spectral part of the quantization noise outside the signal band—i.e. the out-of-band (OOB) quantization noise—can be eliminated. A decimator (or down-sampler) finally reduces the sample rate to f_N, where—assuming an ideal brick-wall filter—neither signal nor quantization noise aliasing occurs. Thus, the inband quantization noise power IBN_{OSR} is reduced by the suppression of its inband spectral density, which allows to increase the effective resolution not only by increasing the bitwidth B of the quantizer, but also by increasing the OSR. Figure 2.5b illustrates the process of decimation with and without the digital low-pass filter and how the combined process of decimation and filtering achieves the low IBN.

The IBN of an oversampled quantizer can be calculated analogously to (2.13), by including the spectral spreading of the quantization noise and subsequent digital

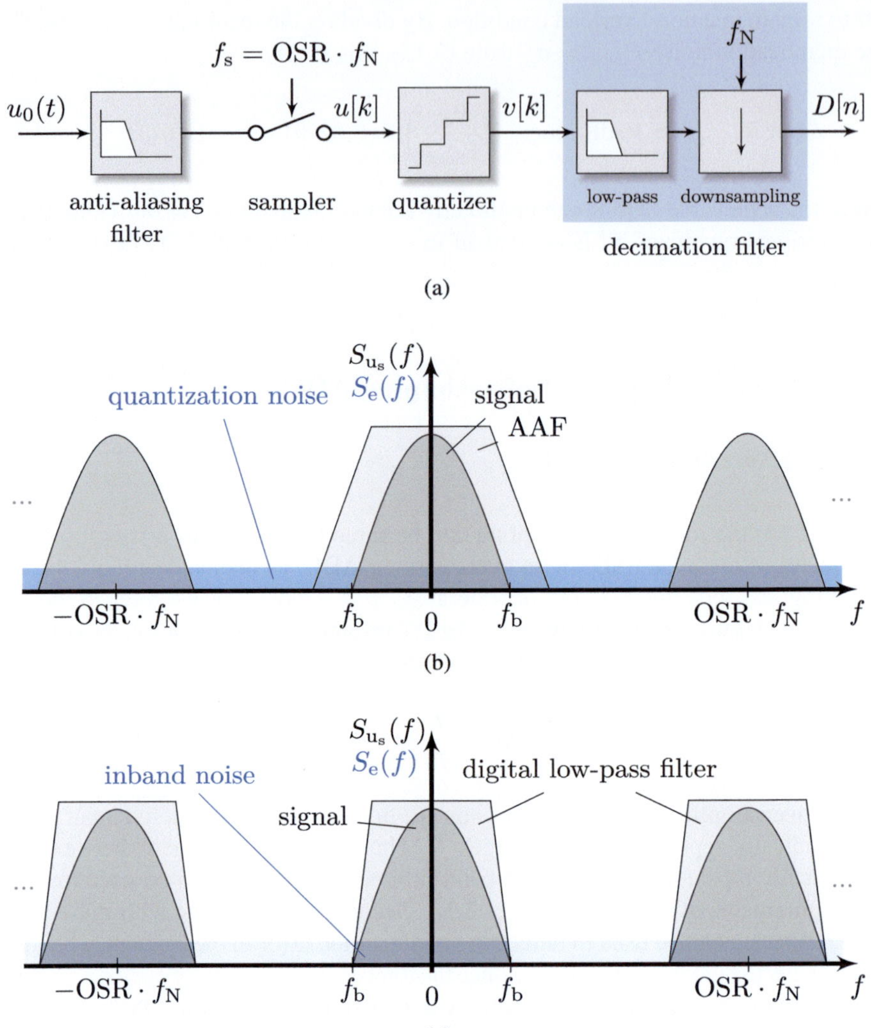

Fig. 2.4 (a) Block diagram of an oversampling ADC employing a relaxed AAF, an oversampled quantizer outputting $v[k]$ at sampling frequency f_S, and a digital low-pass filter with subsequent decimation (decimation filter) outputting $D[n]$ at Nyquist frequency f_N, (b) illustrative PSD with periodic repetitions of the input spectrum at multiples of the sampling frequency and spectrally spread quantization noise by oversampling (blue), and (c) spectrum after the digital low-pass filter (before decimation). ©2021 Wagner. Reprinted, with permission, from Wagner [27].

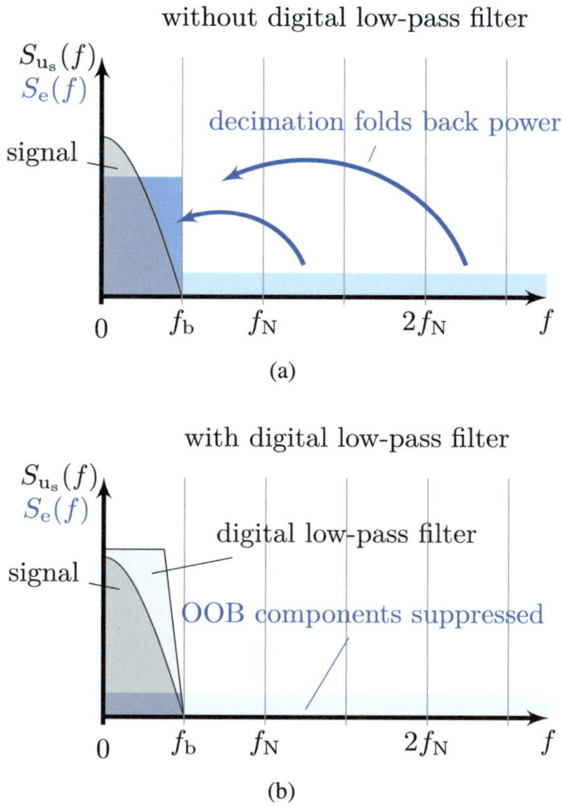

Fig. 2.5 Decimation of an oversampled quantizer output. Spectrally spread quantization noise (a) is not digitally low-pass filtered and completely folds back into the inband during decimation, (b) is low-pass filtered before decimation, OOB noise is eliminated and final IBN is reduced. ©2021 Wagner. Reprinted, with permission, from Wagner [27].

low-pass filtering

$$IBN_{\mathrm{OSR}} = \int_{-f_s/2}^{f_s/2} |H_{LP}(z)|^2 \cdot S_e(f)\,\mathrm{d}f \tag{2.17}$$

with $z{=}e^{s\cdot T_s}{=}e^{(\sigma+jw)} \cdot T_s$. The digital low-pass filter is assumed to be an ideal brick-wall filter

$$H_{\mathrm{brick}}(f) = \begin{cases} 1, & -f_b \leq f \leq f_b \\ 0, & \text{otherwise,} \end{cases} \tag{2.18}$$

which suppresses all OOB quantization noise power [5]. Then, (2.17) can be rewritten as

$$IBN_{\mathrm{OSR}} = \int_{-f_b}^{f_b} S_e(f)\mathrm{d}f = \frac{\Delta^2}{12} \cdot \frac{f_N}{OSR \cdot f_N} = \frac{\Delta^2}{12} \cdot \frac{1}{OSR} \tag{2.19}$$

where we use the quantization noise power spectral density $S_e(f)$ from (2.14). This IBN is used to calculate the SQNR of an oversampled ADC

$$SQNR_{\mathrm{OSR}} = 10\,\mathrm{dB} \cdot \log_{10}\left(\frac{P_{\mathrm{sig}}}{IBN_{\mathrm{OSR}}}\right) = 6.02\,\mathrm{dB} \cdot B + 10\,\mathrm{dB} \cdot \log_{10}(OSR) + 1.76\,\mathrm{dB}. \tag{2.20}$$

For each doubling of the sampling frequency, $SQNR_{\mathrm{OSR}}$ improves by 3 dB, which—according to the SQNR of a Nyquist converter in (2.15)—is equivalent to 0.5 bit in resolution in comparison to a quantizer operating at Nyquist-rate. Consequently, instead of giving the quantizer 1 bit more resolution, also four-times more oversampling can be used.

2.2.2 *Noise-shaping ADCs*

Although oversampling allows to improve the SQNR by 3 dB for each doubling of the sampling rate, unrealistically high sampling frequencies would be required in order to obtain a high-resolution ADC from a low-resolution quantizer. To realize, e.g. a 16 bit audio-band ADC from a single-bit quantizer, we would need a thirty-times doubling of the sampling frequency, meaning about $44\,\mathrm{kS/s} \cdot 2^{30}$, ending up at an unrealistic terahertz range. By expanding the idea of spectrally reduced quantization noise in the inband by oversampling, the concept of noise-shaping becomes an obvious solution.

2.2.2.1 Loop Filter, Signal and Noise Transfer Functions

The ADC concepts presented so far are—in terms of system theory—open-loop configurations. In control systems, given an input and a disturber, the disturber can be reduced towards the output by adding a high gain controller upfront. Thus, the logical step to further reduce the IBN is to realize such a control-loop with the quantization noise as the disturber. This allows to implement two different transfer functions, one for the input signal and one for the quantization noise: the signal transfer function (STF) and the noise transfer function (NTF), respectively. The loop filter of this control system is ideally chosen such that the STF is unity in the inband and, e.g. low at OOB frequencies, and thus forms an ideal (brick-wall) low-pass filter:

$$STF_{\mathrm{ideal}}(f) = \begin{cases} 1, & -f_b \leq f \leq f_b \\ 0, & \text{otherwise.} \end{cases} \tag{2.21}$$

The ideal NTF should eliminate the IBN, while its magnitude OOB can be much larger.

$$NTF_{\text{ideal}}(f) = \begin{cases} 0, & -f_\text{b} \leq f \leq f_\text{b} \\ \gg 0, & \text{otherwise.} \end{cases} \tag{2.22}$$

Thus, ideally, the quantization noise is completely suppressed in the signal band by the NTF and completely shaped into the gap between the signal images. This idea obviously requires a certain oversampling operation such that there is an OOB region to shape the quantization noise into. In contrast to the ideal NTF, the ideal STF does not influence the inband signal.

Such ideal STF and NTF are obviously not realizable. A feasible solution for a controller, however, is an integrator, as we know from control theory that it eliminates DC disturbances and still suppresses low-frequency disturbances. We come back to this idea in Sect. 2.3. With an integrating controller, STF and NTF resemble low- and high-pass filters in practical realizations. The frequency-domain representation

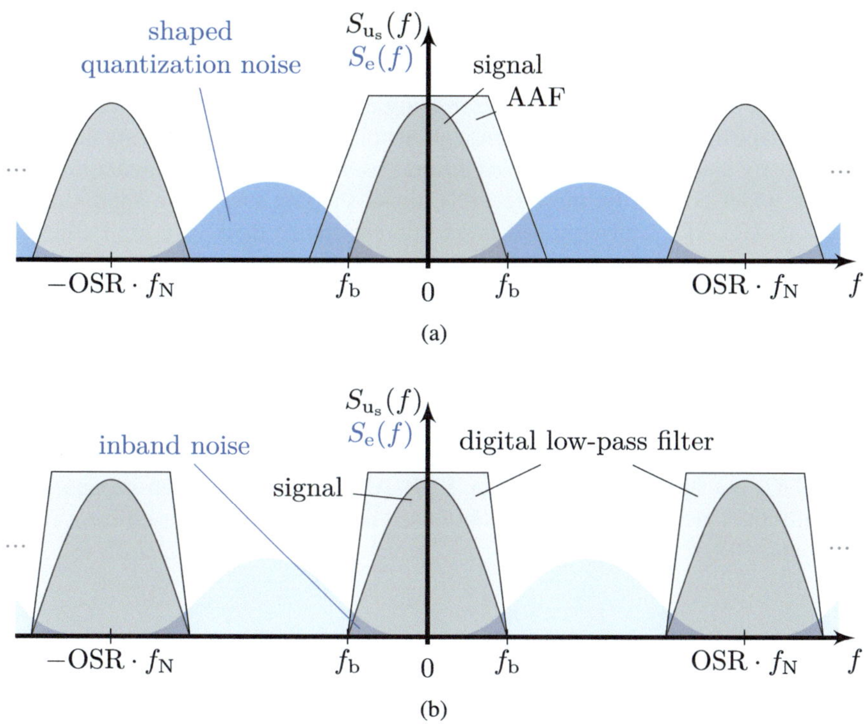

Fig. 2.6 (a) PSD of quantization noise (blue) is shaped out of the inband towards OOB and is (b) removed by a digital low-pass filter. ©2021 Wagner. Reprinted, with permission, from Wagner [27].

of such a noise-shaping process is illustrated in Fig. 2.6a. Similar to an oversampled converter, a digital low-pass decimation filter is required subsequent to a noise-shaped quantizer in order to remove the excessive OOB shaped quantization noise prior to decimation.

Then, similar to (2.19) the IBN of a noise-shaped oversampled converter can be calculated by integrating the noise-shaped quantization error spectrum only over the inband frequency range due to the assumed brick-wall decimation filter:

$$\begin{aligned} IBN_{\text{shaped}} &= \frac{1}{f_s} \cdot \int_{-f_s/2}^{f_s/2} |H_{\text{brick}}(f)|^2 \cdot |NTF(f)|^2 \cdot S_e(f)\,\mathrm{d}f \\ &= \frac{\Delta^2}{12} \cdot \frac{1}{f_s} \cdot \int_{-f_s/2}^{f_s/2} |H_{\text{brick}}(f)|^2 \cdot |NTF(f)|^2\,\mathrm{d}f \\ &= \frac{\Delta^2}{12} \cdot \frac{1}{f_s} \cdot \int_{-f_b}^{f_b} |NTF(f)|^2\,\mathrm{d}f. \end{aligned} \tag{2.23}$$

2.2.2.2 Noise-shaping ADC Architectures

The described principle of oversampling and noise-shaping can practically be used for any kind of low-resolution quantizer using any kind of loop filter, which would allow to shape the low-resolution quantization noise to OOB while leaving the inband signal mostly unaltered. In the state of the art (SoA), only two implementations have achieved wider popularity: more recently, noise-shaping successive approximation register (SAR) ADCs have gained large interest due to their very high efficiency especially in scaled technologies. An overview of noise-shaping SAR ADCs is, e.g. found in [35]. Major innovations included passive rather than active noise-shaping within the capacitive digital-to-analog converter (DAC), higher-order passive noise-shaping, as well as embedding mismatch error shaping, which simultaneously serves as linearization of DAC mismatch. The more traditional noise-shaping ADC has been the DS ADC, where a low resolution quantizer is embedded into a typically active loop filter [1]. The low-resolution quantizer has classically been a low-resolution flash ADC, but also two-step [36] or SAR ADCs [37] have been proposed and especially the latter is meanwhile widely used in many SoA power-efficient implementations [38].

2.3 Delta-Sigma ADCs

2.3.1 Delta-Sigma Modulator and Decimation Filter

In this section, the working principle of free-running DS ADCs is reviewed. They have gained their popularity as they trade analog complexity for time and digital

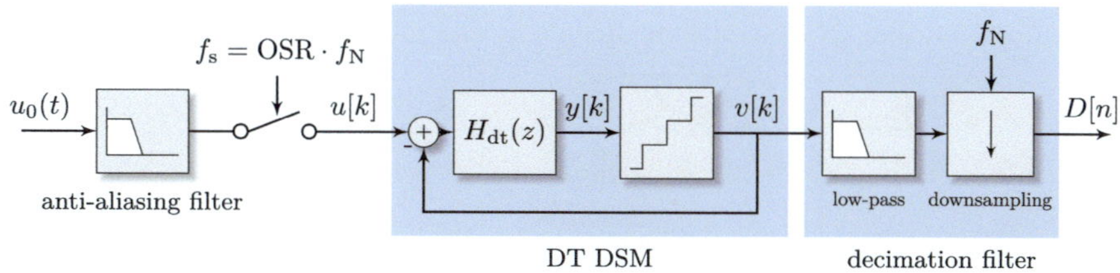

Fig. 2.7 Block diagram of a DTDS ADC implementing a loop filter $H_{dt}(z)$ around a low resolution quantizer. It is preceded by an AAF and sampler, and followed by a decimation filter to remove the OOB quantization noise before subsampling to Nyquist frequency. ©2021 Wagner. Reprinted, with permission, from Wagner [27].

processing, which is very suited for modern complementary metal-oxide-semiconductor (CMOS) technologies due to increased transit frequencies. Proposed in [1], the DSM is based on the above introduced concepts of oversampling and noise-shaping. It mainly consists of the analog DSM as well as the digital decimation filter as indicated in Fig. 2.7. An AAF is used before sampling, but with a relaxed roll-off due to the oversampling used. Even though literature describes also band-pass [2, 39], and high-pass [40, 41] loop-filter implementations, we only review lowpass (LP) DS ADC, as we only need the concept to switch later to the incremental Delta-Sigma modulator (-IDSM). Depending on the realization of the loop filter and thus the position of the sampling operation, DSMs are either referred to as DT DSMs with loop filter $H_{dt}(z)$, cf. Sect. 2.3.1.1, where the sampling operation is performed before the loop filter, or as CT DSMs with loop filter $H_{ct}(s)$. In the latter, the loop filter is realized as CT filter and sampling is only performed at the internal quantizer [42], which will be described in Sect. 2.3.1.2.

The decimation filter, which is a fundamental requirement both for oversampled as well as noise-shaped ADCs, is indicated as a single operational block in Fig. 2.7. In practice, the filtering and decimation operations can be interleaved, e.g. in cascade-of-integrators comb (CIC) filters [43]. The decimation filter is almost always neglected in all publications, even though an efficient implementation with sufficient OOB suppression is required in practice. The few papers that include it on-chip prove that its power consumption is easily in the same range as that of the analog part of the ADC [44, 45].

2.3.1.1 Discrete-Time DS Modulators

A DS ADC with DT loop filter is shown in Fig. 2.7. It is an implementation of a noise-shaping ADC enclosing the quantizer with a feedback loop. The DT loop filter $H_{dt}(z)$ can be implemented in switched-capacitor (SC) technology. In order to calculate the transfer functions, the quantizer is replaced by its linear quantization

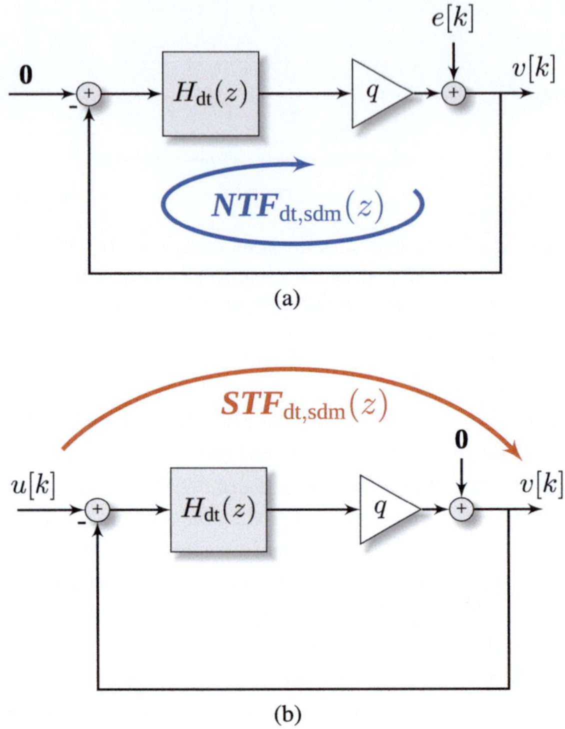

Fig. 2.8 Linearized DTDSM for (a) NTF and (b) STF calculation.

noise model as described in Sect. 2.1.2 and illustrated in Fig. 2.8—thus the system is modeled as a linear time-invariant (LTI) system. This allows to obtain the NTF of a DT free-running DSM as depicted in Fig. 2.8a as

$$NTF_{\mathrm{dt,dsm}}(z) = \frac{1}{1 + q \cdot H_{\mathrm{dt}}(z)} \tag{2.24}$$

and the STF of a DT free-running DSM as depicted in Figure 2.8b as

$$STF_{\mathrm{dt,dsm}}(z) = H_{\mathrm{dt}}(z) \cdot NTF_{\mathrm{dt,dsm}}(z) = \frac{H_{\mathrm{dt}}(z)}{1 + q \cdot H_{\mathrm{dt}}(z)}. \tag{2.25}$$

with the loop-filter function $H_{\mathrm{dt}}(z)$ and the effective quantizer gain q.

First-order DS modulator: As we argued in Sect. 2.2.2, the ideal STF and NTF would be unity and zero within the inband, i.e. at $f \ll f_s$. Comparing this with the NTF and STF in (2.24) and (2.25), an obvious solution is to realize $H_{\mathrm{dt}}(z)$ with very large gain at low frequencies. This leads to the most basic DSM realized with a first-order loop filter using an integrator as $H_{\mathrm{dt}}(z)$:

$$H_{\mathrm{dt}}(z) = \frac{z^{-1}}{1 - z^{-1}} \tag{2.26}$$

which results in the STF and NTF of a first-order DSM as

$$STF_{\mathrm{dt,dsm}}(z) = z^{-1} \quad \text{and} \quad NTF_{\mathrm{dt,dsm}}(z) = \frac{1 - z^{-1}}{q}. \tag{2.27}$$

By applying the discrete-time Fourier transform (DTFT), i.e. substituting $z = e^{j2\pi f T_s}$, the NTF and STF of a first-order DSM can be plotted in the frequency domain, which is shown in Fig. 2.9 [46]. Obviously, the NTF becomes a high-pass and the STF a unity transfer function. Based on (2.23), using (2.27), and assuming $f \ll f_s$, i.e. large OSR, the ideal IBN of the first-order DSM can be approximated as [[4], Sect. 2.5.1]

$$IBN_{\mathrm{shaped}} = \frac{\pi^2}{3} \cdot \frac{\Delta^2}{12} \cdot \frac{1}{q^2} \cdot \left(\frac{1}{OSR}\right)^3. \tag{2.28}$$

In comparison to oversampling alone, the SQNR of a noise-shaped ADC benefits exponentially more from oversampling, i.e. 9 dB improvement for each doubling of the OSR in the case of a first-order loop filter. Redoing our example from Sect. 2.2.2), we try again to realize a 16 bit audio-band ADC from a single-bit quantizer. While a pure oversampling ADC became practically impossible, the first-order DSM achieves the required specification with approximately 10-times doubling of the sampling frequency, i.e. a 44 kS/s output data rate for the DS ADC is achieved with about a 50 MS/s sampling rate of the first-order DSM—which is a very realistic endeavor! Another fact to be noted is that DS ADCs use both principles, oversampling and noise-shaping. While pure oversampling ADCs do work, as outlined in Sect. 2.2.1), a DS ADC can not just use noise-shaping. Inserting OSR=1 into (2.28) reveals that

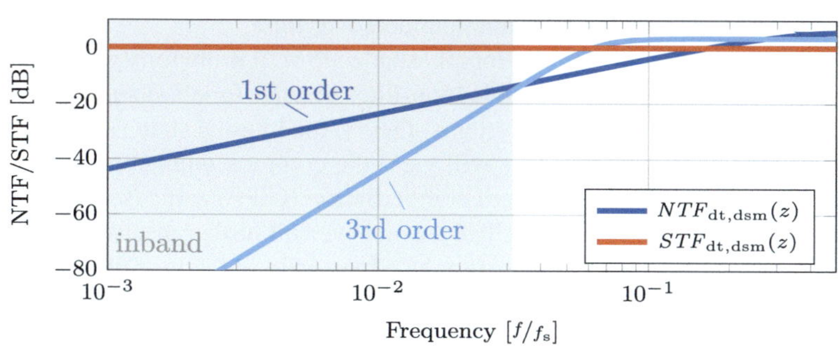

Fig. 2.9 STF for a first-order (red) and NTFs for a first-order (dark blue) and third-order (light blue) DTDSM. ©2021 Wagner. Modified, with permission, from Wagner [27].

the IBN is in fact larger than the IBN of a quantizer alone, cf. (2.13). Thus, a DS ADC will need oversampling in order to benefit from the idea of noise-shaping.

Ideal higher-order DS modulators: In order to improve the performance and IBN of the DSM in (2.28), obvious solutions are to increase the resolution of the underlying quantizer, i.e. its bitwidth B, or to increase the OSR. While we will discuss the increase of the internal bitwidth separately in Sect. 2.3.2.2, a significantly more effective improvement lies in reevaluating our choice of loop filter. As discussed in Sect. 2.2.2, the ideal characteristics of the NTF and STF are to be zero and unity, respectively, within the inband. From Sect. 2.3.1.1, it follows that the loop filter should exhibit high gain at low frequencies. Consequently, our initial choice was to implement an integrator. However, when one integrator exhibits a marked improvement in IBN, it is inevitable that the question will be raised as to whether higher-order integration would not in fact prove superior [47]. Ideally, this means to cascade L integrators in the loop filter. The NTF for such an ideal higher-order DT modulator can be derived similar as described for a first-order DT modulator in (2.27) and results in [4]:

$$NTF_{\text{dt,dsm}}(z) = \tfrac{1}{q}\left(1 - z^{-1}\right)^{L}. \tag{2.29}$$

This leads to the generalized expression for the IBN of an ideal higher-order DSM:

$$IBN_{\text{shaped}} = \frac{\pi^{2L}}{(2L+1)} \cdot \frac{\Delta^2}{12} \cdot \frac{1}{q^2} \cdot \left(\frac{1}{OSR}\right)^{(2L+1)} \tag{2.30}$$

which impressively shows the increased exponential benefit of oversampling for higher-order loop filters.

However, such a chain of integrators is intrinsically unstable. Even when introducing ways for phase compensation, higher-order loop filters remain only conditionally stable. Conditional stability means that the system is only stable for a certain range of input signals following a bounded input, bounded output (BIBO) criterion [4]. This leads to the terminology of a maximum stable amplitude (MSA), which is the highest amplitude of a single sinusoidal input, for which a higher-order DSM remains stable, i.e. all internal state variables remain bounded over time.

In order to achieve such a conditionally stable DSM, it is obviously not possible to just cascade integrators, but a suitable loop-filter transfer function and thus NTF has to be derived. In terms of control systems' theory, the internal state variables need to be controllable or observable. This will yield the common architectures of the cascade-of-integrators with feedforward compensation (CIFF) and cascade-of-integrators with feedback compensation (CIFB), which we will further discuss in Sect. 2.3.2. Then, conditional stability requires to reduce loop-gain in favor of phase margin. In practice, we speak of *scaling* the loop filter. The SoA of stability analysis is summarized in Sect. 2.3.4 and the SoA for the practical loop-filter design process will be presented in Sect. 2.4.

An example of the NTF of a properly scaled third-order DT DSM is depicted in Fig. 2.9 alongside the first-order one. It can be clearly seen that the third-order NTF has a steeper slope of 60 dB per decade in contrast to the 20 dB obtained by the first-order DSM. Even though loop filter scaling for stability will not allow to achieve the ideal IBN of higher-order DSM derived in (2.30), the higher-order suppression of the inband quantization noise spectral density still yields exponentially better integrated IBN for a given OSR.

2.3.1.2 Continuous-Time DSMs

In Sect. 2.3.1.1 DSMs are introduced and derived using DT loop filters. Such DT DSMs are traditionally implemented using SC circuits, where capacitor ratios determine the scaling coefficients and thus the loop-filter transfer function, which becomes very robust over mismatch, process, and temperature. Although switched-current implementations have been introduced, they have never become widely adopted as part of the state of the art [48]. Despite the clear advantage of DT DSM and SC circuits, increased bandwidths are more difficult to realize due to shortened settling times and thus increased power consumption [5]. Also, the whole loop filter of a DT DSM is a sampled system, thus it requires a front-end AAF to avoid aliasing, cf. Fig. 2.7. Actually, the first DSM was implemented with a CT loop filter [1], which is kind of obvious as SC circuits had not been proposed by then. Also, early integrated circuit (IC) implementations of DSM were proposed with CT loop filters [6]. A DS ADC using such a CT loop filter is shown in Fig. 2.10. The sampling operation is moved inside the DSM upfront of the quantizer. This results in a mixed-signal DSM with a CT loop filter $H_{\mathrm{ct}}(s)$ and a sampled quantizer. The loop is closed via a CT DAC in the feedback path.

Many advantages of the CT DS ADC are based on the relocation of the sampling operation inside the modulator: slower signal transitions compared to a DT SC implementations occur, thus the CT integrators demand for reduced dynamic requirements, consume less power, or allow for higher sampling rates. No precise input sample and hold (S/H) is necessary. Also, all sampling related nonidealities are introduced at the quantizer, and are attenuated by the NTF of the CT DSM, which yields one of the most often named advantages of CTDSM, the implicit anti-aliasing filter [4, 5, 49, 50]. Two major disadvantages of CT DSM should also be mentioned: first, the CT nature of signal processing makes all timing related nonidealities much more critical. Especially jitter sensitivity in the CT feedback DAC as well as excess loop delay (ELD) are prominent examples [4]. Second, as RC-time constants match worse than capacitor ratios, precise loop-filter coefficients are difficult to achieve, which makes CT DSM sensitive in architectures, where absolute filter matching is required, as, e.g. in multi-loop DSM [4].

Signal and noise transfer functions of CT DS modulators: The derivation of the STF and NTF of a CT DSM is not as straightforward due to the mixed-signal nature of the loop filter. However, the modulator in Fig. 2.10 can be rearranged to separate the DT and CT parts: by shifting the CT loop filter and the S/H operation over the summation point to upfront the loop and into the feedback loop, resp., the system can be split into a CT part $H_{ct}(s)$ upfront the loop, and an all DT loop-filter part [4, 51–55]. The resulting block diagram is shown in Fig. 2.11a. Therefore, the equivalent DT loop filter

$$H_{dt,eq}(z) = \mathcal{Z}\left\{\mathcal{L}^{-1}\left\{H_{ct}(s)\cdot H_{dac}(s)\cdot q\cdot e^{-\tau_{ELD}\cdot s}\right\}\Big|_{t=nT_s}\right\} \tag{2.31}$$

can be derived [[54], Eq. 3]. It accounts for the behavior of the loop filter $H_{ct}(s)$ and also includes the behavior of the feedback DACs. It is based on the idea that both the feedback signal into the loop filter $v[k]$ and the sampled output of the filter $y[k]$ are in the DT domain, cf. Fig. 2.10. Thus, if the impulse responses are identical at every sampling point $t = nT_s$, the DT and CT filters behave the same. With the equivalent filter, the NTF of the free-running CT modulator can be expressed as in (2.24) by

$$NTF_{ct,dsm}(f) = \frac{1}{1 + q\cdot H_{dt,eq}(e^{j2\pi f})} \tag{2.32}$$

and transformed to the frequency domain by using the DTFT. The STF of the CT DSM can be calculated by concatenating the equivalent DT NTF with $H_{ct}(j2\pi f)$, which is the CT prefilter in Fig. 2.11a.

$$STF_{ct,dsm}(f) = \frac{q\cdot H_{ct}(j2\pi f)}{1 + q\cdot H_{dt,eq}(e^{j2\pi f})}. \tag{2.33}$$

For a first-order CT DSM, the loop filter $H_{ct}(s)$ is an integrator with corner frequency $\omega_I = f_s$ [4]:

$$H_{ct}(s) = \frac{fs}{s}. \tag{2.34}$$

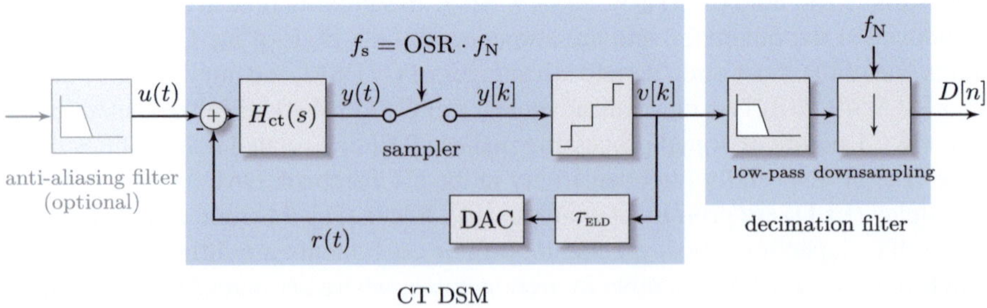

Fig. 2.10 Block diagram of a CTDS ADC with L^{th} order loop filter $H_{ct}(s)$. Also, a typical nonideality of the CT DSM is shown, which is unwanted ELD τ_{ELD} in the feedback loop. ©2021 Wagner. Reprinted, with permission, from Wagner [27].

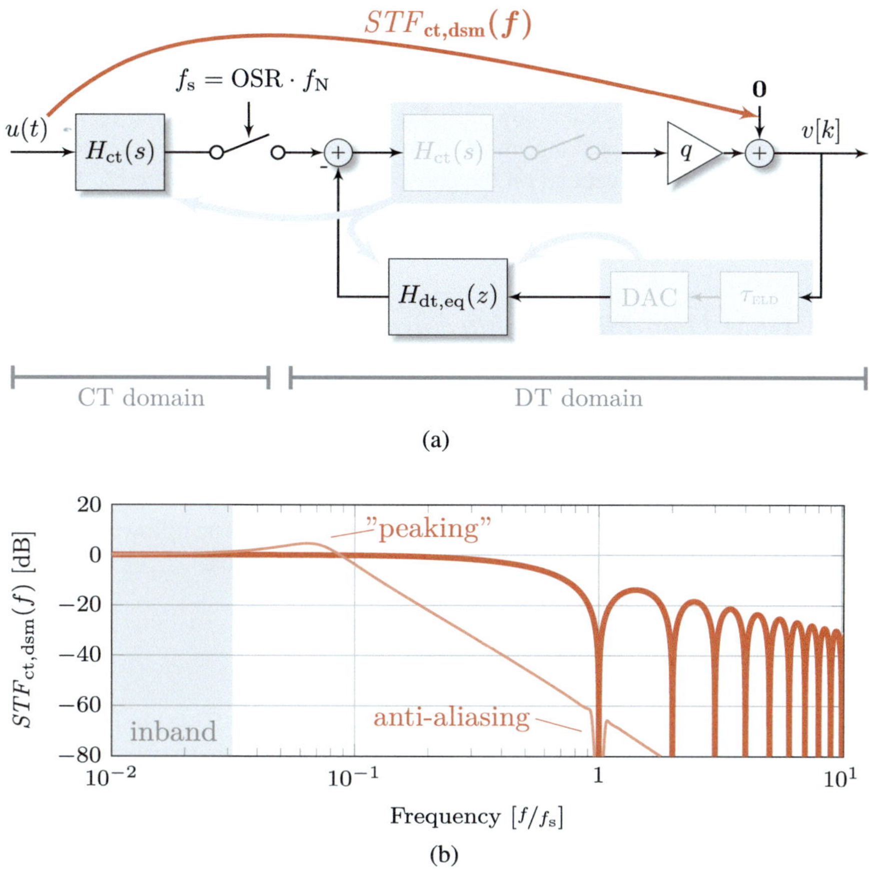

Fig. 2.11 (a) Model of a CT DSM for STF calculation and (b) STFs showing the anti-aliasing behavior for a first-order (thick) and third-order modulator (thin). ©2021 Wagner. Reprinted, with permission, from Wagner [27].

If $H_{dt,eq}(z)$ is made identical to the DT loop filter $H_{dt}(z)$, as, e.g. in (2.26), the resulting NTF of the CT DSM is identical to the DT case in Fig. 2.9 and is therefore not shown. The CT STF, however, is majorly different. Its mixed-signal nature yields the characteristic of a CT low-pass filter multiplied with the periodicity of a DT filter. As seen in Fig. 2.11b, it shows a continuous roll-off and a deep notch around f_s. The exemplary first-order CT STF has an overall sinc-like shape with notches at multiples of f_s, suppressing the aliases in the decimation process. As stated as early as in [47], this behavior is referred to as implicit AAF and has the major advantage that a dedicated AAF is typically not required or highly relaxed. Furthermore, not only is the AAF omitted, but depending on the application, some of the otherwise required front-end or channel-select filtering can be implicitly realized by the CT DSM [56].

With an increasing CT loop-filter order, the implicit STF filtering becomes more prominent, as is shown for an exemplary third-order modulator plotted alongside a first-order one in Fig. 2.11b. Noteworthy, alongside the better OOB attenuation of the STF due to the higher-order loop filter, Fig. 2.11b also indicates that such higher-order STF can result in peaking, most often happening at OOB frequencies. This is known in literature to depend on the chosen loop-filter architecture [2, 4, 5], for which we give an overview next.

2.3.2 DSM Loop-Filter Architectures

When it comes to the implementation of higher-order NTFs, plenty of different modulator topologies and architectures have been proposed throughout literature, ranging from single-loop, multistage noise shaping (MASH), sturdy MASH (SMASH), and other more sophisticated topologies. Realizations can be found in DT and CT domain and are discussed in more detail in literature as, e.g. in [4, 5]. The following is a brief overview to the extent necessary for the remainder of this book.

2.3.2.1 Single-Loop DSM Architectures

The obvious extension of a first-order loop filter is to replace $H_{\mathrm{dt}}(z)$ in Fig. 2.7 by a higher-order filter. We have already discussed in Sect. 2.3.1.1 that a concatenation of more than one integrator results in implicit instability. A feasible higher-order loop filter is the chain of integrators (CoI) filter. There, every state variable is controllable (by the feedback loop) or made visible (towards the output), and the loop-gain is reduced for BIBO stability. Two classic single-loop architectures can be derived from this.

Distributed feedback topology (CIFB): Even before Candy's often cited paper on the second-order modulator [47], in [57] the incorporation of a higher-order loop filter had been proposed, where a cascade of integrators is realized in the forward path of the DSM loop, with each integrator receiving a weighted feedback path via a DAC. The architecture is illustrated in Fig. 2.12 for an exemplary third-order CIFB loop filter. Nonunity forward gains c_i can be used for internal swing scaling, but they are linearly dependent with the feedback weights a_i. In order to achieve a unity STF, the input signal gain $b_1 = a_1$, if a_1, b_1 are either realized as the same SC circuit in a DT DSM or if a_1 is realized as non-return-to-zero (NRZ) DAC in a CT DSM. For other DAC realizations and thus feedback waveforms $r(t)$, the feedback gains must be adapted [4]. Note that in practice each feedback path is realized with a differently scaled DAC$_i$. This CIFB DSM implementation comes with one major advantage, but three distinct disadvantages:

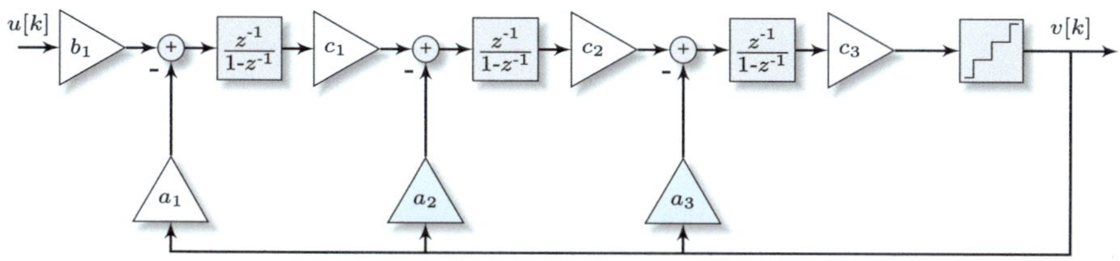

Fig. 2.12 Third-order DT DSM with feedback compensation (light blue) to realize a CIFB architecture. ©2021 Wagner. Modified, with permission, from Wagner [27].

- In addition to the filtered quantization noise, the integrator outputs also contain a significant amount of the input signal [2]. Since the integrator input signals in a stable system are controlled to be zero over time, the feedback signal and each integrator output must be equal over time. Since in a DSM the output, and therefore the DAC feedback, follows the input signal over time, the integrator outputs must carry a significant amount of the input signal. The integrators therefore need to allow significant swings.
- Since the outermost loop and the scaling coefficient realize the highest order of the loop filter, its scaling is usually very small for stability reasons, which places strict requirements on its circuit design, e.g. in terms of noise: small scaling coefficients result in SC integrators with small sampling capacitors or CT integrators with large resistors, both of which cause higher circuit noise contributions.
- In DSMs, errors inserted within the loop filter are effectively suppressed by the preceding part of the loop filter. As a_2 is in practice implemented with a separate DAC, especially for DSM with low OSR, nonidealities introduced by this path are not well attenuated. Thus, for example, DAC nonlinearity is still important for DAC_2.
- Advantageously, the CIFB architecture has a very low tendency to peaking, and Butterworth-characteristic, maximally flat roll-offs are achievable for its STF. This is very advantageous if strong OOB signals are present, such as in receiver scenarios [58], which could overload the whole DSM in case of peaking at frequencies with large OOB interferers.

Feedforward topology (CIFF): While the CIFB makes the internal states controllable via feedback, the CIFF architecture makes the states visible towards the output via forward paths d_i. An exemplary resulting CIFF loop filter is indicated in Fig. 2.13. The forward gains c_i can again be used for internal swing scaling, but they are again linearly dependent on the feedforward weights d_i. The differences to the CIFB DSM are basically the opposite of the above summary:

- As no feedback DAC is inserted into internal nodes, the integrator outputs do mostly process quantization noise and thus swing requirements are highly relaxed.
- The first-order loop is realized through the outermost integrator, and the highest order through the chain including the innermost integrator. This yields that the

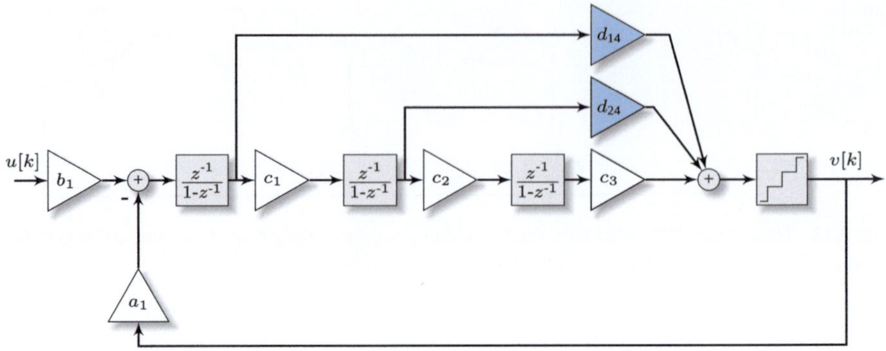

Fig. 2.13 Third-order DT DSM with feedforward compensation (dark blue) to realize a CIFF architecture. ©2021 Wagner. Modified, with permission, from Wagner [27].

scaling coefficient for the outermost integrator is much higher as in the CIFB architecture with advantages for scaling the passive components.

- As there is no internal DAC, only the nonidealities of the outermost DAC matter.
- The major disadvantage of the CIFF architecture is its tendency to a peaking STF which is, e.g. critical in receiver applications [58] with strong OOB signal content.
- Moreover, the CIFF architecture requires a summing point for all forward paths which is often realized with an active adder in the SoA. As this adder is placed inside the loop, its bandwidth and dynamic requirements are typically high. Alternatives are summation at the quantizer [59] or capacitive coupling into the last integrator [60].

In addition, the two architectures in Figs. 2.12 and 2.13 can be identically converted into each other to realize identical loop filters $H_{\mathrm{dt}}(z)$ and NTF [4, 5].

Performance prediction of scaled free-running DSMs: To take into account the scaling coefficients for the estimation of the IBN, (2.30) has to be adapted with the coefficients as in Figs.2.12 or 2.13 to

$$IBN = \frac{\pi^{2N}}{(2L+1)} \cdot \frac{\Delta^2}{12} \cdot \frac{1}{q^2} \cdot \frac{1}{a_1^2 \prod_{i=1}^{L} c_i^2} \cdot \left(\frac{1}{OSR}\right)^{(2L+1)} \tag{2.35}$$

based on the work in [61]. The denominator of the newly added term multiplied by the quantizer gain is the loop gain of the modulator. Using single-bit quantization, the last scaling coefficient is compensated for by the quantizer gain [4], Eq. 2.25]. By increasing it, the suppression of the IBN is higher. However, this can decrease the stability up to instability of the modulator, as discussed in Sect. 2.3.4.

Local resonators: Higher-order loop filters achieve their improved inband noise suppression by realizing more zeros in the NTF by more integrators in the loop filter. In the example in Sect. 2.3.1.1 and Fig. 2.9, all NTF zeros were located at DC, i.e.

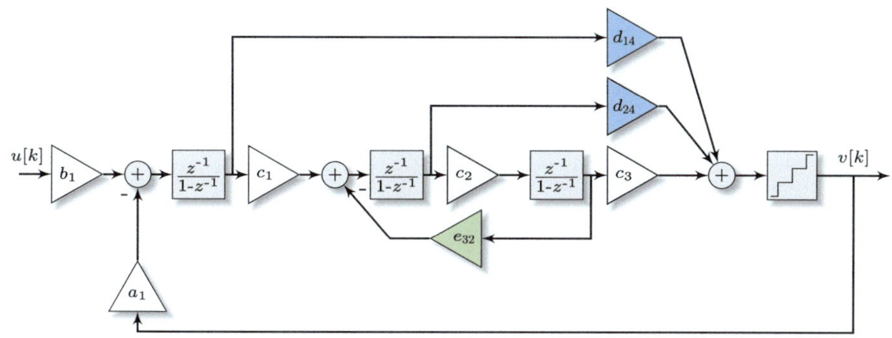

Fig. 2.14 Third-order DT DSM with feedforward compensation (dark blue) and local resonator (green) to realize a CRFF architecture. ©2021 Wagner. Modified, with permission, from Wagner [27].

$f = 0\,\text{Hz}$, $z = 1$. This raises the question, if those zeros can also be optimally placed by spreading them over the inband. This can practically be achieved by combining pairs of integrators to a resonator, which realizes a pair of complex zeros in the NTF and then leads to notches in the inband of the NTF. Such a local resonator can be realized with a local negative-feedback loop around a pair of integrators as illustrated for the CIFF architecture from Figs. 2.13 in 2.14 using the coefficient e_{21}. The architecture is then referred to as cascade-of-resonators with feedforward compensation (CRFF) [2]. Similarly, a CIFB loop filter extended by a local resonator is called cascade-of-resonators with feedback compensation (CRFB). The local feedback can be placed around each pair of integrators, whereby this placement has an impact on the practical circuit realization and STF. Also, for very large OSR, the resonator scaling coefficient tends to very small values which makes its robust realization more difficult [2]. Even though an analytical derivation does not result in a handy expression similar to (2.30), it can be shown numerically that the distributed zeros indeed yield a further improvement of the achievable inband noise suppression [2, 62].

Similar to Fig. 2.9, an exemplary third-order CRFF DSM is simulated and its NTF shown in Fig. 2.15. The three NTF zeroes are optimally spread which places one at DC and the remaining conjugate complex pair of zeros into the inband. Between DC and the resonator frequency, the NTF only shows first-order noise-shaping behavior, as expected by the single DC-zero, while third-order noise shaping is obtained at frequencies higher than the resonator. Overall, the optimal placement of zeros including a resonator minimizes the IBN.

Distributed signal feed-in paths: By realizing a DSM as in Fig. 2.10 or specifically as in Figs. 2.12 or 2.13, the actual loop filter for the signal and noise transfer functions are identical. Thus, if a DSM is designed for a certain NTF, the STF is determined as well. One possibility to overcome this drawback is to include signal feed-in paths, i.e. feedforward coefficients from the input signal to internal integrator output nodes [62]. This is illustrated in Fig. 2.16 for the CIFB architecture from Fig. 2.12, now extended

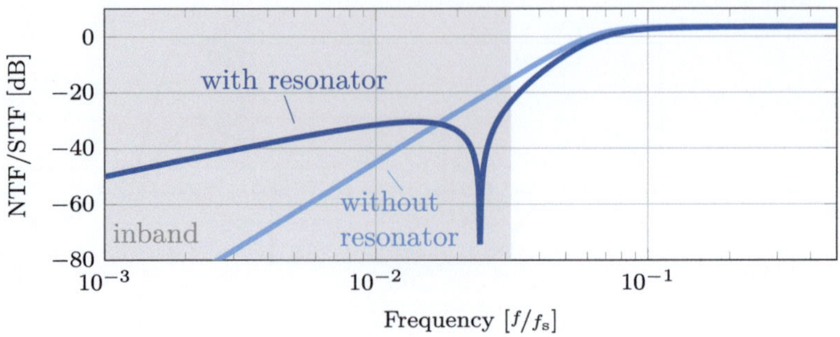

Fig. 2.15 NTFs for a third-order DT DSM with (blue) and without (light blue) resonator. The modulator with resonator shows a notch close to the band edge. ©2021 Wagner. Modified, with permission, from Wagner [27].

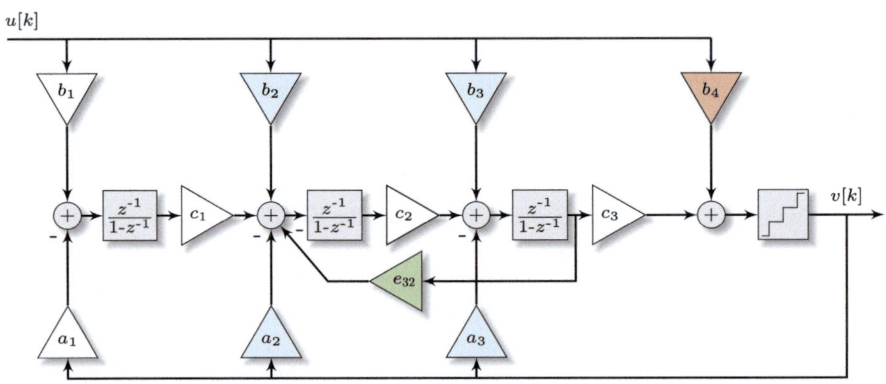

Fig. 2.16 Third-order DT DSM with CRFB architecture and distributed feed-in paths ($b_{2..4}$) to modify the STF.

by a resonator e_{21} and the feed-in paths b_i. The same distributed input paths can also be used in CIFF loop filters. Therewith, the STF becomes somewhat independent from the NTF and both can be separately optimized [2].

This has, e.g. been used in [63] to avoid peaking in a CT CIFF DSM.

Another very prominent implementation of a feed-in path was introduced in [64], where (next to the usual input path b_1) only the direct innermost feed-in path from the input signal to the input of the QTZ (b_4 in Fig. 2.16) was proposed for a DT CIFF DSMs. It has been shown that the DSM loop filter then processes only the quantization noise, since the signal contained in the output of the quantizer is fed in by the forward path and not by the DSM loop filter. This architecture is called the "*low-distortion*" topology. It can be shown that by choosing a DT CIFF architecture including a unity gain scaled low-distortion path, the STF of the DT DSM becomes unity. Interestingly, [64] claimed to achieve unity STF independent of order and OSR of the DSM, but did not mention the internal QTZ bit width and only simulated a multi-bit example. By adding a unity direct path from input $u[k]$ to the quantizer input in Fig. 2.8b, yields

$$STF_{dt,dsm}|_{LD} = q \cdot \frac{1 + H_{dt}(z)}{1 + q \cdot H_{dt}(z)} \tag{2.36}$$

As it was claimed by [64], $STF_{dt,dsm}|_{LD} = 1$ for $q = 1$, i.e. for multi-bit DSM [65], whereas for $q \neq 1$ as in the case for a single-bit DSM, a unity STF is not achieved at all frequencies.

Mixed-CIFF/FB and single-Opamp loop filters: For completeness, it shall be mentioned at this point that higher-order single-loop architectures have also been implemented with other than the shown CIFB or CIFF architectures. To combine advantages from both, a mixed feedback/feedforward (FB/FF) architecture was proposed in [66] and later used manifold, e.g. in [44, 67]. It most often replaces the second feedback path (a_2 and thus DAC_2 in Fig. 2.12) from the CIFB architecture and inserts a single feedforward compensation path instead. Thereby, STF peaking is kept relatively small, while the swing and thus linearity requirements of the first integrator of the CIFB architecture are simultaneously relaxed.

Higher-order loop filters have also been realized without the classical chain-of-integrator implementation. The idea has become prominent under the term of single-opamp resonator (SOR), or single-opamp loop filters. In early SoA, SOR were merely promised as being a low-power solution since the sheer number of amplifiers is reduced. But as especially the first and the last integrator in a loop filter have utmost different specifications [68, 69], it was doubtful if this would truly yield competitive designs. Still, many successful DSM have been presented using SOR, e.g. [70, 71], some even achieving among the best efficiencies [72, 73].

2.3.2.2 Single-bit Versus Multi-bit Architectures

We have seen that the DSM loop filter firstly spreads the quantization noise by oversampling and then high-pass filters it by noise-shaping. While increasing the OSR also increases the overall sampling frequency, and thus the power consumption in the decimation filter, reaching impractically high operating frequencies for wideband operation, increasing the loop-filter order comes at the cost of reduced stability, worse loop gain, and lower MSA. The third option for performance improvement is to reduce the quantization noise in the first place by employing an internal multi-bit quantizer in the DT and CT DSM in Figs. 2.7 or 2.10 [2, 74]. The intrinsic resolution increases proportionally to the number of quantizer levels, because with higher quantizer resolution the quantizer step width Δ and hence the quantization noise power decrease proportionally in accordance to (2.13).

Perhaps even more importantly, employing a multi-bit internal quantizer offers additional advantages. First, multi-bit modulators tend to make higher-order loop filters more stable, as the internal quantizer gain can be approximated to be unity, and the requirements for loop-gain scaling is reduced. Thus, almost ideal higher-order noise-shaping performance is obtained for modulator orders of two to four [75] employing multi-bit quantization. Second, the MSA of multi-bit DSM [76] is typically significantly higher than that of single-bit DSM [34], which directly translates into relaxed input referred thermal noise requirements and thus lower power consumption. Third, as a multi-bit feedback DAC resembles the analog input signal more closely, their difference, which is processed by the loop filter, becomes lower and thus the dynamic requirements of the loop filter can be relaxed. Fourth, when using CT loop filter implementations, the use of NRZ multi-bit feedback DACs also reduces clock jitter sensitivity [4, 5].

A key drawback is that the feedback DAC becomes a significant bottleneck in multi-bit DSM architectures. Even though of much lower resolution, it requires a linearity equal or better than the required overall modulator linearity [77]. This is due to the DAC errors being directly fed into the input of the modulator as can be seen, e.g. in Fig. 2.10. For a single-bit internal quantizer, this problem does not arise because a two-level DAC is intrinsically linear. Besides noise, static offset and gain errors occur here, to which a DSM is highly tolerant [77]. Dynamic errors, such as finite decision time and metastability, can be partially handled by delayed latching and ELD compensation [78].

Concerning this critical problem of DAC nonlinearity, there exists a vast amount of publications dealing with techniques to linearize the feedback DAC in multi-bit DSM, based on design consideration [79], dynamic element matching [80, 81], or digital and analog correction techniques operating in the fore- or background [2, 8, 67, 75, 82–101]. The analog or digital correction techniques mostly rely on finding the underlying DAC nonideality (static and dynamic element mismatch), and then correct those mismatches either in the digital [83, 88, 90, 91, 95, 96, 99] or analog domain [89, 93, 98, 100, 101]. They usually rely on LUTs to store the correction for DAC errors [83, 88, 90, 91, 95, 96, 99]. Such linearization techniques cost more or less complex circuitry, causing an increase in area and power consumption as well as switching activity. The interested reader is referred to the published work.

Concerning the quantizer itself, especially in the case of low speed and large OSR DSM, many SoA publications consider the multi-bit quantizer to be unproblematic and do not even explain its nonidealities. Obviously, for large OSR and higher-order loop filters, the nonlinearity of the quantizer is heavily suppressed. The implementation of the internal multi-bit quantizer can though become complex in wideband, high-speed DSM because of the required short decision time. Also, low OSR and high-linearity designs have employed, e.g. dynamic element matching (DEM) at the quantizer [44, 102]. The internal DSM quantizer has mostly been realized as Flash ADC, but also tracking ADCs [103], (noise-shaped) SAR ADCs [37, 38], and voltage controlled oscillator (VCO)-based quantizers [104] have been used.

2.3.2.3 Multi-Stage Architectures

Increasing the order of the loop filter of single-stage modulators to higher than two results in issues concerning stability, nonideal noise-shaping by reduced loop-gain (scaling), and reduced MSA. As an alternative, cascaded or MASH topologies have been proposed as early as in [105] with the intention to achieve almost ideal high-order noise shaping combined with the stability and MSA of low order, single-stage modulators.

As illustrated in Fig. 2.17, the idea is close to that of two- or multi-step ADCs: the error of the quantization in the first stage is extracted and requantized by a second stage. This can be repeated for more stages. Each stage typically consists of first or second-order DSMs, which assures stability and high MSA. With an appropriate digital recombination of DSM outputs, the quantization noise of all but the last stage can be canceled. Famous examples are the three-stage (1–1–1) MASH DSM [106],

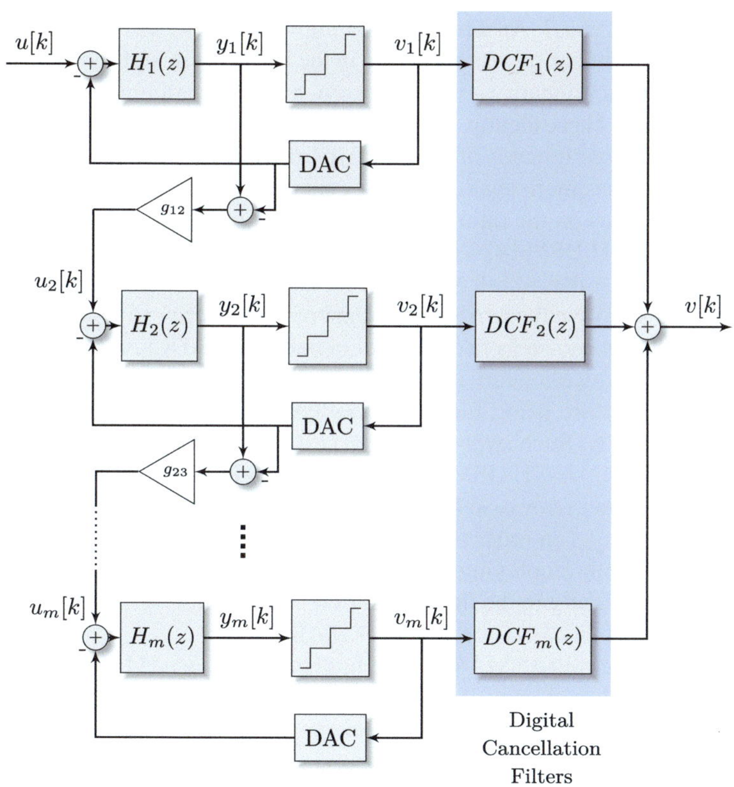

Fig. 2.17 Generalized architecture of MASH DSM with m stages, where DCF_i are digital cancellation filters.

the second-order first-order (SOFO, 2–1) DSM [107], the fourth-order three-stage (2–1–1) DSM in [108], or the Leslie-Singh architecture which employs an Xth-order DSM in the first, but a Nyquist ADC in the second stage and is thus called X-0 MASH DSM [109].

The well-known drawback of such cascaded DSMs is mismatch between the analog loop filters and the digital cancellation filters (DCFs) in Fig. 2.17, which yields leakage of (less-shaped) quantization noise from earlier stages into the overall output. It can be shown that the digital filters need to resemble the analog NTF and STF [2, 4, 5]. This becomes very critical for CT implementations of MASH DSMs, as the employed CT RC-filters are prone to process and temperature induced mismatch [4], and also for wideband MASH DSMs, as also the influence of amplifier dynamic nonidealities must be compensated [7].

Obviously, MASH architectures can be combined with the idea of multi-bit quantization. Besides the intrinsically reduced quantization noise, also higher MSA and more aggressive NTF scaling can be achieved. When multi-bit quantization is employed in all stages of a MASH DSM, the challenge of DAC linearization remains as outlined in Sect. 2.3.2.2. An advantageous implementation was proposed in [110], where multi-bit quantization was only used in the last stage of a MASH DSM. There, the multi-bit DAC nonlinearity is of no concern, as the DCF of higher stages features high-pass filtering. Then, ideally, only the high-order shaped multi-bit quantization noise of the last stage remains after the digital recombination. Disadvantageously, leakage then becomes more challenging, as generally the required analog-digital filter matching depends on the ratio of the IBN of the first stage alone versus the IBN of the overall MASH DSM [4]—thus the lower the order and quantizer bitwidth of the first stage against the overall MASH order and the quantizer bitwidth of the last stage, the more sensitive the MASH is against noise leakage.

Similar to what is known from other two- or multi-step ADCs, where the quantization error (or the residue) of earlier stages is extracted for requantization, the residue can be amplified before being further processed. This is illustrated in Fig. 2.17 by scaling coefficients g_i. Such *inter-stage gain (ISG)* can be implemented between the individual stages of a MASH DSM. But as long as single-bit quantization is used, the extracted quantization error is so large in magnitude such that the ISG is actually an attenuation with $g_i \leq 1$ in order to avoid overloading of the following stages [111]. On the contrary, with employing multi-bit quantization in earlier stages, ISG can actually add significantly to the final resolution [7], but with the disadvantage of the now necessary DAC linearization in earlier MASH stages.

More recently, so called SMASH DSM has been introduced [112]. The main idea is to eliminate the digital cancellation filters of the conventional MASH DSM and thus to eliminate the problem of noise leakage by analog-digital filter mismatch. It does so by feeding the output of the second stage into the loop filter of the first stage. By carefully looking at the resulting architecture, this resembles a single-loop DSM with an internal quantizer being a 0-X noise-shaping quantizer, as illustrated in Fig. 2.18 [113].

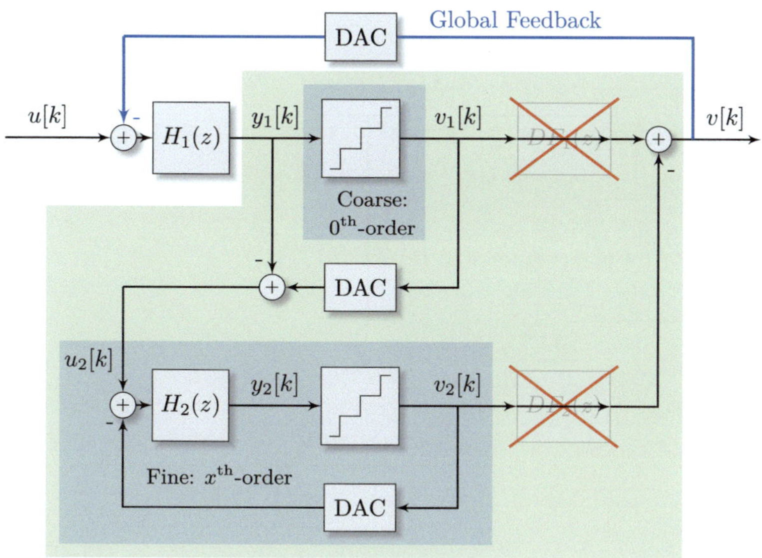

Fig. 2.18 SMASH DSM with two stages. Green: the 0-X quantizer embedded within the first stage single-loop DSM.

A disadvantage of the architecture is that in SoA implementations the required adder is not realized inside the loop, but the adder is split into an out-of-the-loop digital adder, as well as an additional analog adder from the output of the second stage to the overall input. This advantageously omits the in-the-loop digital adder, but disadvantageously any error of the additional DAC contributes again directly to the input and noise as well as linearity need to be concerned. Even more critically, the SMASH architecture thereby becomes highly sensitive to any mismatch between the two feedback DACs [114].

2.3.3 DSM Employing FIR DACs

Figure 2.19 shows the concept of implementing a finite impulse response (FIR) DAC in the feedback of a DSM. The concept was proposed as early as [103, 115, 116] and has recently become popular [117], e.g. for single-bit DT [115, 118, 119], single-bit CT [120], 3-level CT [121], and multi-bit CT loop filters [102, 122].

The FIR filter $F(z)$ reduces the OOB quantization noise of the DSM output signal $v[k]$ before it is being fed back into the loop filter. This process smooths the feedback signal which consequently follows the input signal more closely. The advantage of employing a FIR feedback is twofold: first, it reduces the residue of input minus feedback signal which yields reduced requirements for signal swings in the subsequent

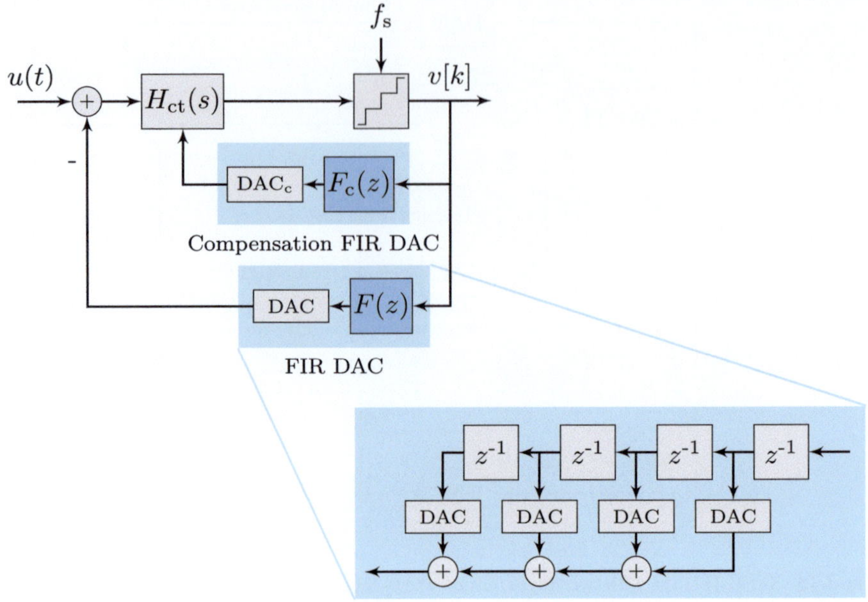

Fig. 2.19 A single-bit DSM with FIR DAC $F(z)$ in the outermost feedback path and compensation FIR DAC $F_c(z)$ to restore the original NTF.

loop filter, and hence relaxed opamp specs for swings. Second, the reduced step size of the FIR-feedback DAC also reduces clock jitter sensitivity. An efficient implementation of the FIR DAC is done in a mixed-signal manner with analog weights and digital tap delays as indicated in Fig. 2.19, where for the sake of simplicity equally weighted FIR taps are shown. Advantageously, element mismatch between individual weights of the FIR taps introduces no nonlinearity, but only (slightly) modify the FIR filter transfer function. This makes the FIR feedback especially advantageous in single-bit DSM. There, an N_{FIR}-tap FIR filter $F(z)$ reduces the step size of the feedback DAC by N_{FIR}, which resembles a $\log_2(N_{FIR})$-bit multi-bit DAC, but without introducing its commonly concerned nonlinearity.

The additional delay caused by the FIR feedback in the outer feedback paths results in a modified NTF and reduces loop stability. In order to avoid that, a compensation FIR DAC $F_c(z)$ is usually employed [117], which is implemented in any of the inner feedback paths into the loop filter. Effectively, the loop filter behind the insertion point of the compensation FIR is restored to its original form, and thus the original NTF is restored as well [115, 123]. The advantages of the FIR DAC are seen in the loop-filter part, which is enclosed between the FIR DAC and the respective compensation FIR DAC. Note that the farther down the loop filter the compensation FIR DAC is realized, the more its nonidealities are suppressed by the loop filter, but the larger and more aggressive its scaling coefficients become. The coefficients of the compensation FIR DAC $F_c(z)$ can be found by impulse-invariant transform (IIT) [117].

It is important to recognize that even though the compensation FIR $F_c(z)$ restores the original NTF, the STF of a DSM with FIR DAC is not restored and is modified depending on the number and actual scaling of FIR taps [120, 124].

2.3.4 DSM Stability Analysis—State of the Art Overview

As we discussed throughout Sects. 2.3.1.1 and 2.3.2.1, higher-order loop filters are only conditionally stable, and one needs to choose appropriate architectures and scaling coefficients to make a compromise between MSA, aggressiveness of noise shaping and IBN. Consequently, stability considerations of the closed-loop DSM is of utmost importance. Loop-filter stability is defined in the SoA as the condition, where all internal state variables remain bounded over time [125] for a bounded input signal (BIBO-criterion). As discussed in Sect. 2.3.2.1, a conditionally stable, higher-order loop filter needs to have all internal state variables being controllable or observable, for which CIFF, CIFB or its variants are the most common implementations. Afterwards, a simple measure to ensure stable operation in a possibly unstable control loop is the reduction of the loop-gain by appropriate filter scaling.

The state of the art for the design of higher-order DSMs and their NTF is based on some classic literature such as [76, 126–128], which focus in particular on the very nonlinear, single-bit converters. Therein, the authors describe ad-hoc rules for the design of stable, higher-order modulators. The procedures are based on the consideration of various NTF norms and the concomitant analysis of the quantizer input. *Lee* argues in [127]—based on simulations—that $\|NTF\|_\infty < 2$ provides a necessary condition for a stable fourth-order modulator under zero-direct current (DC) input, which is why he also postulates that for finite input signals $\|NTF\|_\infty \ll 2$. This led to the often cited *Lee's rule*:

$$\|NTF(z)\|_\infty = \max_{\omega\in[0,\pi]} \left(\left|NTF\left(e^{i\omega}\right)\right|\right) < 1.5, \tag{2.37}$$

meaning that the maximum out-of-band gain (OBG) of the NTF should be smaller than 1.5. Today, this rule is often misinterpreted as a guarantee of stable behavior if $\|NTF(z)\|_\infty = 3$ dB is maintained, but Lee did not claim this in his publication. Unfortunately, his original thesis [129], which is often referenced, is not publicly accessible. In contrast to Lee, [126] used the 2-norm and postulated $\|NTF(z)\|_2^2 < 3$ as criterion for stable behavior. In [128], this was tested by *Schreier* for different modulators under different input conditions, and he found by simulation that Lee's criterion is too conservative for second-order modulators, appears to be approximately correct for third-order modulators, and is however too lax for higher than fourth-order modulators. Schreier found something similar for the 2-norm based rule, which is why he postulated that one cannot prove stability on the infinity or two-norm. One should clarify though that the authors of the original papers did not claim this either. Another example is the use of the 1-norm according to [130, 131], who justifiably introduce a general stability test as $\|NTF(z)\|_1 < 3 - u_{\max}$ (where $u_{\max} = MSA$). However, Schreier again shows in [128] that this condition is so conservative for single-bit converters that none of the modulators he describes as stable by simulation pass the test.

Baird and Fiez [132] proposed the method of root-locus plots known from LTI stability analysis, but adopted it for single-loop, single-bit DSM: the effective quantizer gain q from (2.11) in Sect. 2.1.2, which is also present in the DSM block diagram in Fig. 2.8, is used as the variable gain of the root-locus. It turns out that the location of the closed-loop poles is inside the stable region (unit circle), as long as proper loop-filter scaling is chosen and the effective quantizer gain does not drop below a critical value, which is equivalent to omitting excessive quantizer overload by a too large input signal. An example is found in [4], Fig. 2.19].

A more sophisticated measure for stability and reliability of DSMs was presented by *Risbo* in [125, 133]. It allows to analyze the stability of higher-order, single-bit modulators based on the quasi-linear modeling of a DSM [34] and a more realistic noise model. Risbo thereby introduced the noise power amplification factor $A(q)$, which connects the variance at the input of the quantizer with the variance of the quantization error depending on the quantizer gain. He showed that minimizing $A(q)$ leads to an increased MSA. He further proposed two reliability measures. Even though the method is very powerful, it is almost never referenced or used in more recent SoA, most likely as Lee's rule comes much more *handy*.

For multi-bit converters, whose consideration is simpler because of the much lower correlation between quantization error and stimulus, the more accurate uniform distribution of quantization noise, and the much better description of the quantizer by a linear model, the 1-norm condition of *Kenney and Carley* [130] is usually adopted and leads to very accurate predictions.

Even more recently, further work was shown to assess the stability analysis and optimization for DSM. For example, [134] proposed a stability analysis for single-bit modulators. The solution of a mixed integer nonlinear program (MINLP) based on a mixed-logical dynamical (MLD) system model of a DSM allows an exact stability prediction over a finite time horizon. The work of Zorn et al. is based on a switched system model and allows to optimize the MSA of the DSM by rescaling the internal system dynamics and thus improve the stability of the system [135–137]. Rombouts et al. describe the stability analysis of multi-bit DSMs based on the multiple-input describing function analysis, which allows to predict the overloading level of the internal quantizer [138]. Lota et al. propose the prediction of the MSA for higher-order, single-bit DSMs based on a quasi-linear model with nonlinear feedback control [139], which was extended for the multi-bit case in [140].

Despite such ongoing research, the topic of NTF design and optimization via the above listed ad-hoc rules is highly relevant today, not to say they are virtually the only methods used in practice, despite heuristic optimizers, which will be looked at in the next Sect. 2.4. Even though there is also much literature on nonlinear [141] and quasi-linear [34] DSM analysis, and such analysis provides more realistic results, in particular the internal dynamics for DSMs of higher order and even more when considering nonidealities are still seen as an unsolved problem in the SoA [5].

2.4 Design Flow and Loop-Filter Scaling of DSMs

So far, we have reviewed the functionality of DSM, architectural choices, the necessity to stabilize higher-order modulators by loop-filter scaling, and a SoA overview was given on the theory of conditional stability of DSM. Next we discuss, how we actually find a suitable architecture and its scaling coefficients.

2.4.1 General Design Flow

A typical iterative design process is illustrated in Fig. 2.20, where various steps for the analysis, synthesis, modeling, simulation and optimization are required. These techniques can either be manually executed by a designer or partially/completely

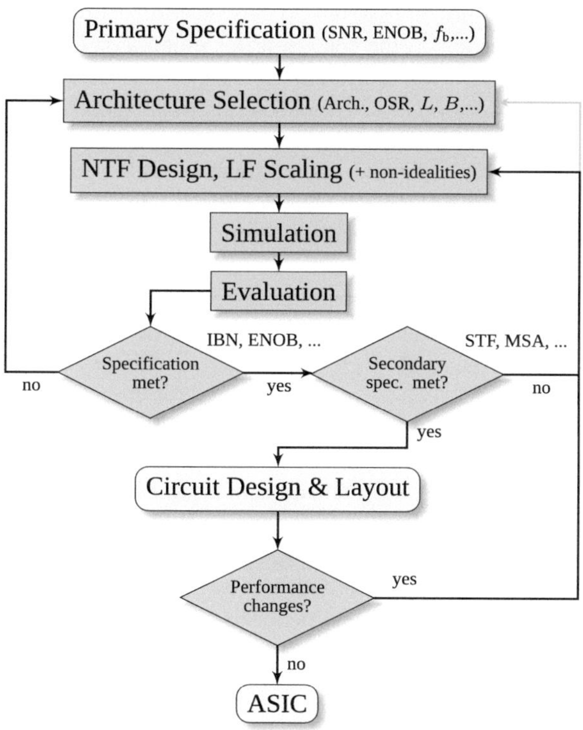

Fig. 2.20 Design process of a DSM from specification to circuit-level as in [142] with L the order of the DSM, B the number of quantizer bits, and OSR the oversampling ratio. ©2021 Wagner. Reprinted, with permission, from Wagner [27].

automated [142]. Nonetheless, the individual design steps are similar in both ways and thus generally described next. Given a required core specification, at the beginning of the design a suitable modulator architecture has to be found. The major parameters are the effective number of bits (ENOB) or SQNR, and signal bandwidth f_b. Experience or literature based decisions yield the selection of a single- or multi-bit, single- or multi-stage, DT or CT implementations, etc. The design space for loop-filter order L, OSR, and internal number of bits B must be explored. This can, e.g. be done using ideal estimations of the achievable performance based on generic equations of NTF and IBN, cf. Sect. 2.3.

2.4.2 Modulator Scaling and Optimization

With a suitable DSM architecture being selected, the loop filter needs to be synthesized on system-level i.e. the scaling coefficients of the loop filter must be found. Over the last decades, many approaches to obtain these scaling coefficients were reported. Even before Candy's paper on the second-order modulator [47, 57] proposed the incorporation of a higher-order loop filter by means of a chain of integrator in the forward path of the DSM loop filter with each integrator receiving a weighted feedback DAC, thus realizing a CIFB architecture. Those early works came with proposed scaling coefficients which had proven conditionally stable operation. Alternatively, coefficients obtained by a grid search were reported in literature as *optimal* scaling coefficients [61, 111, 143] and were subsequently adopted in many practical designs.

Probably the most common approach is based on analytically placing the poles and zeros of the NTF to desired locations [5], whereas the desired locations are however based on empirical findings such as the stability criteria of Lee's rule for single-bit DSM, cf. (2.37), or the design approach by Kenney and Carley [130] for multi-bit DSM. For both, *Schreier's Delta-Sigma Toolbox* is most popular in the DS community [144]. Therein, the NTF is designed for a DT DSM prototype, for CT implementations a subsequent transformation into CT domain is used to give the CT DSM the same NTF. Normalization of loop-filter swings can be done as well. Though, there is no design space for the STF, nor can nonideal behavior be included. For both, additional iterations are required at this abstraction level in order to fine-tune the loop-filter implementation and to maximize the performance of the DSM after including system-level nonidealities, such as, e.g. ELD [145] or finite gain-bandwidth product (GBW) [146].

An alternative approach is to apply optimization on system level, e.g. using heuristic search for the loop-filter synthesis and thereby simultaneously account for non-ideal behavior [147]. For such system level optimization, various approaches have been reported over the last decades, where we only name a few in the following. These system level design tools are intended to either support the designer, or even automate parts of the design process. They integrate diverse depth of nonideal loop-filter modeling and employ diverse optimization algorithms for DT and CT DSMs at individual steps of the design process of Fig. 2.20.

Generally, such DSM design tools aim for minimizing the IBN, maximize the signal-to-noise-and-distortion ratio (SNDR), maximize the MSA, obey constraints like swings within the loop filter, etc., and achieve that by adjusting the scaling coefficients, and thus the NTF. In [148], Doboli et al. present a MINLP based optimization process for DT DSM topologies. In [149], *Anders et al.* describe a design process based on the stability considerations in [125] in order to increase the MSA of CT modulators. The tools by Gielen et al. (DAISY) and de la Rosa et al. (SIMSIDES) [150, 151] report tools to optimize parameters of the individual building blocks based on (MATLAB and Simulink) behavioral models. Recently, several new approaches using artificial neural networks (ANNs) for the high-level design of DSMs have been presented. In [152], an ANN is used together with behavioral model simulations to derive the circuit-level design variables of DSMs for a given set of specifications. In [153] the system level design of DSMs is split into three different steps, employing ANNs in two of them: at first a classification neural network is used to derive an architecture that fits best to the given specifications. Next, a regression neural network is used to derive design variables for this specific architecture. At last, behavioral simulations are performed choosing simulation points close to the design variables derived by the regression network. The best performing design variables after simulation are chosen as the final values. The resulting tool using this approach was named AISAD and presented in [154].

However, none of these tools has attained the popularity of Schreier's Delta-Sigma Toolbox, for a variety of reasons. For example most custom tools were simply not publicly available, and also many were very specialized on specific problems or using specific software. The Delta-Sigma Toolbox is available for everyone having a MATLAB license, and it can be changed, extended and flexibly used by the designer. Parts are available as Python scripts to omit the MATLAB license [155].

Brückner, Wagner, Kässer and Ortmanns proceeded a different approach by introducing a web-based design environment [147]. Originally meant for CT implementations, it supports DT/CT single-loop architectures, allows constraints and includes nonidealities, and comes with the unique advantage to account for STF design. This tool has been extended in the last years to allow band-pass and incremental DSM, various DAC waveforms, FIR-DACs, and many more. As it is based on a highly-parallel graphics processing unit (GPU) based simulation, it delivers optimized DSM scaling in real-time. This makes it very practical for both design exploration as well as fine-tuning of scaling coefficients.

An overview on some SoA design tools is given in Table 2.1.

Table 2.1 Simulation, design, and optimization tools for DSMs and I-DS ADCs. At the time of writing this book actively maintained tools are marked bold. ©2021 Wagner. Modified, with permission, from Wagner [27].

Tool Name	Platform	References	Functionality	Domain	Type	Release
Delta-Sigma Toolbox	MATLAB/C	[76, 130, 144]	sim, syn	DT, (CT)	LP/BP	1988[a]
SIMSIDES	MATLAB/C	[151, 156]	sim, opt	DT, CT	LP/BP	2005
www.sigma-delta.de	C++/ Javascript	[147, 157, 158]	sim, syn, opt	DT/CT	LP/BP/INC	2014
AISAD	MATLAB/ Python	[154, 159]	syn, opt	DT, CT	LP	2024
ASIDES	MATLAB/ HSPICE	[160]	sim, syn, opt	DT	LP	1995
DAISY	C	[161]	sim, syn, opt	DT, CT	LP	2000
DISCO	MATLAB	[162]	sim, syn	CT	LP	2007
HOST	C	[163, 164]	syn, opt	DT	LP/BP	1999
MIDAS	C++	[165]	sim	DT	LP	1992
TOSCA	C	[166]	sim	DT	LP	1993
ZSIM	Fortran	[167]	sim	DT	LP	1990
–	MATLAB/ Spectre	[168]	sim, syn, opt	DT	LP	2007
–	MATLAB	[169, 170]	sim, syn	(CT)	BP	1997
–	MATLAB	[171]	syn	DT	LP	2007
–	MATLAB	[172]	sim	DT	LP/BP	2009

sim = simulation; syn = loopfilter synthesis support; opt = application specific filter optimization
[a]Parts of the toolbox date back as early as 1988

Chapter 3
Theory of Incremental Delta-Sigma ADCs

DS ADCs use oversampling and noise-shaping in order to achieve high resolution. The core idea is to trade circuit accuracy such as component matching, which is required in many other types of ADCs, against time by oversampling, and to use the concept of control theory to reduce the inband quantization noise by spectral noise shaping. In many sensor applications, such as e.g. biomedical front ends, imagers, and multi-sensor readouts, such high-resolution ADCs are required, but they should also be capable of being multiplexed in the time domain between individual channels. Other applications require single-shot conversion without memory, and thus without inter-sample interference. While the highest efficiency for high-resolution ADCs is nowadays achieved by DS ADCs or by noise- and partially mismatch-error shaping SAR ADCs [38], such true sample-to-sample conversion at Nyquist rate is unfortunately not possible with these architectures as noise shaping introduces memory to the system.

A solution to this dilemma is the I-DS ADC which was proposed in [12, 173, 174]: even though it is based on the same principles as a conventional DS ADC, like employing a coarse quantizer embedded into a loop filter and operating with oversampling, however, all memory elements are repeatedly reset after M fast clock cycles, after which a single Nyquist output sample is generated by a digital reconstruction filter. The original authors stated that as a result of the regular reset, the proposed system would be simpler than the oversampling ADC (replacing a complicated low-pass decimation filter by a counter), would not require dithering, and could thus convert DC signals without tonal behavior. On the other hand, it would be restricted to very slowly varying signals [12]. Nowadays, incremental converters are still used for high resolution, but have been shown to operate with input signals in the MHz-range, and have—in various architectures—allowed to achieve best-in-class power efficiency. Thus, it is a mighty converter architecture and worth looking at in detail!

M. Ortmanns et al., *Incremental Delta-Sigma ADCs*,
https://doi.org/10.1007/978-3-032-09468-1_3

Interestingly, Robert et al. [173] reference the work of *van de Plassche* [11] as similar and earlier work having shown their principle. In fact, [10, 11] introduced a CT current-input free-running DSM, and combined it with a digital up/down counter to produce the digital output. That counter was only run over a predefined time and thereafter reset, thus resembling a first-order CoI filter. While the analog DSM was in fact not explicitly reset, it was claimed that due to the current-mode input and the current-mode feedback, the integrating capacitor would remain almost at a virtual ground over time and thus not have significant memory between subsequent samples; but this is only approximately true and only Robert et al. introduced the explicit periodic reset of analog memory in the DSM and also introduced the terminology for *incremental DS ADC* [173].

This chapter will introduce the functionality of the incremental operation and describe it in the time and frequency domain. The analytical derivation of an overall NTF and STF is presented which allows the calculation of performances regarding the aliasing of higher frequency bands by decimation subsequent to the digital reconstruction filter. Various types of digital reconstruction filters will be presented and compared.

3.1 Functionality of Incremental Delta-Sigma ADCs

The basic architecture of an I-DS ADC is simply derived from a conventional, free running DS ADC, cf. e.g. Fig. 2.7, and is schematically depicted in Fig. 3.1a. In order to obtain a memoryless system, one has to make sure that all memory elements within the modulator as well as within the digital filter are periodically erased, as proposed by [12]. The ADC consequently consists of an *I-DSM*, which is a conventional DSM but with all analog integrators being resettable, and which produces M low-resolution output data points at rate f_s. The low-resolution data is then fed into a digital filter, whose internal registers are also periodically reset, and where the overall output is finally decimated to Nyquist rate f_N. We should note that the term *decimation filter* is usually employed for the digital filter in a free-running DS ADC, which consists of a low-pass filter with subsequent decimation as described in Sect. 2.3.1. There, its functionality is to cancel all spectral OOB components before decimating the low-pass filtered high-resolution output bitstream. As the functionality of the digital filter in the I-DS ADC in Fig. 3.1a is different, we use the term *reconstruction filter* throughout this book.

An upfront S/H element at Nyquist rate can be optionally employed in an I-DS ADC as shown in Fig. 3.1a. Its usage or omissions has a significant influence on the signal-transfer behavior of the I-DS ADC [175]. If used, the STF becomes unity by the sampler. If omitted, the STF is influenced by the I-DSM, but dominated by the digital reconstruction filter, which will be further discussed in Sect. 3.3 and in Sect. 4.1.1.3. Note that due to the possible Nyquist-rate S/H upfront the I-DSM, we give the subsequent signal variables $u_n(t)$, $v_n(t)$, and $d_n(t)$ and index n in order to indicate that those operate on a single Nyquist input. Only the overall output variable

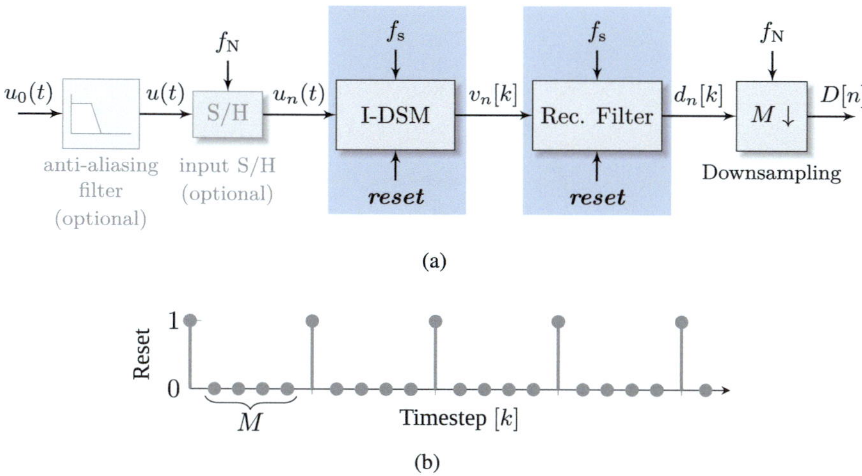

Fig. 3.1 (a) Basic block diagram of an I-DS ADC with an (optional) Nyquist-rateS/H, an oversampled I-DSM, a digital reconstruction filter and a digital output $D[n]$ giving a Nyquist sample corresponding to the analog Nyquist input sample, and (b) illustration of its oversampled operation and reset with finite reset time.

$D[n]$ is chosen the same as for the free-running DSM in Fig. 2.7, because both are high-resolution digital output data at Nyquist rate. Though $D[n]$ of the I-DS ADC is the digital representation of only a single Nyquist input sample, i.e. a true-Nyquist ADC output, while $D[n]$ of the free-running DS ADC is the digital representation of a filtered version of the continuous-time input.

Finally, the preceding AAF shown in Fig. 3.1 can be disregarded when discussing the basic functionality of an I-DSM, similar as we did in the discussion of free-running DSM. It will be shown that it actually has very different requirements in incremental operation. We omit the AAF in the following discussion, but further discuss its requirements in Sect. 4.1.4.

The timing of the reset signal of the I-DS ADC in Fig. 3.1a is illustrated in Fig. 3.1b. For I-DS ADCs, M is often called the OSR, which is in fact only true if the reset length is infinitely short. For finite reset lengths M_{reset}, the OSR for I-DS ADCs can be expressed by

$$\text{OSR} = M + M_{\text{reset}}, \tag{3.1}$$

i.e. the actual number of operational clock cycles M of the I-DSM plus the finite number of reset cycles M_{reset} gives the number of clock cycles needed for one Nyquist conversion and thus the oversampling ratio OSR of the input signal. The reset removes any memory from the system leading to independent samples at Nyquist rate at the output [176]. This finite duration of the reset time is relevant for circuit-level implementations and will be investigated further in Sect. 4.3.5. In the following, unless otherwise stated, we assume for simplicity that the reset happens in an infinitely short amount of time, which makes OSR=M. This is also assumed in most SoA.

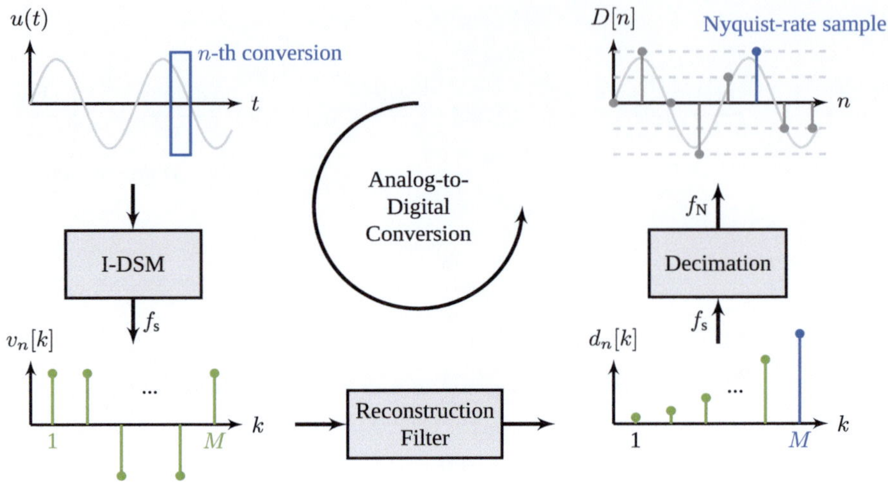

Fig. 3.2 Working principle of a single-bit I-DS ADC. ©2020 IEEE. Modified, with permission, from Vogelmann et al. [177].

Working principle of I-DS ADC: After we have introduced the block diagram of an incremental ADC in Fig. 3.1, next its working principle is explained with the help of Fig. 3.2. An analog signal $u(t)$ is fed to the ADC. It is either sampled at Nyquist rate and thereafter held constant for the remainder of the M operation cycles of the I-DSM, or the S/H is omitted and the CT input is directly fed into the oversampled I-DSM, cf. Fig. 3.1. In the illustration in Fig. 3.2, a CT sine wave is chosen as the input signal. During the course of a single Nyquist conversion, the signal therefore varies slightly, as indicated by the black frame. Figure 3.2 shows a single conversion operation resulting in exactly one Nyquist sample. Without the loss of generality it is assumed that this is the n-th conversion. The I-DSM can practically be any conventional DSM architecture enhanced by a periodic reset. It acts as an encoder that encodes the analog input signal into a low-resolution digital output sequence $v_n[k]$ of length M, where k is the current cycle extending from 1 to M. After $k = M$, an (infinitesimally short) reset takes place, erasing the memory of the modulator. The digital output $v_n[k]$ runs at internal rate f_s, which is OSR=M times faster than the Nyquist-rate sampling frequency f_N. The sequence $v_n[k]$ is fed into a digital reconstruction filter that acts as decoder. This filter processes all M samples of $v_n[k]$ and provides its last output sample $d_n[M]$ as high-resolution, Nyquist-rate output $D[n]$. Then, the analog and digital filters are reset and the conversion of the next Nyquist sample starts. The sequence $D[n]$ is the Nyquist-rate digital representation of the analog input signal.

CT I-DSM implementation: Although the majority of published I-DS ADCs have been designed and published using DT (SC-based loop filters, similar to free-running modulators), I-DSMs can also be realized in the CT domain. Relaxed settling speed requirements for the integrators and implicit signal filtering makes this type of modulator usually interesting [178, 179]. Some state-of-the-art work on CT I-DS ADCs are [180, 181] report on a first-order CT I-DS ADC. In [182], a modified CT I-DS MASH architecture is shown, where a first-order I-DSM is used for both stages. In [183, 184], the first higher-order CT I-DS ADC is published and similar designs are implemented and evaluated in [175, 185]. A design relying on a FIR filter in the feedback path is reported in [186, 187]. In [188], a CT I-DS ADC comprising a feedback DAC with hybrid mismatch error shaping is proposed. The Zoom ADC in [189] is also based on a CT I-DSM. A SMASH I-DSM with CT loop filter has been realized in [24] with 2MS/s Nyquist rate. An extended-counting CT I-DSM with high efficiency was presented in [190].

I-DS ADCs as linear time-variant system: Irrespective of the actual implementation of the DSM, both CT and DT implementations share a similar challenge: free-running DSMs could be approached completely in the frequency domain with the help of STF and NTF of LTI systems. The decimation filter could be approximated as brick-wall filter, and the inband performance could be used for highly accurate performance prediction. On the contrary, the operation of I-DS ADCs makes their analysis utmost different, as the reset in I-DS ADC changes the system from a time-invariant to a time-variant system. In order to work around the analysis of time-variant systems, different approaches can be found in literature: the most frequently used approach is the analysis in the time domain, where the system is considered as LTI within the reset window [176, 191, 192]. This will be reviewed in Sect. 3.2. Another approach regards the I-DS ADC having a finite-length impulse response, and thus analyzes them in the frequency domain as FIR filters [193, 194]. This will be discussed in Sect. 3.3.

3.2 Time-Domain Description of I-DS ADCs

3.2.1 Time-Domain Description of a First-Order I-DS ADC

Figure 3.3 shows a first-order DT I-DS ADC derived from its free-running counterpart in Fig. 2.7. The quantizer is already replaced by the linear model as discussed in Sect. 2.1.2, i.e. by a quantization noise source $e[k]$ and the effective quantizer gain q. The preceding AAF and input S/H are omitted and will be discussed separately in Sect. 4.1.4. The variable n denotes the nth Nyquist-rate sample and $k = 1 \dots M$ is the running variable of the oversampled modulator, where M samples at f_s are used to generate one sample at f_N as illustrated in Fig. 3.2. Due to the finite number M of discrete time steps, the describing equations can be given in a straightforward way in the time domain. With $a_1 = b_1$, the output of the integrator $V_{n,i1}[k]$ in Fig. 3.3 is calculated by

$$V_{n,i1}[k] = b_1 \cdot \sum_{l=0}^{k-1} (u_n[l] - v_n[l] V_{\text{ref}}) \tag{3.2}$$

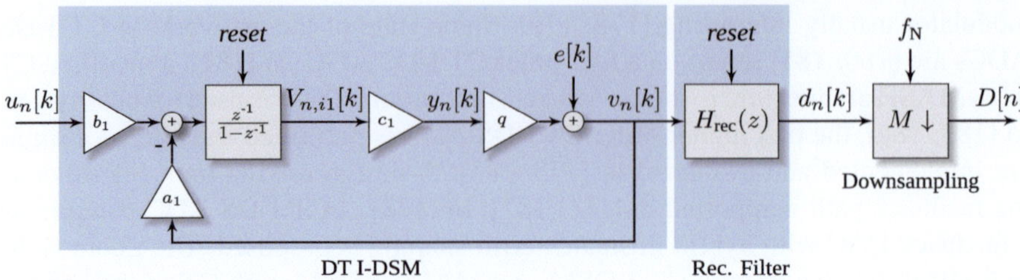

Fig. 3.3 Linear model of a DT first-order I-DS ADC.

where the fed back quantized modulator output is

$$v_n[k] = q \cdot y_n[k] + e[k]. \tag{3.3}$$

The I-DSM generates M oversampled low resolution outputs $v_n[k]$ for $k \in [1, M]$. The input signal $u_n[k]$ is the DT input obtained e.g. by the SC network at the integrator input by sampling the continuous-time input $u_n(t)$ at the oversampling frequency f_s. If an upfront S/H is used to sample $u(t)$ at Nyquist-rate f_N, as optionally illustrated in Fig. 3.1, or alternatively if the input is assumed sufficiently slow, then $u_n(t)$ is constant over a whole conversion cycle, and also the oversampled DT input $u_n[k]$ in Fig. 3.3 is approximately constant.

$$u_n[k] \approx \hat{u}_{const}. \tag{3.4}$$

Consequently, (3.2) can be simplified to

$$V_{n,i1}[k] = b_1 \cdot k \cdot \hat{u}_{const} - b_1 \cdot \sum_{l=0}^{k-1} v_n[l] V_{ref}. \tag{3.5}$$

The core difference of an I-DSM to a free-running DSM is its periodically applied reset as shown in Fig. 3.1b. After the M-th sample, the integrator output $V_{n,i1}[M]$ becomes

$$V_{n,i1}[M] = b_1 \cdot M \cdot \hat{u}_{const} - b_1 \cdot \sum_{l=0}^{M-1} v_n[l] V_{ref}. \tag{3.6}$$

This can be rearranged to

$$\hat{u}_{const} = \frac{V_{ref}}{M} \sum_{l=0}^{M-1} v_n[l] + \frac{V_{n,i1}[M]}{b_1 \cdot M}. \tag{3.7}$$

The first term of (3.7) consists of the scaled summation of the coarsely quantized modulator output samples $v_n[\mathrm{k}]$, which can be interpreted as a digital approximation of the constant input

$$\hat{u}_{const} \approx \frac{V_{\mathrm{ref}}}{M} \sum_{l=0}^{M-1} v_n[l]. \tag{3.8}$$

The second term in (3.7) represents then the residual quantization error after M conversion cycles

$$e_{\mathrm{resid}}[k] = \frac{V_{n,i1}[M]}{b_1 \cdot M} \tag{3.9}$$

in analog form which can be made arbitrarily small for bounded $V_{n,i1}$ and increasing M. The availability of the residual quantization error at the output of the analog integrator scaled by M can be used advantageously in advanced architectures in order to refine the output quantization error and thus increase the resolution; this is referred to as the *extended-counting architecture* and will be presented in Sect. 5.

The summation operation of the coarsely quantized I-DSM output $v_n[k]$ in (3.8) can be realized by a digital integrator. Consequently, a feasible reconstruction filter for the first-order I-DSM in Fig. 3.1b is a digital integrator. Then, the last oversampled output of the integrating digital reconstruction filter represents the analog input as in (3.8), and the subsequent decimator creates one high-resolution digital output sample of the I-DS ADC at Nyquist-rate f_{N}.

$$D[n] = \frac{V_{\mathrm{ref}}}{M} \sum_{l=0}^{M-1} v_n[l]. \tag{3.10}$$

This sample $D[n]$ represents the digital, time- and value-discrete approximation of the analog input during the n-th Nyquist period as indicated in Fig. 3.2, and it is independent of previous samples due to the reset before the incremental operation. Thus, true Nyquist-ADC behavior is obtained.

3.2.2 Time-Domain Description of a Third-Order CIFF I-DS ADC

The same calculation principle can be applied to higher-order loop filters. For the frequently used CIFF architecture this is e.g. described in detail in [5]. The equations become quite cumbersome depending on the architecture and order and in the following the derivation of [16, 176] is shown for the third-order CIFF I-DSM in Fig. 3.4.

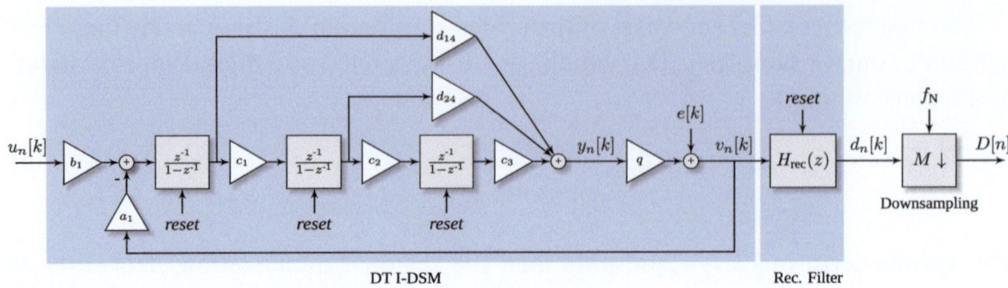

Fig. 3.4 Third-order CIFF I-DSM with reconstruction filter.

Similar as in the derivation toward (3.6) for the first-order I-DSM, using $a_1 = b_1$, the outputs of the three integrators of the third-order CIFF I-DSM in Fig. 3.4 can be found at the last of M operation cycles as:

$$V_{n,i1}[M] = b_1 \cdot \sum_{k=0}^{M-1} (u_n[k] - v_n[k] \cdot V_{\text{ref}}) \tag{3.11}$$

$$V_{n,i2}[M] = c_1 \cdot \sum_{l=0}^{M-1} V_{n,i1}[l] = b_1 c_1 \cdot \sum_{l=0}^{M-1} \sum_{k=0}^{l-1} (u_n[k] - v_n[k] \cdot V_{\text{ref}}) \tag{3.12}$$

$$V_{n,i3}[M] = c_2 \cdot \sum_{m=0}^{M-1} V_{n,i2}[m] = b_1 c_1 c_2 \cdot \sum_{m=0}^{M-1} \sum_{l=0}^{m-1} \sum_{k=0}^{l-1} (u_n[k] - v_n[k] \cdot V_{\text{ref}}) \,. \tag{3.13}$$

For stable operation, we can assume that the output of the last integrator $V_{n,i3}$ is bounded by $\pm V_{\text{ref}}$. This assumption can be expressed by the following inequality:

$$- V_{\text{ref}} \leq b_1 c_1 c_2 \cdot \sum_{m=0}^{M-1} \sum_{l=0}^{m-1} \sum_{k=0}^{l-1} (u_n[k] - v_n[k] \cdot V_{\text{ref}}) \leq V_{\text{ref}}. \tag{3.14}$$

Without loss of generality and as done for the first-order analysis in (3.4), a constant input $u_n[k] = \hat{u}_{const}$ is assumed during the n-th conversion cycle, which simplifies the equations to

$$\frac{-V_{\text{ref}}}{b_1 c_1 c_2 \cdot k_{\text{Norm}}} \leq \hat{u}_{const} - \frac{1}{k_{\text{Norm}}} \cdot \sum_{m=0}^{M-1} \sum_{l=0}^{m-1} \sum_{k=0}^{l-1} v_n[k] \cdot V_{\text{ref}} \leq \frac{V_{\text{ref}}}{b_1 c_1 c_2 \cdot k_{\text{Norm}}} \tag{3.15}$$

with k_{Norm} being

$$k_{\text{Norm}} = \sum_{m=0}^{M-1} \sum_{l=0}^{m-1} \sum_{k=0}^{l-1} 1 = \frac{M \cdot (M-1) \cdot (M-2)}{3!}.$$

We see that the unknown (constant) input signal $\hat{u}_{const}$ and the rhs sum are limited by an interval, which can be made arbitrarily small by increasing M. Thus, an increasingly accurate estimate for the input signal is achieved by an arbitrarily large M as

$$\hat{u}_{const} \approx \frac{V_{\text{ref}}}{k_{\text{Norm}}} \cdot \sum_{m=0}^{M-1} \sum_{l=0}^{m-1} \sum_{k=0}^{l-1} v_n[k]. \tag{3.16}$$

Similar as found for the first-order I-DSM in (3.10), the structure of a feasible reconstruction filter in the case of the third-order I-DSM from Fig. 3.4 can be identified with the help of (3.16). It is a third-order CoI filter, which triple-sums the output sequence of the I-DSM v_n and scales it by k_{Norm}.

Also in the third-order I-DSM, the remaining quantization error is available in the analog domain. This is easily derived from (3.13), which—for a constant input signal—is rewritten to

$$\frac{V_{n,i3}[M]}{b_1 c_1 c_2} = k_{\text{Norm}} \hat{u}_{const} - \sum_{m=0}^{M-1} \sum_{l=0}^{m-1} \sum_{k=0}^{l-1} v_n[k] V_{\text{ref}}. \tag{3.17}$$

Rearranging this yields

$$\hat{u}_{const} = \frac{V_{\text{ref}}}{k_{\text{Norm}}} \sum_{m=0}^{M-1} \sum_{l=0}^{m-1} \sum_{k=0}^{l-1} v_n[k] + \frac{1}{k_{\text{Norm}}} \cdot \frac{V_{n,i3}[M]}{b_1 c_1 c_2} \tag{3.18}$$

which shows that the (constant) analog input signal is represented by its digital approximation in (3.16) and a scaled version of the output of the last integrator at the last (M^{th}) sampling point. This scaled last integrator output is thus the analog representation of the residual quantization error, whose further quantization and digital cancellation could be used to refine the overall digital output. Again, this is known as *extended counting* and will be described further in Chap. 5.

3.2.3 Time-Domain Description of a Second-Order CIFB I-DS ADC

As discussed throughout Sect. 2.3.2, the loop filter of a DSM can be realized by various architectures like CIFF, CIFB, or including resonators. We will discuss in Sect. 3.3.3 that I-DSM ADCs do not take advantage of STF shaping of the CIFB architecture, and further in Sect. 4.1.2 that resonator zeros have no advantage in I-DSM ADCs. This is why most SoA I-DSM are realized as CIFF architectures. Nonetheless, the same principle of time-domain derivation as used above for a first- and third-order CIFF I-DSM can also be employed for other architectures in order

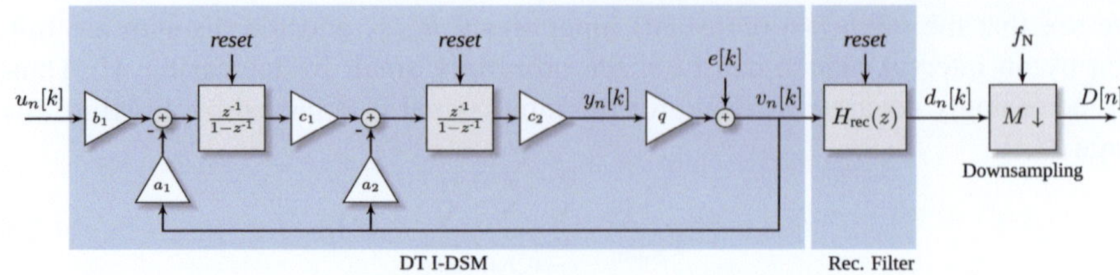

Fig. 3.5 Linearized block diagram of a second-order CIFB I-DS ADC.

to derive suitable reconstruction filters. As an example, this is performed for the second-order CIFB I-DSM in Fig. 3.5.

Assuming $a_1 = b_1$ leads to the following equations:

$$V_{n,i1}[M] = a_1 \cdot \sum_{k=0}^{M-1} (u_n[k] - v_n[k] \cdot V_{\text{ref}}) \tag{3.19}$$

$$\begin{aligned} V_{n,i2}[M] &= \sum_{l=0}^{M-1} \left(c_1 \cdot V_{n,i1}[l] - a_2 \cdot v_n[l] \cdot V_{\text{ref}}\right) \\ &= \sum_{l=0}^{M-1} \left(c_1 \cdot a_1 \cdot \sum_{k=0}^{l-1} (u_n[k] - v_n[k] \cdot V_{\text{ref}}) - a_2 \cdot v_n[l] \cdot V_{\text{ref}}\right). \end{aligned} \tag{3.20}$$

Again without loss of generality, a constant input $u_n[k] = \hat{u}_{const}$ is assumed during the nth conversion cycle. This yields

$$V_{n,i2}[M] = \sum_{l=0}^{M-1} \left(c_1 \cdot a_1 \cdot \sum_{k=0}^{l-1} \left(\hat{u}_{const} - v_n[k] \cdot V_{\text{ref}}\right) - a_2 \cdot v_n[l] \cdot V_{\text{ref}}\right) \tag{3.21}$$

which can be rewritten to

$$\begin{aligned} V_{n,i2}[M] =& c_1 \cdot a_1 \cdot \frac{M \cdot (M-1)}{2!} \cdot \hat{u}_{const} - \\ & \sum_{l=0}^{M-1} \left(c_1 \cdot a_1 \cdot \sum_{k=0}^{l-1} v_n[k] \cdot_{\text{ref}} + a_2 \cdot v_n[l] \cdot V_{\text{ref}}\right). \end{aligned} \tag{3.22}$$

For stable operation, we can again assume that the output of the last integrator $V_{n,i2}$ is bounded by $\pm V_{\text{ref}}$. Using

$$k_{\text{Norm2}} = c_1 \cdot a_1 \cdot \frac{M \cdot (M-1)}{2!}$$

this assumption can be expressed with the following inequality:

$$\frac{-V_{\text{ref}}}{k_{\text{norm2}}} \leq \hat{u}_{const} - \frac{V_{\text{ref}}}{k_{\text{norm2}}} \cdot \sum_{l=0}^{M-1} \left(c_1 a_1 \cdot \sum_{k=0}^{l-1} v_n[k] + a_2 \cdot v_n[l] \right) \leq \frac{V_{\text{ref}}}{k_{\text{norm2}}}. \quad (3.23)$$

As the boundaries can again be made arbitrarily small with increasing M, the constant input signal can be approximated with

$$\hat{u}_{const} \approx \frac{V_{\text{ref}}}{k_{\text{norm2}}} \cdot \left(c_1 a_1 \sum_{l=0}^{M-1} \sum_{k=0}^{l-1} v_n[k] + a_2 \sum_{l=0}^{M-1} v_n[l] \right). \quad (3.24)$$

This shows that the reconstruction filter—derived by the time-domain analysis—is in this case not a simple CoI filter of the same order as the I-DSM, but a scaled combination of a first- and a second-order CoI. We will further discuss such differences in reconstruction filters in Sect. 3.5.

Also in the case of this second-order CIFB I-DSM, the remaining quantization error is available in the analog domain and can be used for the mentioned *extended-counting technique*. This is derived from (3.22), which we can rewrite as

$$\hat{u}_{const} = \frac{V_{\text{ref}}}{k_{\text{norm2}}} \cdot \left(c_1 a_1 \sum_{l=0}^{M-1} \sum_{k=0}^{l-1} v_n[k] + a_2 \sum_{l=0}^{M-1} v_n[l] \right) + \frac{V_{n,i2}[M]}{k_{\text{norm2}}} \quad (3.25)$$

where the lhs term is the digital estimate of the input signal after the reconstruction filter, and the rhs term is the residual quantization error present as a scaled version of the analog output of the second integrator at the last clock cycle, i.e. $V_{n,i2}[M]$.

3.2.4 Time-Domain Description of CT I-DS ADCs

Also for CT I-DS ADC the time-domain analysis can be performed. For the example of a first-order CT I-DSM, the sum in (3.2) needs to be replaced by an integral

$$y_n(t) = \int_0^t (u_n[\tau] - r[\tau]) \, \mathrm{d}\tau \quad (3.26)$$

with r(t) describing the waveform of the DAC feedback signal according to the quantizer output. The sampled quantizer output is then defined as

$$v_n[k] = q \cdot y_n(t)|_{t=kT_s} + e[k] \quad (3.27)$$

and the digital representation of the input signal can be analogously obtained by (3.10).

3.3 Frequency-Domain Description of I-DS ADCs

3.3.1 Overall Signal and Noise Transfer Functions *in I-DS ADCs*

Free-running DS ADCs are described in the frequency domain based on their transfer functions. Thereby, only the transfer characteristic of the DSM is investigated, which is obtained with the help of the linearized, time-invariant system model, cf. Sects. 2.3.1.1 and 2.3.1.2 for DT and CT DSM, rendering the DSM into an LTI system. The influence of the digital low-pass filter is usually neglected, as it is assumed to be an ideal brick-wall filter. This eliminates all OOB components before they can alias into the inband due to subsampling by the decimator.

This approach is not applicable to I-DS ADCs, as the periodic reset of all memory in the analog loop- and digital reconstruction filter makes the incremental ADC a LTV system. Therefore, as outlined throughout Sect. 3.2, incremental ADCs are described in the time domain in most SoA. However, this time-domain derivation has to be carried out for each architecture and combination of I-DSM and respective reconstruction filter $H_{\text{rec}}(z)$. Moreover, including nonideal behavior is hardly possible as e.g. discussed in [179]. Consequently, a frequency-domain analysis of I-DS ADCs would be very advantageous, as it could derive the performance of any architecture and include nonidealities, similar as known for free-running DSM. The signal transfer characteristic of an I-DS ADC without upfront S/H has been reported as early as [195]. The measured signal transfer characteristic of a CT I-DS ADC was e.g. reported by Mokhtar et al. [186]. Thereby—in contrast to free-running modulators—the signal transfer characteristic of an I-DS ADC measures how an applied input signal with known power and frequency, e.g. between $DC \ldots f_s/2$, is seen at the output of the I-DS ADC after the decimating reconstruction filter. Even though OOB input signals are folded back into the inband, we can plot their frequency dependent attenuation over their original frequency range, knowing that they will not be found at their original position in the ouput spectrum, but aliased into the inband at the decimated I-DS ADC output $D[n]$.

Illustration by an exemplary simulation of spectra at different positions of the ADC: For clarification of this procedure, an example is simulated in Fig. 3.6 for an exemplary third-order CIFF DSM, in free-running (left) and incremental operation (right). The free-running DSM is combined with a brick-wall decimation filter, while the I-DSM is followed by a CoI3 reconstruction filter, cf. Sect. 3.2.2. The spectra of the signals at different stages of the ADCs are plotted. Two simulations are performed with sinusoidal input signals of equal amplitude but once inband (blue) and once OOB

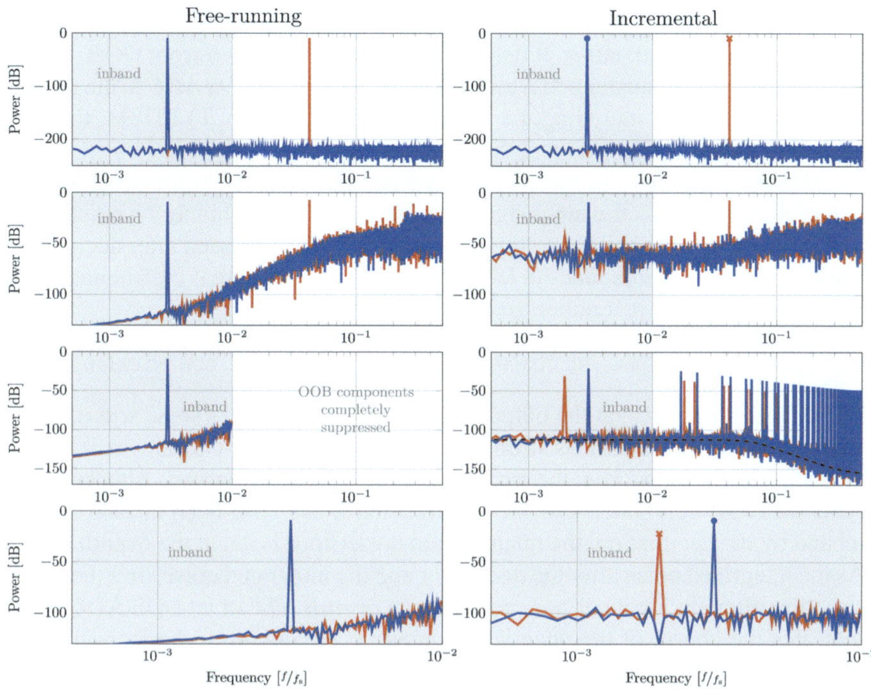

Fig. 3.6 Spectra at different stages of a DS ADC using two different sinusoidal inputs (simulated independently): one inband sinusoidal signal (blue) and one OOB sinusoidal signal (red). The spectra of the free-running DS ADC are shown on the left and the spectra of the incremental DS ADC on the right. Top row: the spectra at the ADC input. Second row: the spectra at the modulator output. Third row: the undecimated output spectra after the digital filter. Bottom row: the decimated output spectra of the ADC.

(red), and the spectra are plotted on top of each other. In the top row of Fig. 3.6, the spectra of the modulator input signals are shown (with a white numerical simulation noise floor). The second row shows the spectra at the modulator outputs. The third row shows the spectra after the digital filters but before decimation, and the last row shows the spectra of the decimated outputs. The following observations can be made

- Without surprise, the left side shows the typical behavior known from DSM: the modulator output shows shaped quantization noise together with the input signals filtered by the STF of the DSM, which is approximately one in this case. The digital low-pass (brick-wall) filter eliminates all OOB signals and quantization noise (QN). The decimator leaves the inband spectrum ideally unchanged, only sampled at a new rate.
- The right side shows completely different behavior. At the overall output (bottom), the inband input signal (blue, o-mark) is seen identically as in the free-running case, but the original OOB signal (red, x-mark) is seen in the inband as well, slightly attenuated.

- Consequently, we see that an input signal has frequency dependent gain before it ends up in the inband, either at its original position or aliased from OOB.
- We call this signal transfer behavior the *overall STF of an I-DS ADC* in the remainder of the book, i.e. including the influence of the (DT or CT) I-DSM, the decimating reconstruction filter, and the time-variance due to the reset.
- We also note that this *overall STF* cannot be seen after the reconstruction filter before decimation, as the input signal power is spread into higher Nyquist zones, and the final signal power in a single-frequency bin is only seen after decimation.
- Finally, we should note that the I-DSM output does not reveal the attenuation of the OOB signal and it is rather useless to look at.

For the quantization noise, the observations for the I-DS ADC can be extended:

- The quantization noise at the output of the I-DSM still shows some sort of noise-shaping, though by far not as strong as for the free-running DSM.
- The quantization noise after the reconstruction filter shows low-pass filter behavior.
- After all OOB components of the quantization noise have been aliased into the inband by the decimation, the quantization noise-floor is flat at the overall output.
- As the integrated noise after the decimator and the integrated noise after the reconstruction filter are identical, we can define an *overall NTF* of an I-DS ADC which indicates the course of the quantization noise at the output of the reconstruction filter, indicated as dashed line in Fig. 3.6, knowing that after decimation the whole overall NTF-shaped QN is found aliased and flat in the inband.

The definition of the *overall signal or noise transfer functions for incremental ADCs* are illustrated in the block diagram in Fig. 3.7. The reader should therefore bear in mind that, when simulating these transfer functions, the respective OOB spectral contents were aliased into the first Nyquist zone. However, before this occurred, they experienced the respective overall transfer function.

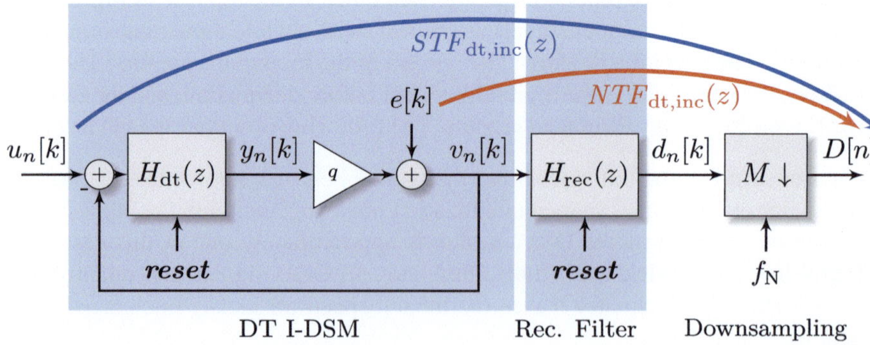

Fig. 3.7 Concept of overall transfer functions illustrated in the block diagram of a DT I-DS ADC. ©2021 Wagner. Modified, with permission, from Wagner [27].

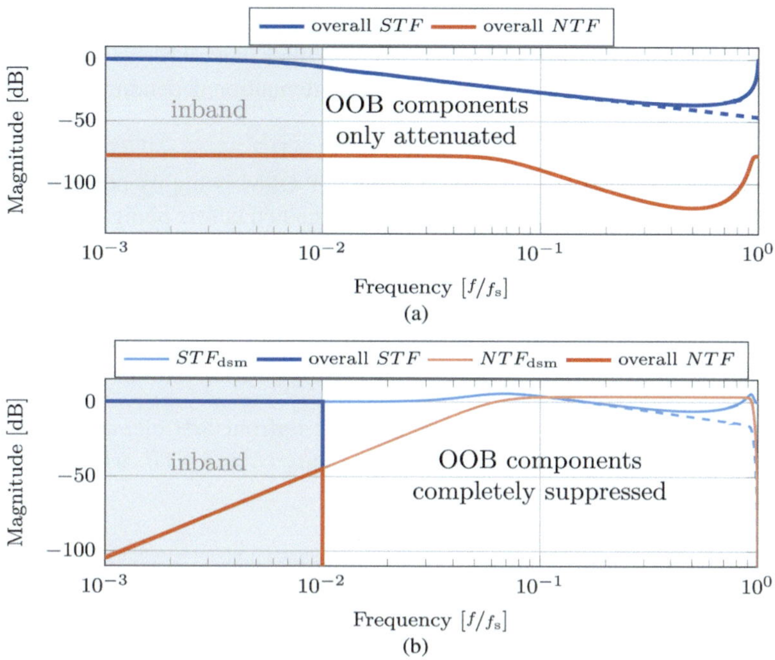

Fig. 3.8 (a) Overall STF and NTF of an exemplary third-order DT (solid) and CT (dashed) I-DS ADC without upfront S/H combined with a third-order CoI reconstruction filter at M=100. (b) STF and NTF of a free-running DT (solid) and CT (dashed) DSM combined with an ideal brick-wall low-pass filter. Both simulated at an OSR of 100 based on the same third-order DSM architecture and scaling (Schreier Toolbox with an OBG of 1.5). ©2021 Wagner. Modified, with permission, from Wagner [27].

Exemplary simulation of the overall STF and NTF: As an example and another comparison to free-running DSMs, the overall STF and NTF of an exemplary third-order I-DS ADC without front-end S/H are plotted in Fig. 3.8 and compared to the STF and NTF of an identically built free-running DSM combined with its brick-wall low-pass filter.

From this illustrative example we can note that

- Neither the course of the overall STF nor NTF of an I-DS ADC can be explained with the course of the transfer functions of the free-running DSM alone.
- The shape of the overall STF resembles that of its third-order CoI reconstruction filter.
- While the overall STF of the DT I-DS ADC (solid) is periodic with f_s and thus plotting only from $DC \ldots f_s/2$ would be enough, the overall STF of the CT I-DS ADC (dashed) shows a continuous roll-off for higher frequencies and is plotted from $DC \ldots f_s/2$ to illustrate the difference.

- In the I-DS ADC, OOB signals are—in contrast to free-running DS ADCs—not further filtered by the (brick-wall) decimation filter, but all higher Nyquist bands are folded into the inband after some modest attenuation depending on their OOB frequency location.
- Even though the overall STF of the CT I-DS ADC has a roll-off toward f_s, an implicit AAF as known from free-running CT DSM is highly compromised, as e.g. the second Nyquist band is almost not attenuated before being aliased into the inband—this is further discussed in Sect. 4.1.4.
- The overall NTF of an I-DS ADC does not show the usual high-pass filter behavior. Higher frequency components are not further attenuated by a brick-wall low-pass filter, but higher frequency quantization noise is only shaped by the overall NTF and then the complete band from DC..$f_s/2$ is folded back into the inband.

Finally, we should note that by including an upfront S/H element to the I-DS ADC, the simulation example would yield a unity overall STF, while the overall NTF would not change in Fig. 3.8.

3.3.2 Description of I-DS ADCs as a Finite Impulse Response Filter

In order to describe an I-DS ADC in the frequency domain, in this section its representation as a filter with FIR is revised as introduced in [192, 194, 196]. The FIR model considers the incremental DS ADC as an LTI system for a finite number of samples M. This principle can be applied to any DT transfer function with periodically applied reset and can therefore be used for both, the overall NTF and overall STF calculations. We start with the Z-transform of the overall transfer function of a free-running DT DSM concatenated with its subsequent digital filter to obtain

$$TF_{\mathrm{dt,free}}(z) = TF_{\mathrm{dt,dsm}}(z) \cdot h_{\mathrm{rec}}(z) \tag{3.28}$$

where $TF_{\mathrm{dt,dsm}}(z)$ can be the STF or NTF of the free-running DT DSM from (2.24) and (2.25), and $h_{\mathrm{rec}}(z)$ is the transfer function of its reconstruction filter, both still without reset. In the time domain, this yields a convolution based on the impulse responses of the two concatenated sub-systems:

$$h_{\mathrm{free}}[k] = h_{\mathrm{dsm}}[k] * h_{\mathrm{rec}}[k] = \sum_{l=-\infty}^{\infty} h_{\mathrm{dsm}}[l-k] \cdot h_{\mathrm{rec}}[l]. \tag{3.29}$$

Without the periodic reset, the overall impulse response $h_{\mathrm{free}}[k]$ of the ADC is infinite due to the infinite impulse response $h_{\mathrm{dsm}}[k]$ of the modulator. By truncating the convolution sum in (3.29), the impulse response of an equivalent FIR filter can be obtained for $k \in [1, M]$ [196]:

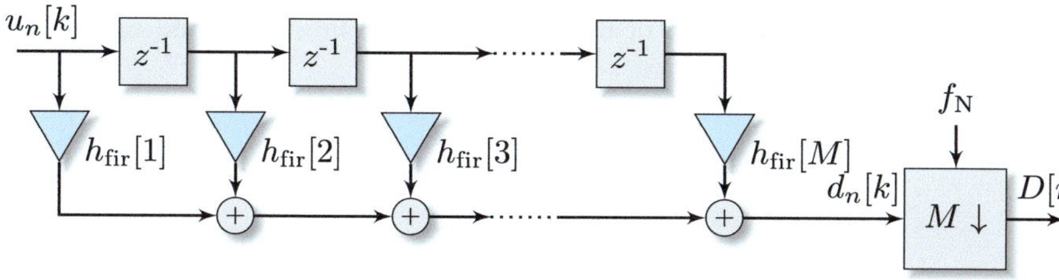

Fig. 3.9 FIR filter model of an I-DS ADC illustrated for an oversampling ratio of M. ©2019 IEEE. Modified, with permission, from Wagner et al. [194].

$$h_{fir}[k] = \sum_{l=1}^{M} h_{\text{dsm}}[l-k] \cdot h_{\text{rec}}[l]. \tag{3.30}$$

The finite length accounts for the reset, as it is the case in an I-DS ADC. A block diagram that represents this FIR model is illustrated in Fig. 3.9. It can be adapted to be used for both the input signal $h_{\text{fir,stf}}[k]$ and the quantization noise $h_{\text{fir,ntf}}[k]$.

3.3.3 *Overall STF and NTF Calculation using the FIR Model*

Based on the FIR model, the frequency response of I-DS ADCs can be derived. This was introduced in [196] for a first-order DT I-DS ADC with CoI reconstruction filter. In the following, the approach is generalized to higher-order modulators with arbitrary reconstruction filters as proposed in [194].

3.3.3.1 Calculation of the Overall Transfer Functions for DT I-DS ADCs

Based on the FIR model of an I-DS ADC, (3.30) can be rewritten as

$$h_{\text{fir}}[k] = h_{\text{free}}[k] \cdot \text{rect}_M[k] = \mathcal{Z}^{-1}\left\{TF_{\text{dt,free}}(z)\right\}[k] \cdot \text{rect}_M[k] \tag{3.31}$$

where $TF_{\text{dt,free}}(z)$ is the combination of the STF or NTF of the free-running DSM with the reconstruction filter transfer function, cf. (3.28), and $h_{\text{free}}[k]$ is its impulse response, cf. (3.29). $\text{rect}_M[k]$ is a DT rectangular window function with length M, defined as

$$\text{rect}_M[k] = \begin{cases} 1, & 1 \leq k \leq M \\ 0, & \text{otherwise.} \end{cases} \tag{3.32}$$

The Z-domain representation of the overall transfer function based on the FIR model in Fig. 3.9 is then found as

$$TF_{\mathrm{dt,inc}}[z] = \sum_{l=1}^{M} h_{\mathrm{fir}}[l] \cdot z^{-l}. \tag{3.33}$$

With applying the discrete-time Fourier transform (DTFT), the overall transfer function in the frequency domain can be obtained by substituting $z = e^{j2\pi f/f_s}$

$$TF_{\mathrm{dt,inc}}(f) = \sum_{l=1}^{M} \mathcal{Z}^{-1}\left\{TF_{\mathrm{dt,free}}(z)\right\}[l] \cdot e^{-j2\pi f/f_s l} \tag{3.34}$$

which allows to calculate the overall transfer functions $STF_{\mathrm{dt,inc}}(f)$ or $NTF_{\mathrm{dt,inc}}(f)$ in the frequency domain based. Thereby, $TF_{\mathrm{dt,free}}(z)$ in (3.28), which is the combination of the free-running modulator transfer function $TF_{\mathrm{dt,dsm}}(z)$ and the reconstruction filter transfer function, is either chosen as STF or NTF.

Deriving the overall transfer functions solely in the frequency domain: Instead of approaching the problem via the time-domain finite impulse response in (3.31), we can alternatively derive the overall DT transfer functions $TF_{\mathrm{dt,inc}}(f)$ of incremental ADCs using the FIR model completely in the frequency domain [194]. This will also be the method of choice when we look at CT I-DSM in Sect. 3.3.3.2. By transforming (3.31) to the frequency domain via the DTFT yields the overall transfer function

$$TF_{\mathrm{dt,inc}}(f) = TF_{\mathrm{dt,free}}\left(\mathrm{e}^{j2\pi f/f_s}\right) \circledast X_{rect,M}(f) \tag{3.35}$$

with

$$X_{\mathrm{rect},M}(f) = \mathrm{DTFT}\left\{\mathrm{rect}_M[k]\right\} = \frac{\sin\left(M\pi f/f_s\right)}{\sin\left(\pi f/f_s\right)} \cdot e^{-j\pi f/f_s \cdot M}. \tag{3.36}$$

The $\circledast$-operator stands for the CT circular convolution—the frequency-domain correspondence of the multiplication in (3.31)—which requires solving the convolution integral

$$X(f) \circledast Y(f) = \int_{f_o}^{f_o+f_s} X(\nu) \cdot Y(f-\nu)\, d\nu. \tag{3.37}$$

3.3.3.2 Calculation of the Overall Transfer Functions for CT I-DS ADCs

Calculation of the NTF for CT I-DS ADC: For CT I-DS ADCs, the calculation of the overall transfer function can be similarly approached. In Sect. 2.3.1.2 it was described that the NTF of a free-running CT DSM can be derived based on the equivalent loop-filter response of a DT DSM. The derivation can be done using the impulse-invariant transformation as described in (2.31), or using the lifting method. The latter was originally described in [197] and later adopted to DSMs in [198], and it allows a straightforward transformation of a sampled CT system to an equivalent DT loop filter including nonidealities such as finite gain, finite GBW and ELD. The derived equivalent DT NTF of a CT DSM ($NTF_{\mathrm{ct,dsm}}(z)$, cf. (2.32)) can be inserted into the impulse response of the FIR model of an I-DS ADC in (3.31). Then, the overall transfer function in the frequency domain in (3.34) becomes

$$NTF_{\mathrm{ct,inc}}(f) = \sum_{l=1}^{M} \mathcal{Z}^{-1}\left\{NTF_{\mathrm{ct,free}}(z)\right\}[l] \cdot e^{-j2\pi f/f_s l} \tag{3.38}$$

where the combined transfer function of the free-running modulator and reconstruction filter is

$$NTF_{\mathrm{ct,free}}(z) = NTF_{\mathrm{ct,dsm}}(z) \cdot h_{\mathrm{rec}}(z)$$

which is equivalent to the DT case in (3.28).

Calculation of the STF for CT I-DS ADC: The calculation of the overall STF of a CT I-DS ADC has to be approached differently. In contrast to the NTF, where the CT and DT DSM can behave identically using a sampled-input-sampled-output equivalence, e.g. by generating a CT loop filter based on the impulse invariant transformation from its DT counterpart, the STF of a CT DSM behaves differently, as it results in a mixed CT/DT transfer function [4, 5]. Thus, using the approach via the Z-domain and impulse response is not possible due to this mixed-signal nature of the CT ADC, which has to be simultaneously described in the Z- and Laplace-domain as was already seen for the free-running case in (2.33).

The derivation of the overall transfer function completely in the frequency domain, as introduced above for the DT I-DS ADC, cf. (3.35), however, can be adapted for a CT I-DSM. Therefore, we use a different reset window function, i.e. the CT rectangular function:

$$\mathrm{rect}_T = \begin{cases} 1, & |t| < \frac{T_{\mathrm{M}}}{2} \\ 0, & |t| > \frac{T_{\mathrm{M}}}{2} \end{cases} \tag{3.39}$$

with $T_{\mathrm{M}} = \frac{M}{f_s}$. This reset window function can be transformed to the frequency domain yielding

$$X_{\mathrm{rect},T} = \mathcal{F}\left\{\mathrm{rect}_T\right\} = \mathrm{sinc}_T(f) = \frac{\sin(T_{\mathrm{M}}\pi f)}{T_{\mathrm{M}}\pi f}. \tag{3.40}$$

For causal systems, the CT rectangular function, which is centered around zero, needs to be shifted by $T_{\mathrm{M}}/2$, which yields

$$X'_{\mathrm{rect},T}(f) = X_{\mathrm{rect},T}(f) \cdot e^{-j\pi f T_{\mathrm{M}}}. \tag{3.41}$$

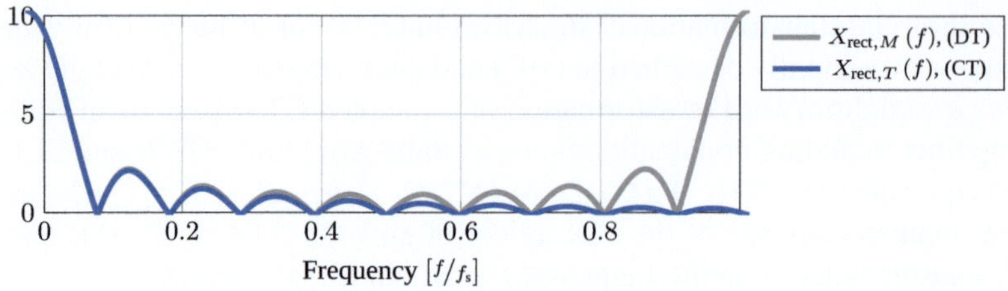

Fig. 3.10 Illustration of Fourier transformation of DT and CT rectangular functions ($X_{\text{rect},M}(f)$ and $X'_{rect,T}(f)$) for $M = 10$. ©2019 IEEE. Modified, with permission, from Wagner et al. [194].

For the sake of illustration, the frequency response of this CT rectangular window $X'_{\text{rect},T}(f)$ is plotted alongside the frequency response of its DT counterpart $X_{\text{rect},M}(f)$ from (3.36) in Fig. 3.10. This visualizes the CT nature as the magnitude decreases with increasing frequency. In contrast, the frequency response of the DT rectangular window function $X_{\text{rect},M}(f)$ is expectedly periodic with f_s, i.e. mirrored at $0.5 f_s$.

In the DT derivation, the multiplication with the reset window yielded a circular convolution in the frequency domain, cf. (3.35). Similarly accounting for the reset in the CT domain, the multiplication with the CT reset window in the time domain corresponds to a convolution in the frequency domain, which lets us derive the overall STF of an incremental converter as

$$STF_{\text{ct,inc}}(f) = STF_{\text{ct,tree}}(f) * X'_{\text{rect},T} \tag{3.42}$$

where $STF_{\text{ct,tree}}(f)$ is the combined transfer function $STF_{\text{ct,dsm}}(f) \cdot H_{\text{rec}}(z)(e^{j2\pi f})$ of the free-running modulator and the reconstruction filter. Therein, the convolution operator in (3.42) represents the integral

$$X(f) * Y(f) = \int_{-\infty}^{\infty} X(\nu) \cdot Y(f - \nu)\, d\nu. \tag{3.43}$$

3.3.3.3 Example for Overall STF and NTF Calculations

The derivation of the overall transfer functions is illustrated by a comparison of calculation and simulation for an exemplary I-DS ADC. Therefore, a DT and a CT third-order single-bit CIFF I-DS ADC combined with a third-order CoI reconstruction filter is used, which is shown in Fig. 3.11. The scaling coefficients of this I-DS ADC are obtained from the free-running third-order CIFF DSM equivalent using the Delta-Sigma Toolbox [144], where the OBG is set to 1.5 (cf. Lee's rule in Sect. 2.3.4, and all integrator outputs are scaled to $\pm\, FS$. The derived coefficients for both the DT and CT modulator are given in Table 3.1.

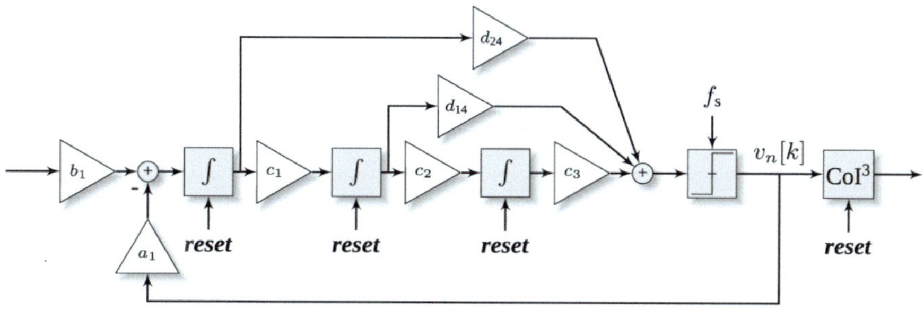

Fig. 3.11 Block diagram of an exemplary DT or CT third-order single-bit CIFF I-DS ADC using a third-order CoI reconstruction filter. The coefficients are derived using the Schreier Toolbox with an OBG of 1.5 and all integrator outputs scaled to ±FS.

Table 3.1 Scaling coefficients for the I-DSM in Fig. 3.11 for OBG=1.5 and all internal swings normalized to $\pm FS$ using [144].

	b_1	a_1	c_1	c_2	c_3	d_{14}	d_{24}
DT	0.37	−0.37	0.26	0.12	3.8	2.1	2.9
CT	0.41	−0.41	0.42	0.13	1.9	1.6	1.4

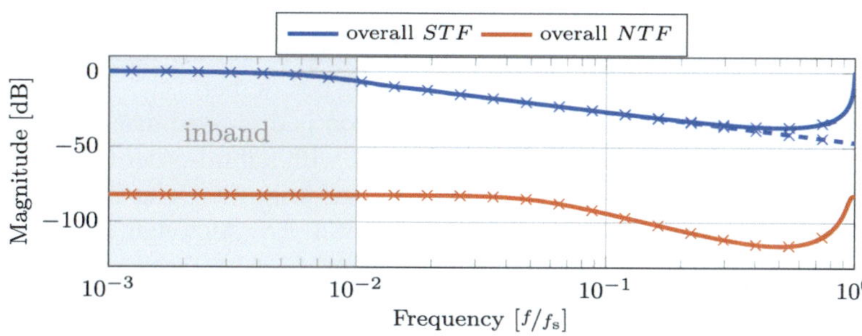

Fig. 3.12 Calculated overall STFs (blue) and NTF (red) of a DT and CT third-order single-bit CIFF I-DS ADC using a third-order CoI reconstruction filter and an oversampling of $M = 100$. The calculations are verified by the simulation points (x-marks). The scaling coefficients are derived using the Schreier Toolbox with an OBG of 1.5 and all integrator outputs scaled to 1 FS. The simulations are done at −6 dBFS. The calculations use a quantizer gain $q \approx 1.7$. ©2025 IEEE. Modified, with permission, from Ismail et al. [199].

The overall STF (blue) and NTF (red) of the exemplary DT (solid) and CT (dashed) I-DS ADCs are calculated using (3.34), (3.42), and (3.38) and are shown in Fig. 3.12 for an oversampling of $M = 100$. The calculations are verified by simulation points (x-marks) and excellent matching is achieved. For the DT I-DS ADC, it can be seen that the STF is repetitive with a periodicity around f_s. For the CT I-DS ADC, the overall NTF is identical to that of the DT counterpart, since both loop filters are derived from the same free-running prototype. In contrast, the overall STF of a

CT I-DS ADC behaves different as was already seen in the simulation in Fig. 3.8, and it features a continuous roll-off for higher frequencies without periodicity around f_S. The calculation results are again confirmed by the simulation points.

3.3.4 Effective Quantizer Gain in Incremental Operation

In all analyses on DSMs, the calculation of STF and NTF has to take into account the effective quantizer gain from its linear model. In the free-running case, we have defined the NTF in (2.24) depending on loop filter and effective quantizer gain q. As the overall $NTF_{\text{inc}}(f)$ for I-DS ADCs is defined upon the $NTF_{\text{free}}(z)$ in (3.34) and (3.38), the effective quantizer gain q must be known, too. For a free-running DSM, it is known that the effective quantizer gain q depends on the modulator architecture, internal bit-width, and input signal power [2, 5]. For I-DS ADCs, this has rarely been addressed in literature [200]. The question is if the applied periodic reset changes the statistics of the derivation of the effective quantizer gain which is obtained from transient simulation using (2.11) in free-running DSM [3].

Due to the reset in incremental operation, the I-DS ADC has a settling behavior at the beginning of each Nyquist conversion. This transient settling might also have an influence on the effective quantizer gain. Therefore, at first the dependency of the effective quantizer gain on M is analyzed. The influence of the input amplitude is discussed and compared to that of its free-running counterpart.

For this, the behavior for incremental operation is discussed using the example of the third-order CIFF single-bit DT I-DS ADC with a third-order CoI filter from Fig. 3.11.Modulator coefficients are again obtained with the Delta-Sigma Toolbox with OBG = 1.5 scaled for a maximum swing of 1 FS. Although the question is discussed with the help of the exemplary I-DS ADC, it has been proven that the resulting statements are valid for a large number of I-DS ADCs.

Dependency on M: The simulation results are averaged over a plentitude of simulation runs with varying input frequency and phase. The input amplitude is fixed to -10 dBFS, and the quantizer gain is simulated for variable number of cycles M per Nyquist sample. The results are shown in Fig. 3.13a and compared with the effective quantizer gain of the same modulator in free-running operation.

If the quantizer input and output of all M occurring samples within every Nyquist conversion are considered for the calculation of q according to (2.11), it can be seen that q shows a strong dependency on M (cf. Fig. 3.13a, dotted blue). While for a small number of samples M the deviation between the quantizer gain of the incremental modulator and the value of the free-running case is quite significant, it becomes smaller for large M. This can be explained by the influence of the transient settling of the incremental system directly after reset: its influence becomes stronger for low M. By omitting just the first two samples (cf. Fig. 3.13a —dashed blue), the calculated quantizer gain becomes already largely independent from the number of cycles M. If the first 5 samples of the incremental operation are neglected, no visible

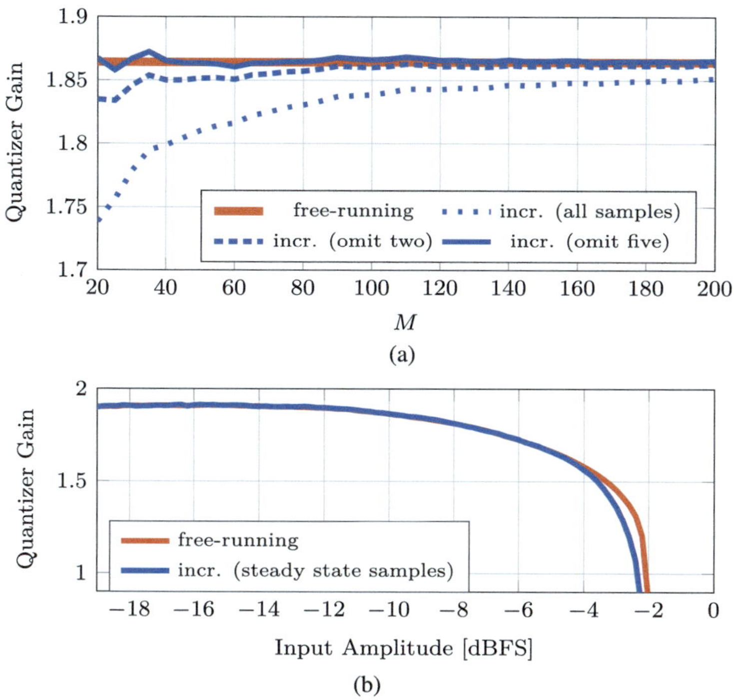

Fig. 3.13 Quantizer gain of the exemplary third-order single-bit I-DS ADC in CIFF architecture from Fig. 3.11 averaged over several simulations, (a) plotted over number of operation cycles M. Free-running operation (red) vs. incremental operation using all samples (dotted blue), all but the first two samples (dashed blue), and all but the first five samples (solid blue) for q calculation. (b) plotted over input signal amplitude. Free-running operation (red) versus incremental operation (blue). ©2023 IEEE. Modified, with permission, from Kaesser et al. [200].

deviation from the free-running case can be seen anymore. Similar behavior could be observed for other modulator orders and architectures.

In free-running DSMs, the effective gain of a multi-bit quantizer can be approximated to be unity [5], as long as no overload occurs or the quantizer input does not become too small such that only the inner QTZ level is always triggered in a single-bit manner. By analyzing the gain of a multi-bit quantizer similarly to the single-bit case for I-DSM, this assumption can be confirmed for I-DSMs.

While the actual number of five first cycles of the shown example will change for different architectures and parameters of the I-DSM, it can be generally stated that by omitting the first few cycles of every Nyquist period and consequently only use samples from steady state operation, the effective quantizer gain of an I-DSM is identical to that of its free-running counterpart.

Dependency on amplitude: The same result can be found for the behavior of the effective quantizer gain over amplitude, for which a simulation is shown in Fig. 3.13b using M=150. Again, results averaged over different input frequencies and phases are shown. The initial samples after reset were omitted to exclude the influence of the initial settling of the I-DSM after reset. It can be seen that the behavior in I-DSM (solid blue) is similar to the behavior of the quantizer gain in the free-running case (solid red).

In conclusion, the effective quantizer gain q of I-DSM can either be taken using simulations from its free-running counterpart or by neglecting the first few samples of each Nyquist conversion during transient settling and consequently only using the steady state samples of the simulations in incremental operation [200].

3.4 Comparison of Performance Predictions for I-DS ADCs

The performance of free-running DS ADCs is determined in the frequency domain with the NTF of the modulator and by using an ideal brick-wall filter for the decimation filter. This leads to the integration of the IBN as shown in (2.23). Using a low-frequency approximation for the modulator NTF and solving the integral leads to a handy approximation as it was shown in (2.35). Since this is not directly applicable for I-DS ADCs due to the periodic reset, the lack of an LTI system, and the inapplicable low-frequency approximation of the loop filter, the SoA found other solutions to predict the performance of I-DS ADCs, both based on the time-domain and frequency-domain considerations outlined in Sects. 3.2 and 3.3, which are discussed and compared next.

3.4.1 Performance Prediction based on Time-Domain Description

The probably most comprehensive transient analysis for DT single-loop I-DS ADCscan be found in [16]. Multiple I-DSM architectures are taken into account. Moreover, also other than CoI reconstruction filters are employed. As one example, [16] derives a theoretical estimate for the performance of the often used third-order, single-bit I-DSM with third-order CoI reconstruction filter [5, 176], cf. Fig. 3.4, for which the LSB is obtained by

$$LSB_{\text{ciff}|\text{markus}} = \frac{V_{\text{FS}}}{a_1 c_1 c_2} \cdot \frac{3!}{(M-2)(M-1)M} \tag{3.44}$$

with the high-level scaling coefficients a_1, c_1, c_2 and assuming a constant input. This equation can easily be generalized for an arbitrary L -th order I-DSM with

$$LSB_{\text{inc}|\text{time}} \approx \Delta \cdot \frac{1}{a_1 \prod_{i=1}^{L} c_i} \cdot \frac{L!}{M^L} \tag{3.45}$$

where Δ is the step width of the internal B-bit quantizer of the I-DSM. Note that the coefficient c_i was not included in the original equation of [16], since only single-bit QTZ were considered where the last scaling coefficient is compensated by the quantizer gain; it is added for completeness in (3.45). In effect, the forward coefficients of the highest order path together with the CoI filter determine the resolution. The IBN can then be calculated with

$$IBN_{\text{inc|time}} = \frac{LSB^2_{\text{inc|time}}}{12} \approx \frac{\Delta^2}{12} \cdot \frac{1}{\left(a_1 \prod_{i=1}^{L} c_i\right)^2} \cdot \left(\frac{L!}{M^L}\right)^2. \tag{3.46}$$

In principle, all subsequent work providing performance estimates for I-DS ADCs is based on the same time-domain considerations and yields similar results. For example, in the work of [178, 183], the same third-order CIFF architecture was analyzed for a CT I-DS ADC, and the ENOB was estimated for its idealized DT equivalent. In [14], the performance prediction of [16] was generalized for different modulator orders L and quantizer levels but without taking the loop-filter scaling into consideration. In [191], the performance of I-DS ADCs at low M was analyzed based on purely differentiating, free-running NTFs and the considerations were adapted for multi-stage designs. Nonetheless, the calculations ignored loop-filter scaling completely.

Derivation of ENOB of unscaled I-DS ADCs: For the sake of simplicity, in the following derivation, we assume an unscaled I-DSM, i.e. the scaling coefficients in (3.45)–(3.46) become unity. Then the IBN from (3.46) simplifies to

$$IBN_{\text{inc|unscaled}} \approx \frac{\Delta^2}{12} \cdot \left(\frac{L!}{M^L}\right)^2. \tag{3.47}$$

We have reviewed the linear quantizer model in Sect. 2.1.2.2, and it is text-book knowledge that a quantizer with step-width Δ_Q features a quantization noise

$$N_{\text{Q}} \approx \frac{\Delta_Q^2}{12}. \tag{3.48}$$

When we replace $\Delta_Q = {}^{FS}/_{(2^{ENOB}-1)}$, ENOB being the effective number of bits of a more general ADC definition and FS being the full-scale voltage, this becomes

$$N_{\text{ADC}} \approx \frac{\left(\frac{FS}{2^{ENOB}-1}\right)^2}{12}. \tag{3.49}$$

After replacing Δ in (3.47) by the actual step-width of the internal B-bit quantizer of the I-DSM, i.e. $\Delta = {FS}/{(2^B-1)}$, we equate (3.49) with (3.47), which results in

$$\frac{FS}{2^{ENOB}-1} = \frac{FS}{2^B-1} \cdot \frac{L!}{M^L} \tag{3.50}$$

$$\Leftrightarrow \quad ENOB_{\mathrm{inc|time}} = log_2\left((2^B-1)\cdot\frac{M^L}{L!}\right) \tag{3.51}$$

which is the equivalent ENOB of an I-DS ADC with CoI reconstruction filter, oversampling ratio of M, order L, and internal bitwidth B.

Note that this derivation does not include loop-filter scaling of the I-DSM, nor does it account for a reduced MSA. Both assumptions are acceptable if a medium-order multi-bit I-DSM is chosen but have to be adapted for scaled or single-bit I-DSMs.

3.4.2 Performance Prediction based on Frequency-Domain Description

While most SoA for the performance prediction of I-DS ADCs is based on time-domain considerations, the accuracy of the prediction is limited. Especially, when loop-filter scaling is required, time-domain predictions largely vary from transient simulations. Simulations on the other hand, while obviously giving the most precise performance estimation especially when nonidealities are included, lack any analytical insight and moreover are very time consuming. The frequency-domain description with its calculation of the overall NTF and STF outlined by Sect. 3.3.3 allows to assess the performance estimation in combination with the white noise quantizer model, as is known from free-running DSM, cf. Sect. 2.2.2.

For the calculation of the IBN of a free-running DS ADC, cf. (2.23) and e.g. (2.30), the modulator $NTF_{\mathrm{dsm}}(f)$, the quantization noise spectral density and the decimation filter function were used. Thereby, the integration over $\pm f_s$ could be simplified by the assumption of a brick-wall decimation filter leading to an integration over $\pm f_b$. In contrast, for a I-DS ADC, the overall NTF must be integrated over the whole sampled spectrum, as no further brick-wall filter is present. This yields

$$\begin{aligned} IBN_{\mathrm{inc|freq}} &= \frac{1}{f_s}\int_{-f_s/2}^{f_s/2} |NTF_{\mathrm{inc}}(f)|^2 \cdot S_e(f)\,\mathrm{d}f \\ &= \frac{\Delta^2}{12\cdot f_s}\cdot\int_{-f_s/2}^{f_s/2} |NTF_{\mathrm{inc}}(f)|^2\,\mathrm{d}f \end{aligned} \tag{3.52}$$

with $NTF_{\text{inc}}(f)$ being the overall $NTF_{\text{dt,inc}}(f)$ in (3.35) for a DT, or $NTF_{\text{ct,inc}}$ in (3.38) for a CT I-DS ADC. Having derived the integrated inband noise IBN_{inc} the SQNR can be estimated by

$$SQNR_{\text{inc|freq}} = 10 \cdot \log_{10}\left(\frac{P_{\text{in}}}{IBN_{\text{inc}}}\right) \tag{3.53}$$

with the input signal power P_{in} [201].

While this approach yields very accurate performance predictions, even under the influence of nonidealities included into the loop-filter NTF, unfortunately, the frequency-domain approach described in [201] lacks a closed analytically derived expression for this performance prediction, which would be similarly accurate and handy as (2.35) in the case of free-running DSM.

3.4.3 Performance Prediction based on Approximate Frequency-Domain Description

An analytical expression similar as derived for free-running DSM in (2.35), which is both accurate and handy, has been introduced in [202]. The approach is based on similar considerations as in the free-running case. A low-frequency approximation is used for the free-running NTF of the DSM, but instead of a brick-wall filter, a CoI reconstruction filter is used

$$H_{\text{rec}}(z) = \left(\frac{1}{1-z^{-1}}\right)^{L_d} \cdot G, \tag{3.54}$$

where its gain G is

$$G = \frac{L_d!}{\prod_{i=0}^{L_d-1}(M+i)} \approx \frac{L_d!}{M^{L_d}}. \tag{3.55}$$

Solving the simplified integral to calculate the IBN in case the modulator order equals the CoI order ($L = L_d$) leads to

$$\begin{aligned} IBN_{\text{inc|approx}} &\approx \frac{\Delta^2}{12} \cdot \frac{1}{f_s} \int_{-f_s/2}^{f_s/2} \left| NTF_{dsm,lf}(z) \cdot H_{\text{rec}}(z) \right|^2 \mathrm{d}f \\ &\approx \frac{\Delta^2}{12} \cdot \left(\frac{G}{q \cdot a_1 \prod_{i=1}^{L} c_i}\right)^2 \frac{1}{f_s} \int_{-f_s/2}^{f_s/2} \left| \left(\frac{1}{1-z^{-1}}\right)^{L_d - L} \right|^2 \mathrm{d}f \\ &\approx \frac{\Delta^2}{12} \cdot \left(\frac{G}{q \cdot a_1 \prod_{i=1}^{L} c_i}\right)^2. \end{aligned} \tag{3.56}$$

However, it was shown in [202] that using this low-frequency approximation of the NTF of the free-running DSM introduces an error once the loop filter is scaled. To account for this error a correction factor $^{OBG_{\mathrm{sc}}}/_{OBG_{\mathrm{unsc}}}$ was introduced, where

$$OBG_{\mathrm{unsc}} = NTF_{\mathrm{dsm}}(z = -1) = 2^L \tag{3.57}$$

is the OBG of an ideal high-pass filtering NTF of the free-running DSM when the loop filter is not scaled, and OBG_{sc} is the OBG of the NTF of the scaled loop filter. With this, the IBN can be estimated with

$$IBN_{\mathrm{inc|approx}} \approx \frac{\Delta^2}{12} \cdot \frac{1}{\left(q \cdot a_1 \prod_{i=1}^{L} c_i\right)^2} \cdot \left(\frac{L!}{M^L}\right)^2 \cdot \frac{OBG_{\mathrm{sc}}}{2^L}. \tag{3.58}$$

An intuitive explanation of the correction factor can be given by looking at Fig. 3.14. There, the overall NTFs of a third-order I-DS ADC with a CoI3 reconstruction filter, $M = 50$, 4-bit quantizer, and different loop-filter scalings yielding in different OBG are depicted. Without the correction factor $^{OBG_{\mathrm{sc}}}/_{OBG_{\mathrm{unsc}}}$, (3.56) basically integrates the DC value of the NTF over the whole frequency range $0 \ldots f_s/2$ leading to a pessimistic prediction. The use of the correction factor is inspired by the concept of an effective noise bandwidth in filters. Therefore, the overall NTF in Fig. 3.14 is approximated as a brick-wall filter with a flat response (inband gain) from DC up to an effective bandwidth, beyond which it is approximated as zero. Notably, the effective noise bandwidth of these NTFs can be estimated using the expression $^{OBG_{\mathrm{sc}}}/_{OBG_{\mathrm{unsc}}}$ [202]. Thereby note how the correction is not needed for the unscaled loop filter with $OBG = OBG_{max} = 8$ for the third-order DSM, where the magnitude of the NTF in Fig. 3.14 is flat and thus no correction for the high-frequency drop in the NTF is needed.

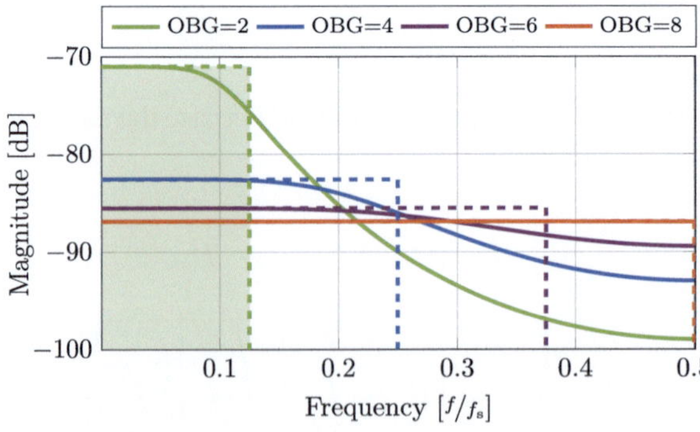

Fig. 3.14 NTFs of a third-order I-DS ADC with a CoI reconstruction filter, $M = 50$, 4-bit quantizer and different loop-filter scalings. The dashed lines illustrate the idea of an effective noise bandwidth using the DC values of the NTFs up to $\frac{f_s}{2} \cdot {}^{OBG_{\mathrm{sc}}}/_{OBG_{\mathrm{unsc}}}$, cf. (3.58). Licensed under CC BY 4.0. Reprinted, with permission, from Kaesser et al. [202].

Derivation of I-DS ADC ENOB including loop-filter scaling: Observing the difference between (3.46) and (3.58), the derivation of the ENOB of an I-DS ADC performed in (3.51) can be extended by the loop-filter scaling to give

$$ENOB_{\text{inc|approx}} = log_2\Bigg[\underbrace{\underbrace{\underbrace{(2^B-1)\cdot\frac{M^L}{L!}}_{(3.51)}\cdot\left(q\cdot a_1\sum_{i=1}^{L}c_i\right)}_{\text{Incl. Scaling as in (3.46)}}\cdot\sqrt{\frac{2^L}{OBG_{\text{sc}}}}}_{\text{Incl. OOB effect, cf. (3.58)}}\Bigg] \tag{3.59}$$

3.4.4 *Comparison of Performance Predictions*

Having reviewed three approaches to predicting the performance of I-DS ADCs, we will now compare them with various simulation results. As a first example, the previously used third-order single-bit DT CIFF I-DS ADC from Fig. 3.11 (scaled for OBG=1.5) is simulated using $M = 100$. In Fig. 3.15, a comparison is shown for the simulated SQNR (black) and its predicted values based on the exact frequency-domain calculation in (3.52) (red), the expression found via time-domain derivations in (3.46) (blue), and the approximated frequency-domain approach from (3.58). The simulation achieves a maximum SQNR of 91dB at an MSA of −3.5 dBFS. The frequency-domain calculation (red) tracks the simulation perfectly and only lacks prediction of the MSA. The prediction using the time-domain analysis largely underestimates the performance and is approximately 15 dB below the simulated

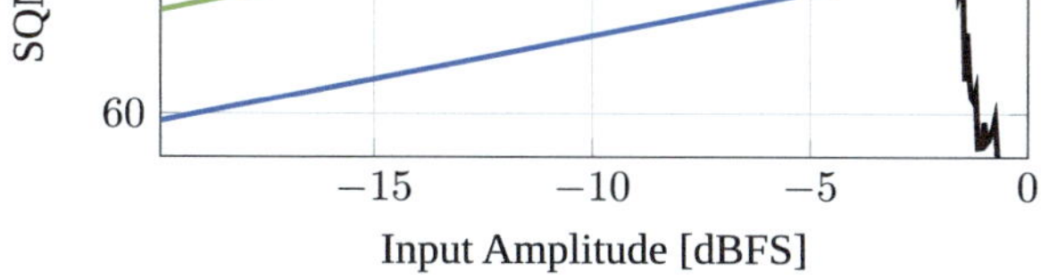

Fig. 3.15 Simulated and predicted SQNR of a third-order single-bit DT CIFF I-DSM scaled for an OBG=1.5 using the Schreier Toolbox and at $M = 100$. Black: simulation. Blue: prediction using (3.46). Red: prediction using (3.52). Green: prediction using (3.58). Licensed under CC BY 4.0. Modified, with permission, from Kaesser et al. [202].

value. The prediction using the approximated frequency-domain approach slightly underestimates the performance and deviates only about 3dB below the simulated values in this exemplary case.

To analyze the behavior of the predictions at different oversampling ratios, the simulated and predicted IBN of the exemplary third-order single-bit I-DS ADC is plotted over M in Fig. 3.16. It can be seen that every approximation nicely tracks the course of the simulation over the oversampling M. However, the time-domain prediction again has a large offset, while the other two predictions only have a slight deviation from the simulation.

Having analyzed the exemplary single-bit I-DS ADC for a loop-filter scaling of OBG=1.5, the performance prediction is extended by an example, where the OBG and thus loop-filter scaling can be varied under otherwise constant conditions. Due to stability, the possible variation of scaling in DSMs using a single-bit quantizer is limited. Therefore, a multi-bit I-DSM with 4-bit quantizer is used. In Fig. 3.17, the

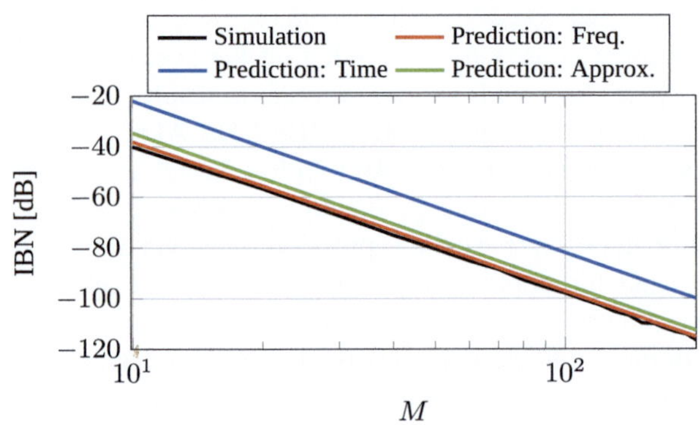

Fig. 3.16 Simulated and predicted IBN of a third-order single-bit DT CIFF I-DSM scaled for an OBG=1.5 using the Schreier Toolbox. Licensed under CC BY 4.0. Modified, with permission, from Kaesser et al. [202].

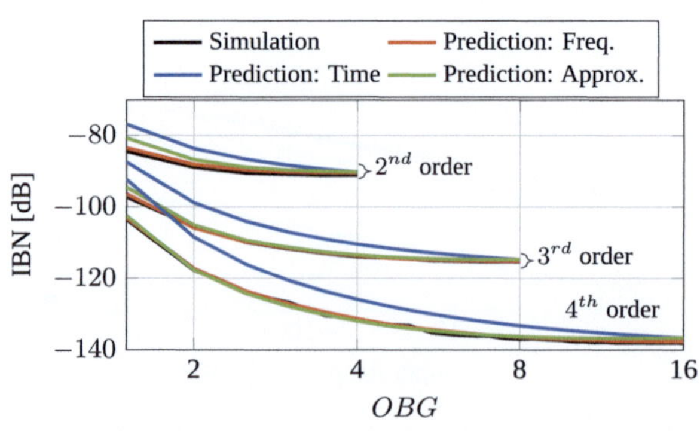

Fig. 3.17 Simulated and predicted IBN of a second- to fourth-order DT CIFF I-DSM with 4-bit quantizer at $M = 50$. The coefficients are derived using the Schreier Toolbox [144]. Licensed under CC BY 4.0. Modified, with permission, from Kaesser et al. [202].

simulated and predicted IBN of a second- to fourth-order DT CIFF I-DSM with 4-bit quantizer at M=50 is shown. The coefficients are derived for varying constraints of the OBG using the Schreier toolbox [144]. Again, it can be seen that the exact frequency-domain calculation (red) and the frequency-domain approximation (green) nicely follow the course of the simulation over various OBG values. The time-domain approximation only show a good fit if the loop filter is unscaled and OBG=$OBG_{\max}$ is not limited. Also note that the frequency-domain approximation becomes slightly worse for lower order and very small OBG, which in fact is not a suitable design choice and is thus of little practical relevance.

3.5 Reconstruction Filters for I-DS ADCs

In Sect. 3.2 the working principle of I-DS ADCs was introduced. Based on time-domain description suitable digital reconstruction filters were derived. A first-order CoI was found for a first-order DT I-DSM in (3.8), a third-order CoI for a third-order DT I-DSM with CIFF architecture in (3.16), and a combination of a second-order and a first-order CoI was derived with the same approach to be a suitable filter for a second-order DT I-DSM with CIFB architecture in (3.24). Also, in the derivation of the frequency-domain description, we have assumed and simulated I-DSM combined with CoI reconstruction filters of the same order until now.

Though, various implementations of reconstruction filters for I-DS ADCs have been proposed in the SoA, especially in combination with higher-order modulators. For a more general analysis, the incremental modulator can thus be regarded as an encoder that encodes the analog input signal $y_n(t)$ into a coarsely-quantized sequence $d_n[k]$. Then, the reconstruction filter works as decoder that intents to give the most precise digital estimate $D[n]$ for the input signal and adds otherwise useful behavior. In the following, a short overview of SoA reconstruction filters is given and subsequently, various implementations are discussed in detail.

3.5.1 Overview on Reconstruction Filter Classes for I-DS ADCs

Linear filters: Most SoA reconstruction filters $H_{\text{rec}}(z)$ in I-DS ADCs are linear filters, similar as in the free-running case. As for free-running lowpass (LP) DSMs, the linear filter needs to have low-pass characteristic, however, the transfer characteristic is altered due to the periodic reset. This is why sophisticated and concatenated filter implementations—as used for free-running DS ADCs—cannot be applied for incremental operation, as they are not implementable due to the periodic reset. In order to generalize the CoI examples from the previous sections, we generalize the reconstruction for arbitrary filter shape. Then, the output at Nyquist rate can be obtained by

$$D[n] = \sum_{k=1}^{M} W_{\text{lin}}[k] \cdot v_n, \qquad W_{\text{lin}}[k] = h_{\text{rec}}[M-k] \tag{3.60}$$

with $W_{\text{lin}}[k]$ being the filter weights of the respective reconstruction filter and $h_{\text{rec}}[k]$ the filter impulse response. Using CoI reconstruction filters with an order equal to the order of the I-DSM has been proposed early in literature [174] and their implementation was later studied many times [16]. As early as [13] it was described that the order of the CoI filter can be higher than that of the I-DSM [176]. Both will be further revised in Sect. 3.5.2.1. Wagner et al. [201] also introduced a modified higher-order CoI reconstruction filter.

Sinc reconstruction filters have been used in the SoA as they implement STF notches which can be advantageously used to suppress disturbers such as e.g. line noise [16, 203]. Also, [204] showed that a sinc reconstruction filter has significant advantages concerning the number of gates and thus physical chip area. They will be shortly revised in Sect. 3.5.3. Another linear filter option is a matched digital filter [205, 206], which resembles the filter transfer function of the modulator. It will be revised in Sect. 3.5.4.

Algorithmic filters: Also algorithmic filters can be used as decoders for I-DSM bit streams. In [207, 208], the Zoomer algorithm was employed on a first-order free-running DSM with DC input. This was then adapted to I-DS ADCs in [209], where it was proven to be the optimal decoder for given initial conditions, which then allowed to define an upper performance bound for reconstruction filters of I-DSMs. Even though this optimal decoder was actually implemented for a second-order I-DS ADC [210], its main use is theoretical analysis as in actual implementation the algorithm requires quite complex calculations, making its circuit implementation very power and area hungry. Also, it requires a strictly constant input during operation, which mandates using a S/H upfront the I-DSM as well as very low circuit noise for functionality.

Another nonlinear algorithm was proposed in [211] for decoding a first-order, single-bit I-DSM which could even outperform the Zoomer algorithm. At first, this comes surprising as the Zoomer algorithm was claimed optimal. But while the optimality of the Zoomer algorithm is based on the assumed initial condition, [211] stated that those were indeed not optimal: the Zoomer algorithm assumes a maximum quantization error for the first sample, while the algorithm proposed in [211] assumes the quantization error as zero. The proposed reconstruction algorithm was realized in [212]. McIlrath [213] extended the work by considering thermal noise (for free-running DSMs with constant inputs). In [214], another algorithm was then proposed which was adapted to work for linearly increasing inputs. In [215], this algorithm was used on a second-order MASH I-DSM with constant input. More recently, also [216] considered constant inputs but under the influence of noise. This work modified the Zoomer algorithm to allow thermal noise, but it sacrificed quantization noise performance instead.

After this short overview of various reconstruction filters in the SoA, next their implementation, core properties, advantages, and disadvantages are elaborated in more detail.

3.5.2 CoI Reconstruction Filter

3.5.2.1 CoI Reconstruction Filter of the Same Order as the I-DSM: $L_d = L$

The most common linear reconstruction filter is the CoI filter [5, 176] due to its simplicity and superior quantization noise performance. It consists of a chain of resettable digital integrators as shown in Fig. 3.18. This leads to the free-running filter transfer function

$$H_{\mathrm{coi}}(z) = \left(\frac{1}{1 - z^{-1}}\right)^{L_d} \cdot z^{-d} \tag{3.61}$$

of a digital CoI with L_d being its order and $d = \{0 \ldots L_d\}$ being the delay of the CoI filter.

The first M CoI filter weights, which are used to build a Nyquist-rate output sample of an I-DS ADC before it is being reset as defined in (3.60), can be written as

$$\mathrm{w}_{\mathrm{lin|CoI}}[k, L_d, d] = \begin{cases} \frac{1}{(L_d-1)!} \prod_{i=1}^{L_d-1} (M - d - k + i) & , 1 \le k \le M - d \\ 0 & , \text{else.} \end{cases} \tag{3.62}$$

As derived throughout Sects. 3.2.1 and 3.2.2 for the first- and third-order CIFF, the order of the CoI reconstruction filter is usually chosen the same as the analog I-DSM [5]. However, the SoA has reported filters with higher order than the modulator [13, 16, 176, 201], which will be discussed in detail in Sect. 3.5.2.2.

Overall STF: The transfer function according to (3.61) of CoI filters from first to fourth order are shown in Fig. 3.19 for an exemplary reset after M=150. The first-order CoI filter provides notches at multiples of f_{N}, but also shows a distinct drop

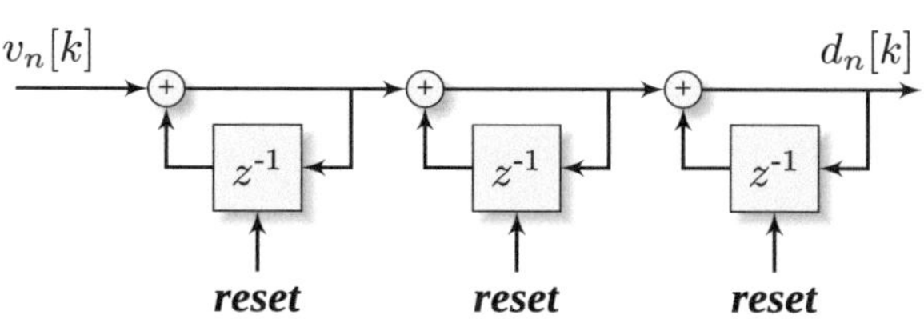

Fig. 3.18 Block diagram of an exemplary resetting delay-free ($d = 0$) third-order ($L_d = 3$) CoI reconstruction filter. *Note* Decimation by M to Nyquist-rate digital output $D[n]$ is not shown. ©2021 Wagner. Modified, with permission, from Wagner [27].

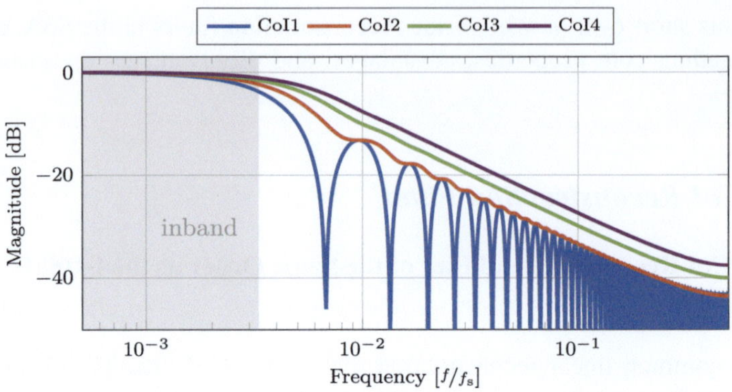

Fig. 3.19 Transfer functions of first- to fourth-order CoI filters according to (3.61) with an exemplary reset after $M = 150$.

toward the bandedge. As we already mentioned, the overall STF of I-DS ADC is mostly defined by the reconstruction filter, and the notches can be advantageously used to reduce interfering signals at certain frequencies. Disadvantageously, the drop at the band edge also causes signal power loss, i.e. an inband frequency dependent SNR will result. By increasing the CoI filter order the drop at the bandedge is advantageously reduced. However, also the OOB suppression is reduced. Considering input referred white noise and using higher-order CoI reconstruction filters, the OOB noise is consequently less suppressed before being folded into the inband by decimation to Nyquist rate. This fact also leads to an increased noise in higher-order incremental converters, when (white OOB) circuit noise is considered, which is known in the SoA as *noise penalty factor (NPF)* [192, 217]. The NPF will be further analyzed in Sect. 4.3.1.

Overall NTF: The shape of the resetted higher-order CoI reconstruction filters in Fig. 3.19 is not straightforward when viewed with the expectation of a free-running DSM operation. There, a low-pass filter is spectrally multiplied with a high-pass filtered quantization noise spectrum, and we expect a higher-order digital low-pass filter to generate more suppression. In the incremental operation, we have to regard the overall transfer functions as introduced in Sect. 3.3.1. The overall NTFs for DT single-bit I-DSMs of first to fourth order concatenated with CoI reconstruction filters of the same order are shown in Fig. 3.20. The loop filters of each modulator are scaled to have an OBG of 1.5 using the DS Toolbox [144]. The oversampling is set to $M = 150$. The superior suppression of quantization noise by the overall higher-order NTFs can be seen.

Filter weights: The normalized filter weights (cf. (3.62)) of CoI filters from first to fourth order are shown in Fig. 3.21, again plotted for an exemplary $M = 150$. Normalization is done by dividing the calculated weights by the sum of all weights. This illustrates how every quantizer decision at time step $k = 1 \ldots M$ is weighted

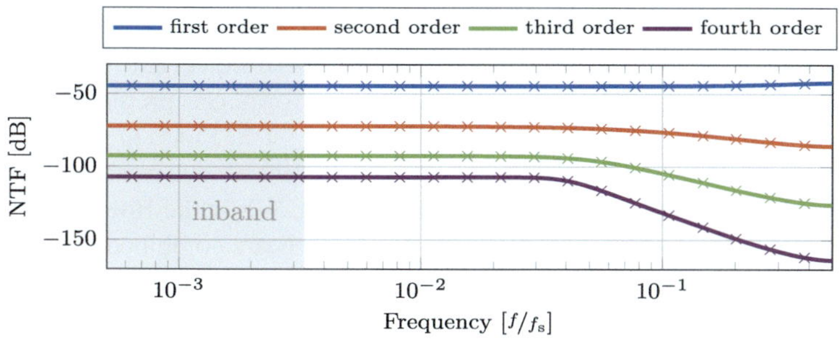

Fig. 3.20 Overall NTF for first- to fourth-order I-DSM combined with CoI filters of equal order.

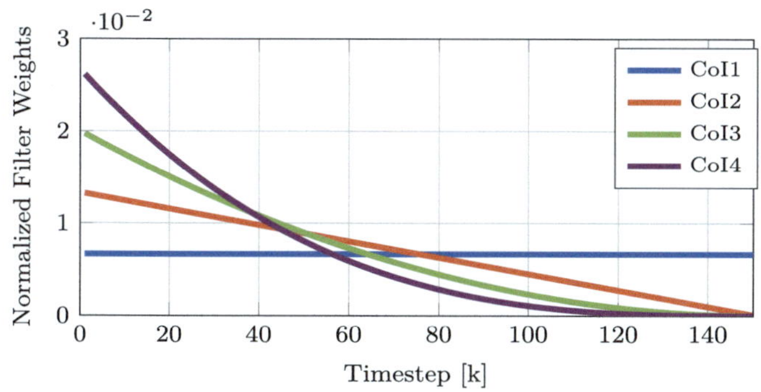

Fig. 3.21 Normalized filter weights of first- to fourth-order CoI filters vs. timestep $k = 1 \dots M$ for an exemplary $M = 150$ according to (3.62).

relatively toward the decimated digital output. Interestingly, in higher-order filters early samples are much higher weighted than later samples. This is because the earlier the decision is taken over by the first digital integrator, the more often it will be integrated by subsequent stages. This property can be exploited on circuit level, as nonidealities, which appear later during a conversion cycle, are less important, which can be used for dynamic nonideality suppression. Also, input signal samples are most dominantly weighted during early cycles, which is useful for the introduction of the so-called RP. Both techniques are further analyzed in Sect. 5.4.

3.5.2.2 CoI Reconstruction Filters with $L_d = L + 1$

CoI reconstruction filters with orders L_d higher by one than the I-DSM loop filter order L have already been reported in [13]. However, the resulting performance differences have not been analyzed a lot and have only been shown in the time domain for direct current (DC) operation [16, 176]. Consequently, the influence of increasing the order of the CoI reconstruction filter by one over the order of the I-DSM is illustrated next.

Overall NTF and QN performance: Wagner et al. [201] analyzed the achievable performances of I-DSM combined with CoI filters of equal and increased order. This is shown in Fig. 3.22a, where the SQNR performance predictions based on the overall NTF and IBN calculations are shown for I-DSM with first- to third-order loop filters. The achievable SQNR is compared to the achievable performance of free-running DSMs of same order, OSR and loop filters. While the free-running DSMs are combined with ideal brick-wall filters, the I-DSMs are combined with CoI filters of the same order or one order higher. Each modulator uses a CIFF architecture, a single-bit quantizer and is scaled for an OBG of 1.5 using the DS toolbox [144]. The difference between the achievable SQNR of the I-DS ADCs from their free-running counterparts is then plotted against the oversampling ratio M.

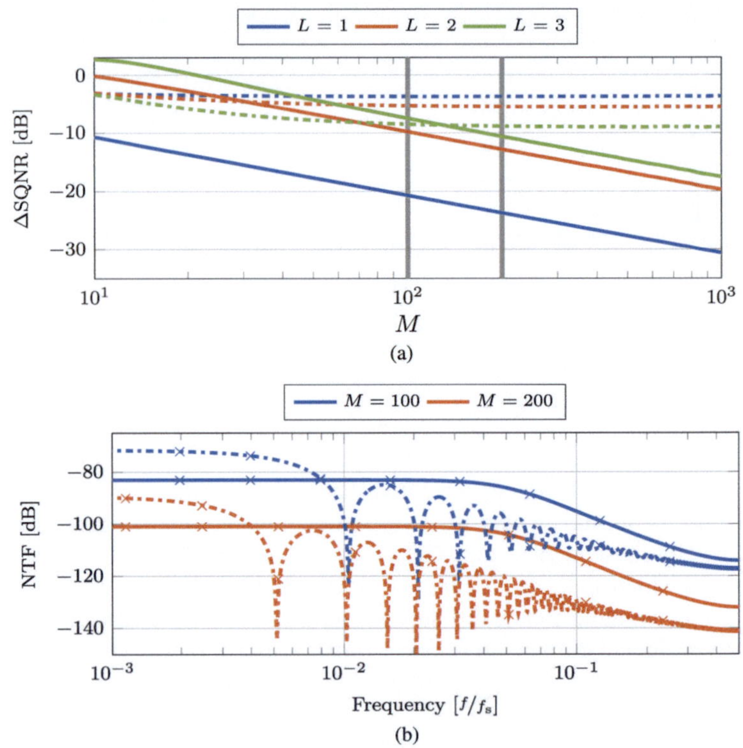

Fig. 3.22 (a) Comparison of performance difference ΔSQNR of I-DS ADCs versus free-running DS ADCs from first- to third-order modulator loop filters. Each modulator uses a CIFF architecture, a single-bit quantizer and is scaled for an OBG of 1.5 using the DS toolbox. Free-running DSMs are combined with brick-wall decimation filters, incremental DSMs are combined with CoI filters of the same order (solid) and one order more (dash-dotted). (b) Calculated NTFs of the third-order I-DS ADC with third-order CoI filter (solid) and fourth-order CoI filter (dash-dotted), verified by simulation points (x-marks). ©2020 IEEE. Modified, with permission, from Wagner et al. [201].

It can be observed that for I-DS ADCs with equal orders of the digital CoI and the analog loop filter (solid lines in Fig. 3.22a), the performance loss compared to free-running DSMs becomes worse for larger M. Noteworthy, at very low oversampling M, higher-order I-DS ADCs can even outperform their free-running counterparts concerning quantization noise suppression (cf. $\Delta SQNR \geq 0dB$ in Fig. 3.22a) [191].

The effect of the increased order of the digital CoI filter $L_d = L+1$ on the achievable SQNR performance is seen when comparing the dash-dotted lines vs. the solid lines in Fig. 3.22a. First, the performance drop of the incremental vs. free-running operation becomes largely constant over M, and only at very low M a remaining dependency is seen. Though, at these very low M, no incrementally operating DS ADC ever outperforms the free-running DSM, which is in contrast to the case with equal digital and analog orders (solid lines). Second, it can be seen that for the increased digital CoI filter order, the performance drop of incremental operation is relaxed.

To explain this behavior, in Fig. 3.22b the corresponding overall NTFs are exemplarily shown for $M = 100$ and $M = 200$ for a third-order I-DSM combined with third- (solid) or fourth-order (dashed-dotted) CoI reconstruction filters. For both M, the NTFs for the modified CoI reconstruction filters feature a sinc-like shape with notches distributed over the whole frequency band. As the width of the (inband) main lobe varies with M, the accumulated IBN consequently reduces exponentially. This leads to the observed rather constant performance difference with the free-running ADC, where the brick-wall decimation filter performs similarly with changed filtering dependent on M.

Overall STF and MSA: As the overall STF of I-DS ADCs is largely defined by the reconstruction filter, the observations made around Fig. 3.19 can be used to explain the effect of using CoI reconstruction filters of one order higher than the I-DSM. The band edge attenuation is reduced, but also the OOB suppression drops. This results in less wideband (white input referred) noise attenuation and thus worse NPF as mentioned in Sect. 3.5.2.1, which will be further investigated in Sect. 4.3.1.

Further, it can be observed that using CoI reconstruction filters of higher order than the I-DSM also the MSA [218] (cf. Sect. 4.1.1.2). This can be explained by knowing that the quantizer of the I-DSM starts overloading close to the MSA. Thereby, the quantizer introduces excessive quantization errors (not quantization noise, but nonlinear quantization errors) during the transient phase of each Nyquist conversion. These nonlinear quantization errors are then filtered differently depending on the used reconstruction filter. In case of a higher-order CoI, they are not efficiently filtered and appear at the output. Therefore, the performance close to the MSA drops and consequently the amplitude, at which the peak SNDR is seen, is reduced the same as if a higher-order CoI is used as reconstruction filter compared to a CoI of the same order as the I-DSM. Thus, the potentially gained SQNR using a higher-order CoI is partially traded with a decreased maximum useful input amplitude. For this reason, as will be discussed in Sect. 3.6, a co-synthesis of higher-order CoI and I-DSM is highly advisable to find I-DSM scaling coefficients with optimal overall performance.

Increasing $L_d > L + 1$: If the order of the CoI reconstruction filter is even further increased, it can be shown that the SQNR is not further increased. On the contrary, the SQNR is decreased compared to a filter of the order $L_d = L + 1$. Therefore, a further increase of the filter order only gives disadvantages and is not useful.

Modified CoI reconstruction filter: As seen in Fig. 3.22a, the performance improvement of a higher-order CoI reconstruction filter is only seen at medium to large OSR. As an alternative, [201] reported a modified CoI reconstruction filter, where a lossy integrator realizes the additional order. This modified CoI reconstruction filter shows performance improvement already for lower OSR. However, the lossy integrator's digital realization is complex; therefore, the modified CoI filter has not been used in the SoA.

3.5.2.3 Achievable Performances of I-DSM using CoI Reconstruction Filters

In order to give an overview of achievable performances when using the shown variants of CoI reconstruction filters for various orders and various OSR, Fig. 3.23 shows the results of achievable SQNR plotted vs. OSR for $L = 1 \ldots 4$ for the conventional CoI with $L_d = L$ and for CoI with increased order $L_d = L + 1$. The I-DSM employs a CIFF architecture and scaling coefficients are obtained using the DS toolbox, using OBG of 1.5 in case of the single-bit quantizer in Fig. 3.23a, and using OBG of 3 in case of the 4-bit quantizer in Fig. 3.23b. The SQNR is calculated using the NTFs of

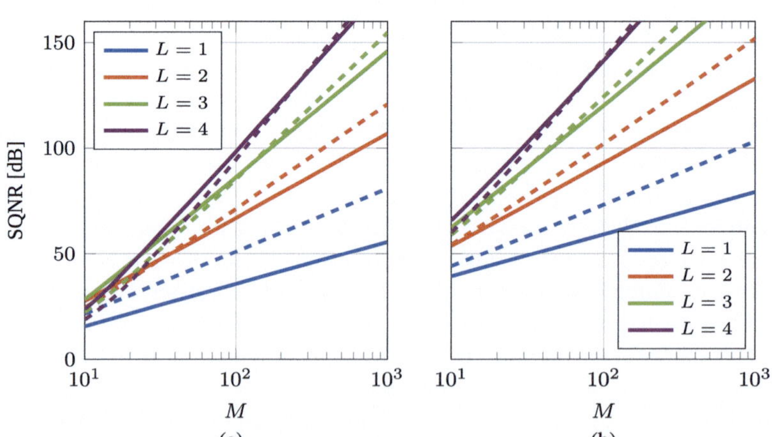

Fig. 3.23 Achievable SQNR of I-DSMs of various orders, calculated for CIFF architectures with an input signal of −6 dBFS. (a) Single-bit quantizer with loop-filter coefficients scaled for an OBG of 1.5. (b) 4-bit quantizer with loop-filter coefficients scaled for an OBG of 3. Solid: $L_d = L$. Dashed: $L_d = L + 1$.

the I-DS ADCs and using an input signal of −6dBFS. Note that the MSA is higher than the chosen input signal, but the overall tendency of achievable performances can be clearly identified.

It can be seen that the higher-order reconstruction filters show their main benefit for low-order I-DSM. Also, as is known for free-running DSM, the performance gains from increasing I-DSM order become smaller the higher the order, due to less aggressive scaling which is required for a stable operation.

3.5.3 Sinc Reconstruction Filter

Sinc filters have also been used as reconstruction filters in I-DS ADCs. They offer the possibility to generate notches in the STF, which can be used, for example, to suppress line noise or other disturbances [16, 203, 219]. While this is similarly possible with a first-order CoI reconstruction filter, cf. Fig. 3.19, this feature is lost for higher-order CoI. Then, sinc filters can be employed. However, this advantage of sinc filters leads to a compromise when it comes to quantization noise performance and SQNR, and I-DS ADCs using sinc reconstruction filters require higher OSR for the same resolution as when using CoI filters. A L_d-th-order sinc filter can be generally expressed as a concatenation of L_d integrators and delayed differentiators as shown in Fig. 3.24, which yields a free-running transfer function

$$H_{\text{sinc}}(z) = \prod_{l=1}^{L_d} \frac{1 - z^{-D_l}}{1 - z^{-1}} \tag{3.63}$$

with L_d being the digital filter order and D_l being the individual decimation factors for the differentiators of the sinc filter. For an exemplary first-order filter, we derive the DTFT

$$H_{\text{sinc}|1}(z) = \frac{1 - z^{-D_1}}{1 - z^{-1}} \xrightarrow{z=e^{j\cdot 2\pi f/f_S}} \frac{sin(\pi D_1 f/f_S)}{sin(\pi f/f_S)} \cdot e^{-j\cdot\pi(D_1-1)f/f_S} \tag{3.64}$$

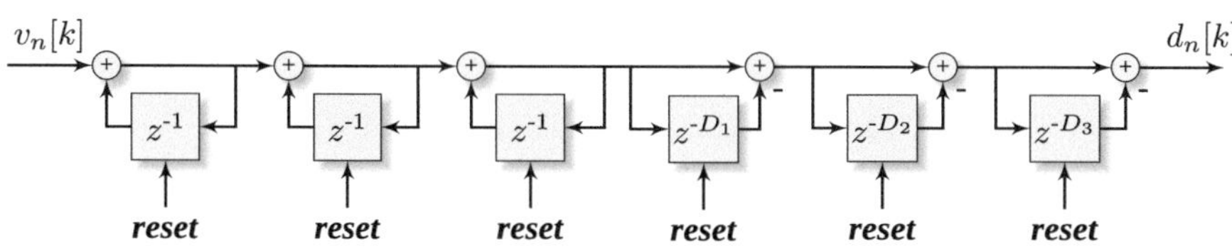

Fig. 3.24 Block diagram of an exemplary third-order sinc filter. ©2021 Wagner. Modified, with permission, from Wagner [27].

which leads to the magnitude of the frequency response

$$\left|H_{\mathrm{sinc}|1}(f)\right| = D_1 \left|\frac{sinc(D_1 f/f_S)}{sinc(f/f_S)}\right|, \tag{3.65}$$

which obviously gives this filter its name. The weights of a first-order sinc filter with length D_1 are then simply found as [3]:

$$w_{lin|\mathrm{sinc}|1}(k) = \begin{cases} 1, & \text{if } (1 \leq k \leq D_1) \\ 0 & \text{, otherwise} \end{cases} \tag{3.66}$$

With $M = D_1$, this becomes identical to a first-order CoI filter. A very hardware efficient implementation of an L_d-th-order sinc filter can be achieved with the CIC structure [43]:

$$H_{L,D}(z) = \left(\frac{1 - z^{-D}}{1 - z^{-1}}\right)^L \tag{3.67}$$

for which an exemplary third-order CIC implementation is shown in Fig. 3.25. There, the individual decimation factors D_l from Fig. 3.24 are identically set as $D_l = D$. Then, in contrast to the general sinc filter in Fig. 3.24, the CIC structure runs the comb part at a reduced sampling frequency f_s/D.

Filter weights: The normalized filter weights of sinc filters from first to third order are shown in Fig. 3.26 for an exemplary M=150. The decimation factors are set to $D = \frac{M}{L_d}$, i.e. 150, 75 and 50 for L_d = 1, 2, and 3, respectively. The weights illustrate how every quantizer decision at timestep $k = 1 \ldots M$ is relatively weighted toward the decimated digital output.

With the chosen $D = M$, the weights of the first-order sinc are identical to the weights of the first-order CoI. Further, it can be seen that the weights of the

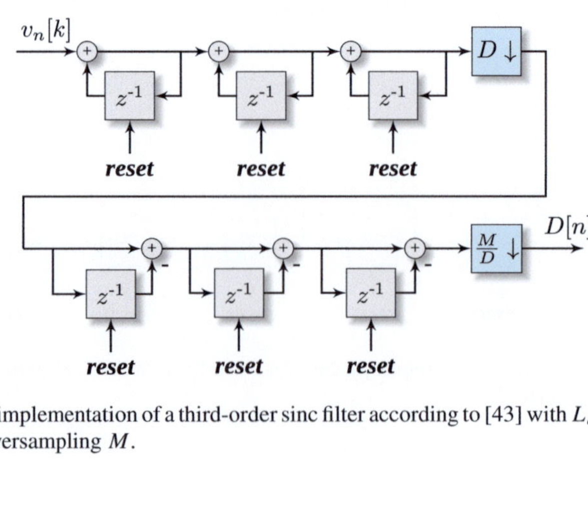

Fig. 3.25 CIC implementation of a third-order sinc filter according to [43] with $L_d = 3$, decimation factor D and oversampling M.

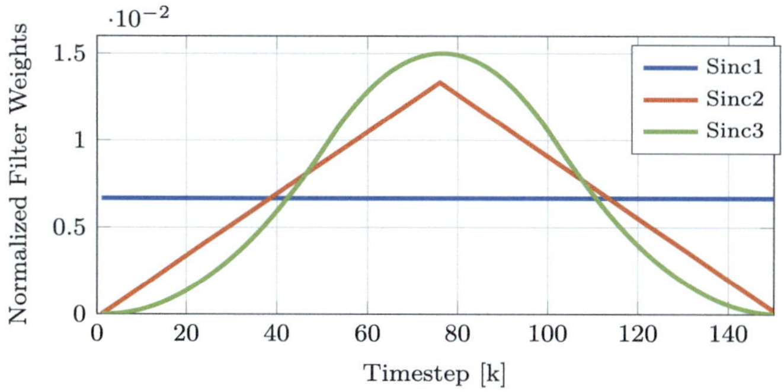

Fig. 3.26 Normalized filter weights of first- to third-order sinc filters vs. timestep $k = 1 \dots M$ for an exemplary $M = 150$ and $D = \frac{M}{L_d}$ (150, 75 and 50).

higher-order sinc filters are approximately symmetrical. They become exactly symmetric if the following condition is met [220]:

$$M = \left(\sum_{l=1}^{L_d} D_l \right) - (L_d - 1)\,. \tag{3.68}$$

If a CIC implementation with $D = D_l$ is used, it is not possible to fulfill this condition for every possible oversampling M. If the decimation factors D_l are chosen such that the rhs of (3.68) is larger than M, symmetry is lost since the differentiation starts too late and the early samples are weighted higher than the late ones. This loss in symmetry then prevents the sinc shaped transfer function and its property to suppress e.g. line noise. If the rhs of (3.68) is smaller than M, the differentiation starts too early which leads to zero weights for the early samples. The symmetry of the remaining weights is then maintained, but the early zero weighted samples lead to an effective reduction of the oversampling and therefore to a loss of SQNR. In practice however, this fact plays only a minor role. All implementations known to the authors that use a sinc reconstruction filter use high oversampling. Therefore, even if the decimation factors are chosen such that the rhs of (3.68) is less than M, the loss in SQNR is negligible due to the high oversampling. If the decimation factors D_l are chosen such that the rhs of (3.68) is greater than M, there is only a small change in the notches of the transfer function due to the high oversampling, and its suppression property is still preserved.

Overall STF: As the overall STF of I-DS ADC is dominated by the reconstruction filter, the resulting STF is basically the transfer function of the reset filter in (3.63) or (3.67). In the CIC filter transfer function, the periodic notches implicitly appear at integer multiples of f_s/D. Given a sampling frequency f_s, D needs adjusting for a certain disturber location such that the periodic notches coincide with the disturber.

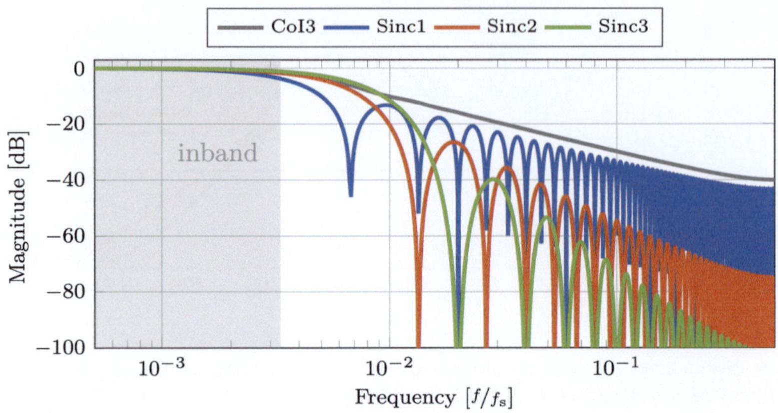

Fig. 3.27 Transfer function of first- to third-order CIC filters for an exemplary OSR of $M = 150$ and decimation factors $D = M/L_d$, which leads to $D = 150$, $D = 75$ and $D = 50$. For comparison reasons the transfer function of the CoI3 is shown, too.

The transfer functions of first- to third-order sinc reconstruction filters are shown in Fig. 3.27, where $M = 150$ and $D = \frac{M}{L_d}$ in Fig. 3.25 has been chosen, which leads to $D = 150$, 75 and 50. The resulting filter transfer functions are compared to a third-order CoI filter. The chosen parameters lead to rhs of (3.68) being too small, i.e. even though early samples are zero weighted, the remaining weights are still symmetrical leading to sinc transfer functions featuring its characteristic notches in Fig. 3.27. Since the decimation factors D are chosen dependent on the filter orders L_d, it can be seen that the notches of each filter are placed at different frequency positions. For minor adjustments of the notch positions the values of D can be slightly varied for each filter order. Though, this is traded with the disadvantage of a decrease in SQNR, as mentioned in the last paragraph.

Alternatively, the more general sinc filter in Fig. 3.24 can be implemented, as it allows to compromise filter implementation complexity, STF design, and achievable SQNR. This is exemplarily shown by choosing individual decimation factors D_l and is simulated in Fig. 3.28. It can be seen that wider notches can be realized in this example.

Overall NTF: The achievable quantization noise performance of I-DSM combined with sinc reconstruction filters can be determined by means of the overall NTF. This is shown for a third-order CIC implementation with $M = 150$ and $D = M/3 = 50$ (red) in Fig. 3.29 [201] and is compared to the overall NTF of a third-order CoI reconstruction filter (blue). Thereby, the I-DSM is chosen again as the exemplary third-order single-bit CIFF architecture scaled for an OBG of 1.5. For the NTF, the sinc filter shows a worse course and thus QN suppression compared to a CoI. In the shown example, using the CIC reconstruction filter achieves about 26 dB worse SQNR, i.e. instead of $SQNR_{\mathrm{CoI}} \approx 99dB$ only $SQNR_{CIC} \approx 73dB$ at −6 dBFS

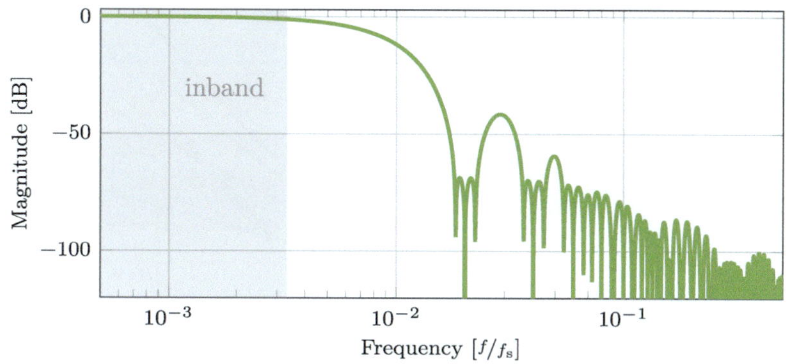

Fig. 3.28 Transfer function of a **general** third-order sinc filter for an exemplary OSR of $M = 150$, $L_d = 3$, and individual decimation factors $D_1 = 45$, $D_2 = 50$, $D_3 = 55$.

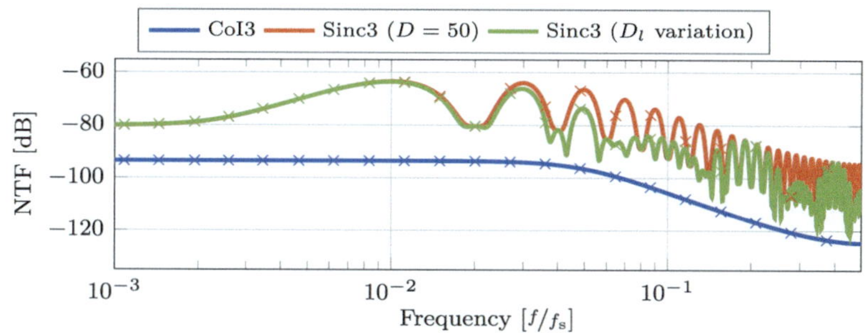

Fig. 3.29 NTFs of the exemplary third-order single-bit CIFF I-DS ADC with a third-order sinc (red: $D = 50$, green: $D_1 = 45$, $D_2 = 50$, $D_3 = 55$) in comparison to a third-order CoI (blue) at an OSR of 150 and the loop filter coefficients scaled for an OBG of 1.5.

input signal. This example illustrates that I-DS ADCs employing sinc reconstruction filters usually require much higher OSR to achieve similar SQNR (Fig. 3.29).

For completeness, the simulation example is extended using the general sinc-filter implementation in Fig. 3.24 and choosing $D_1 = 45$, $D_2 = 55$, $D_3 = 50$ as in Fig. 3.28, where it gave a certain degree of freedom for the STF design at the cost of a more complex implementation. The course of the NTF is also much worse than the CoI filter, and similar to the CIC using $D = 50$. The simulation also reveals similar performance with $SQNR_{\text{sinc}} \approx 76dB$ compared to $SQNR_{\text{CIC}} \approx 73dB$ at the simulated −6 dBFS input signal.

Achievable performances of I-DSM using sinc reconstruction filters: Similar as for the CoI reconstruction filter in Sect. 3.5.2.3, next we summarize achievable performances for various orders of I-DSM and various oversampling ratios using sinc reconstruction filters in order to provide an overview.

Figure 3.30 shows the absolute achievable SQNR plotted vs. OSR for $L = 1 \ldots 4$ for sinc (CIC) filter with $L_d = L$ (dashed), sinc filter with increased order $L_d =$

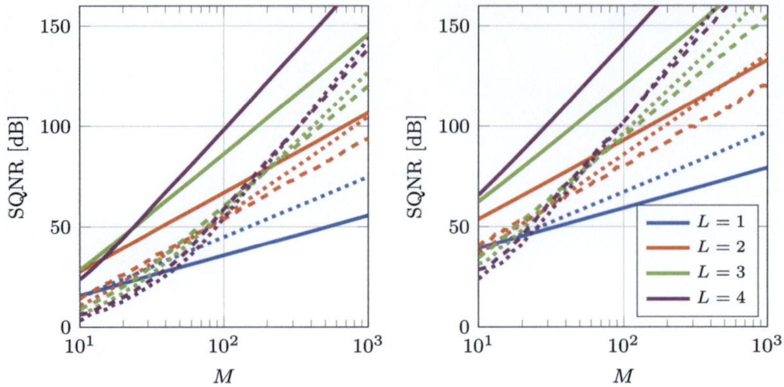

(a) (b)

Fig. 3.30 Achievable SQNR of I-DSM of various orders combined with sinc reconstruction filters, calculated for CIFF architectures with an input signal of -6 dBFS. (a) Single-bit quantizer with loop filter coefficients scaled for an OBG of 1.5 and (b) 4-bit quantizer with loop-filter coefficients scaled for an OBG of 3. Solid: CoI with $L_d = L$. Dashed: sinc filter with $L_d = L$. Dotted: sinc filter with $L_d = L + 1$.

$L + 1$ (dotted). For comparision reasons, the results for CoI reconstruction filters with $L_d = L$ from Fig. 3.23 are shown as well. The exemplary I-DSM employ CIFF architectures. Loop-filter coefficients are derived using the DS toolbox [144] using an OBG of 1.5 for the single-bit I-DSM (left) and OBG of 3 for the 4-bit I-DSM (right). The SQNR values are calculated using the overall NTFs of the I-DS ADCs and assuming an input signal of -6 dBFS. The decimation factors D_l are chosen to fulfill the requirement from (3.68) and to be as similar as possible under this given condition.

It can be seen that the use of a sinc filter leads to a major loss in SQNR compared to the CoI filter. Illustratively taking the example of a single-bit I-DS ADC achieving SQNR= 100 dB, Fig. 3.30a shows that this is achieved with $M \leq 200$ using a CoI reconstruction filter, while $M \approx 500$ is needed for the sinc filter.

Similar as in the case of CoI reconstruction filters, depending on NTF aggressiveness and oversampling M, a higher-order sinc reconstruction filter can lead to slightly more SQNR compared to the sinc filter with same order as the I-DSM. The tradeoffs are similar as for the higher-order CoI filters in Sect. 3.5.2.2, i.e. higher NPF as will be further outlined in Sect. 4.3.1, and more implementation complexity (power and area).

Concluding, it can be stated that while sinc filters can be beneficial for the overall STF by introducing notches which can be used to filter disturbers like line noise, this advantage is traded with highly compromised quantization noise suppression.

3.5.4 Matched Filter

Another filter option described in literature is a matched digital filter [205, 206]. Here, the reconstruction filter resembles (part) of the analog loop filter.

The matched filters were derived using time-domain description for a first-order I-DSM in Sect. 3.2.1, for a third-order CIFF I-DSM in Sect. 3.2.2, and for a second-order CIFB I-DSM in Sect. 3.2.3. By using those filters and as seen in the respective time-domain derivations, the residual quantization error at the output of the overall ADC is replicated at the output of the last integrator of the analog loop filter. This fact can then be used to further refine the residual quantization error by further quantizing and processing the analog representation of the quantization error and subtracting it from the digital output of the reconstruction filter, a process which is called *extended counting* [184, 205], which will be looked at in more detail in Sect. 5.2.

As shown for the third-order CIFF I-DSM in Sect. 3.2.2, the matched filter can turn out to be a simple CoI filter of the same order. Especially for CIFB loop filters, the matched filter though differs from the CoI filter. This was derived for a second-order CIFB I-DSM in Sect. 3.2.3.

The resulting overall NTF is shown next for the illustrative example of a third-order single-bit I-DS ADC, but this time using an I-DSM in CIFB configuration. A third-order CoI reconstruction filter is compared to the matched reconstruction filter. An oversampling of $M = 150$ is chosen and the loop-filter coefficients are scaled using the DS toolbox [144] for an OBG of 1.5. The matched filter gives an overall NTF with a relatively flat course and thus the quantization noise suppression is inferior—especially at OOB frequencies. In the shown example, an SQNR of 98 dB is achieved for the CoI Filter and only 86 dB for the matched filter at −6 dBFS input signal (Fig. 3.31). Concluding, using a matched reconstruction filter can be as easy as using the SoA CoI, but can also yield more complex filters, which then offer worse SQNR. In the latter case this makes only sense if the matched filter is needed for *extended-counting* architectures, as will be discussed in Sect. 5.2.

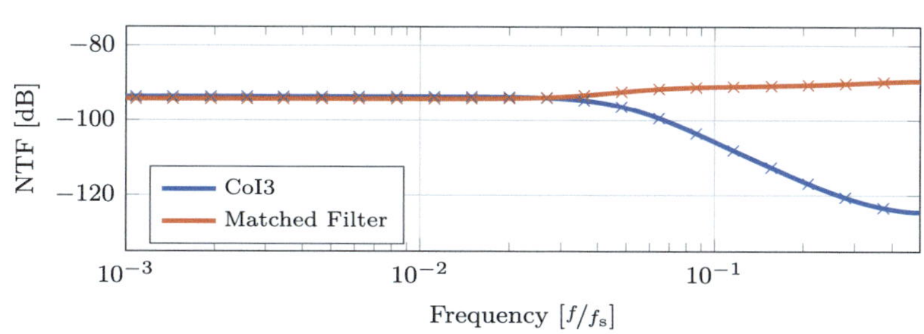

Fig. 3.31 NTFs of a third-order single-bit CIFB I-DS ADC with matched filter (red) in comparison to a third-order CoI (blue) at an OSR of 150. ©2021 Wagner. Modified, with permission, from Wagner [27].

3.5.5 Optimized Linear Filter

Finally, the *optimized linear filter* is reviewed as proposed in [192]. It basically is a FIR filter with optimized weights. In contrast to other optimized filters, the authors of [192] not only account for quantization noise, but also for circuit noise, and a successful implementation of such a reconstruction filter was reported in [221] for a DT I-DS ADC. References [178, 183, 222] employed the same idea for CT I-DS ADCs, however, the optimization of the filter was not based on analysis but on simulation. Therefore, it could also include loop-filter nonidealities such as finite DC gain and GBW.

To give an idea of the achievable performance of this reconstruction filter, the exemplary I-DS architecture is simulated again, and a third-order CIFF I-DS ADC with a 5-level internal quantizer at an OSR of 230 is used. The loop-filter coefficients are scaled using the DS toolbox [144] for an OBG of 3. The oversampling and the internal 5-level quantizer are chosen the same as in the original work in [192]. Furthermore, the same $C_{\text{in}} = 2\text{pF}$ and the same parameter $\gamma = 5$ as in [192] are chosen which leads to the same noise-power ratio

$$\rho = \frac{\Delta^2/6}{\gamma k^T/C_{\text{in}}} \approx 4.02 \cdot 10^6. \tag{3.69}$$

This is the ratio of the mean-square values of the quantization and thermal noises entering the modulator [192, Eq. 13] and it is required to calculate the optimized filter weights.

The resulting normalized weights of the optimal filter are plotted in Fig. 3.32 (red) and compared to the normalized filter weights of a third-order CoI reconstruction filter. It can be seen that the initial samples are approximately equally weighted by the optimal filter, resulting in better thermal noise suppression (less NPF) compared to the CoI filter. The I-DSM combined with the CoI reconstruction filter achieves

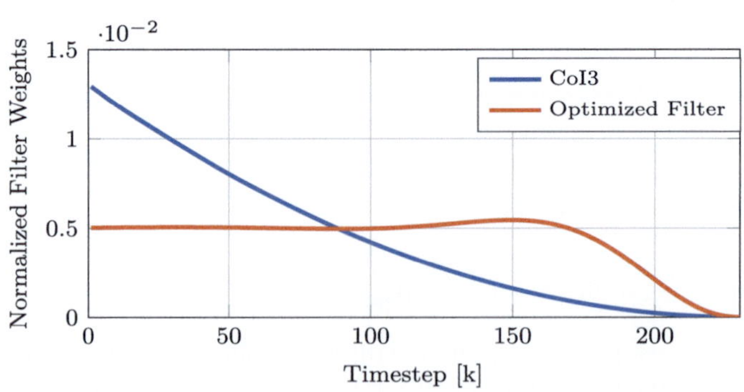

Fig. 3.32 Normalized filter weights over the timestep [k] of the optimized linear filter (red) for an exemplary third-order multi-bit CIFF I-DS ADC in comparison to a third-order CoI (blue).

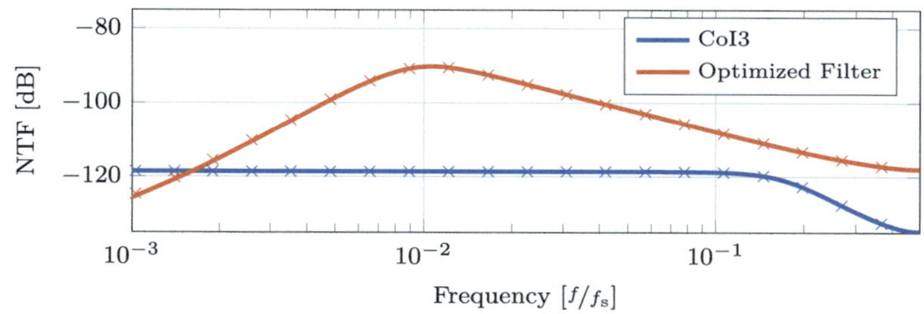

Fig. 3.33 NTFs of a third-order multi-bit CIFF I-DS ADC with optimized linear filter (red) compared to a third-order CoI (blue) at an OSR of 230. The loop-filter coefficients are scaled using the Schreier toolbox for an OBG of 3. The oversampling and the internal 5-level quantizer are chosen to be the same as in [192].

SQNR= 130 dB. In contrast, the exemplary I-DSM combined with the optimal linear filter only achieves SQNR= 112 dB. While this comes surprising for an *optimal* filter, the thermal noise is suppressed approximately 60% more in case of the optimized filter compared to the CoI.

The drop in SQNR performance can also be seen when looking at the overall NTF of the exemplary ADCs illustrated in Fig. 3.33. The optimized linear filter (red) suppresses the quantization noise less than the ADC using the CoI reconstruction filter (blue). Another disadvantage of the optimized filter is its more complex implementation compared to the CoI filter. Concluding, the better thermal noise performance of this filter is bought by a worse quantization noise performance and a more complex implementation.

3.5.6 Algorithmic Filter

Algorithmic filters have been proposed as a reconstruction filter for I-DS ADCs, and an overview of the state of the art was summarized in Sect. 3.5.1. In order to present the idea of algorithmic filters, in the following the *optimal algorithmic filter* as introduced by [209] is reviewed. It was derived with the intention to find a reconstruction algorithm with the optimization goal to maximize the SQNR for a given OSR and was based on the work of [207]. Next, the derivation of this optimal algorithmic filter is shown. Thereby, the modulator is regarded as an encoder that encodes an analog input signal $u_n[k]$ into a binary output sequence $v_n[k]$. The reconstruction filter then receives the I-DSM output sequence $v_n[k]$ as an input and works as a decoder giving the optimal digital estimate $D[n]$ of the analog input $u_n[k]$. For the derivation of the optimal algorithmic filter according to [209], various assumptions are made. First, it is assumed that the analog input signal remains constant $u_n[k] = \hat{u}_{const}$ during one Nyquist conversion period. This can be achieved by arbitrarily slow or DC input signals or by employing a front-end S/H. Second, thermal noise is assumed to be much lower than quantization noise. Both conditions are very strict and limiting as power efficient designs are usually thermal noise limited, while an explicit S/H adds even more power consumption.

For the sake of simplicity, the optimal filter is derived for a first-order I-DSM [223]. The output of the first integrator $V_{n,i1}$ at any time instant $k = 1 \ldots M$ can be described as

$$V_{n,i1}[k] = k \cdot \hat{u}_{const} - \sum_{l=0}^{k-1} v_n[l]. \tag{3.70}$$

The quantizer decides on the sign of $V_{n,i1}[k] \lesseqgtr 0$. Thus, we can rearrange (3.70) to yield

$$\hat{u}_{const} \lesseqgtr \left(\sum_{l=0}^{k-1} v_n[l] \right) / k. \tag{3.71}$$

Thereby, each comparison defines an upper and a lower bound for the input signal. A transition lattice L_m is then defined which is used to find the transition points

$$L_k = \left\{ (p,q) : p, q \in \mathbb{Z}^+, 1 \leq q \leq p \leq k \right\} \tag{3.72}$$

where—for the first-order example—possible transition points are given as

$$S_k = \left\{ \frac{q}{p} : (p,q) \in L_k \right\}. \tag{3.73}$$

At the last step of a Nyquist conversion cycle, the optimal filter calculates the mean of the obtained quantization. The required transition tree becomes more complicated for higher-order I-DSM. Then, the output of a filter, which matches the I-DSM loop filter, is used as predictor in combination with the Zoomer algorithm [207]. This is illustrated in Algorithm 1.

Algorithm 1 Zoomer Algorithm [223]

procedure ZOOMER($y_n[k]$, v_n[k])
  $UB \leftarrow 1$
  $LB \leftarrow$ -1
  for k $\leftarrow$ 1, M **do**
    boundary $\leftarrow$ $y_n[k]$ /weights[k]
    if $v_n[k] \geq 0$ **then**
      LB $\leftarrow$ max(LB, boundary)
    else
      UB $\leftarrow$ min(UB, boundary)
    end if
  end for
  return (UB+LB)/2
end procedure

Thereby, $weights[k]$ are the (impulse response) weights of the I-DSM loop filter from its signal input to the quantizer input. Furthermore, $y_n[k]$ basically represents the predictor and is calculated by using the loop filter from the quantizer output $v_n[k]$ to the quantizer input. The shown algorithm is only valid for modulators with single-bit quantizers, but can be adapted to multi-bit quantizers as stated in [209].

Achievable performances of the I-DSM using an optimal algorithmic reconstruction filter: As the algorithmic optimal filter requires an upfront S/H, the STF will be unity. The overall NTF cannot be calculated for the algorithmic filter. Thus, a performance comparison cannot be done by calculating the estimated SQNR using the NTF, but has to be simulated in the time domain and evaluated based on the simulated spectra. The performance of the optimal algorithmic filter is compared to the performance obtained with CoI reconstruction filters in Fig. 3.34. There, the achievable SQNR is shown over the oversampling M and for different modulator orders L. The simulations are done for a single-bit CIFF modulator scaled using the DS toolbox and using an OBG of 1.5 and an input signal of −6 dBFS. A S/H is used at the input to enable the optimal filter to work. The results are shown as solid lines for the optimal algorithmic reconstruction filter and as dashed lines for the CoI Filter with $L_d = L$.

It can be seen that especially at higher oversampling M and for lower-order I-DSM a large improvement in SQNR can be achieved, e.g. an increase of SQNR by 32 dB from 109 dB to 141 dB for the second-order modulator at $M = 1000$. Even though this is an impressive improvement, the optimal filter requires a quasi constant input signal (using a S/H), and a quantization-noise-dominated I-DSM, where circuit noise must be negligible. The usage of the optimal filter in the presence of dominant circuit noise is an open research problem as its resolution drops dramatically. Also, the implementation complexity of the optimal algorithmic reconstruction filter is rather large. Thus, the optimal filter can rather be seen as a means of predicting the theoretically achievable resolution of an I-DS ADC, but remains impractical for circuit level designs.

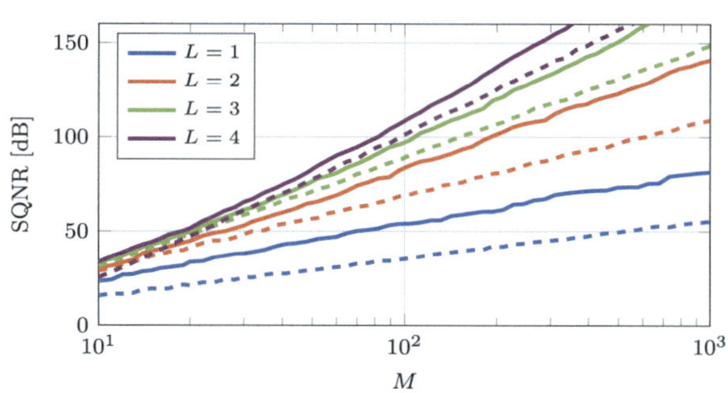

Fig. 3.34 Achievable SQNR of I-DSM of various orders, simulated for single-bit CIFF architectures with an input signal of −6 dBFS and with loop-filter coefficients scaled for an OBG of 1.5. Solid: Optimal algorithmic filter. Dashed: CoI with $L_d = L$.

3.6 System-Level Design of I-DS ADCs

The system level design of free-running DSMs has been described in Sect. 2.4.2. It consists of

- selecting a feasible architecture like single-loop, multi-loop (MASH), SMASH, etc.,
- choosing a feasible loop-filter implementation such as CIFB, CIFF, etc.,
- determining a suitable loop-filter order, OSR and internal bitwidth for the target specification,
- finding scaling coefficients of all loop-filter parameters with the goal of

  1. bounded-input conditional stability,
  2. large MSA, and thereby
  3. maximized SQNR, while
  4. keeping constraints like STF shape and limited swing of internal states,
  5. compensation of the loop filter against the influence of nonidealities like ELD, GBW, etc.

Therewith, a DSM design can be iteratively found to meet a target design specification. Possibly hundreds of papers are found in literature on stability, scaling, and architectural design of free-running DSMs, as well as on design automation for it, from which only a fraction could be listed in Sects. 2.3.4 and 2.4.2. The co-design of the DSM with its mandatory decimation filter is usually not considered in the SoA, as an ideal brick-wall filter is assumed. In practice, the low-pass decimation filter can be designed such that its influence on the overall DSM ADC performance is low.

On the contrary, there is only a very low amount of literature dedicated to the architectural design and loop-filter scaling of incremental DS ADC. The empirical study in [224] suggested that the applied stability criteria derived for free-running DSMs are unnecessarily restricting the design space of I-DSM. Though, as reviewed in Sect. 2.3.4, also free-running DSM are often scaled less aggressively by the common scaling design rules as they could be before becoming unstable. Mohamed and Anders [225] derived a stability prediction, which allows to calculate the internal states of an I-DSM over a finite time horizon (M), which is able to prove if an assumed input amplitude still leads to bounded state and thus stable operation. It considers state trajectories in a mixed-logical dynamic system, and uses an optimization problem to find a worst-case input signal for BIBO stability. However, the work does not further discuss the differences between free-running and incremental operation, and furthermore only investigated I-DSMs with loop filters scaled using the DS toolbox [144], which is based on Lee's $|H|_\infty$ rule or Carley's $|H|_1$ norm, cf. Sect. 2.3.4, both of which are ad-hoc design rules for free-running DSMs.

Consequently, most SoA I-DS ADCs are designed by taking a free-running DSM scaled using state of the art loop filter design methods, adding a periodic reset to the loop filters internal state variables, which then realizes incremental operation of the I-DSM [5, 183], and combining the I-DSM with one of the common reconstruction filters like CoI, sinc or matched.

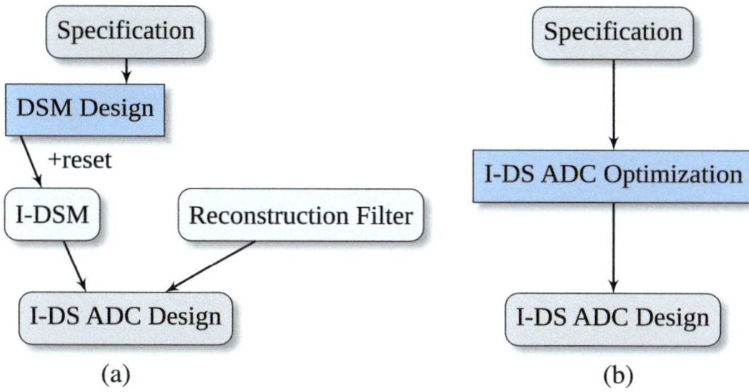

Fig. 3.35 Design flow (a) reusing scaling coefficients of a free-running DSM combined with reset and reconstruction filter versus (b) optimizing the overall I-DS ADC. ©2018 IEEE. Modified, with permission, from Wagner et al. [226].

This conventional design approach for an I-DS ADC is illustrated in Fig. 3.35a, where the DSM is scaled first as a free-running version and the reset and reconstruction filter are added in an independent step to realize incremental operation. While this usually yields functional I-DS ADCs, the question if any of this is *optimal*, and how optimality is thereby defined, is mostly not regarded. Moreover, constraints like internal swings, MSA, and bounded input, bounded output (BIBO) stability can change in incremental operation, but remain unseen by this approach.

Therefore, Fig. 3.35b illustrates the idea of combined optimization of the overall I-DS ADC. Similar as SoA has proposed optimization procedures for the loop-filter scaling of free-running DSMs, partly including nonidealities and constraints into the design flow, cf. Sect. 2.4.2, this can obviously also be done for incremental operation [157].

Illustrative example: The fact that the conventional design approach can be a suboptimal solution is illustrated by an example. First, we scale a free-running third-order CIFF using the DS Toolbox with OSR= 50, a 4-bit quantizer, and the NTF constrained to $|H|_\infty < 2.5$. This yields a free-running DSM with an SQNR of approximateley 112 dB, MSA= -1 dBFS, smooth NTF shape, and limited ($\pm FS$) integrator swings. The scaling coefficients are shown in Table 3.2. By using those coefficients on an incremental DSM and using a third-order CoI reconstruction filter, the SQNR drops to around 104 dB at a reduced MSA of -2 dBFS and all internal swings in the I-DSM increase over those of the free-running design, meaning they would need further hand-tuning to remain within the $\pm FS$ range. The SQNR sweep is shown in Fig. 3.36 (solid blue). When employing a fourth-order CoI instead, but using the same scaling coefficients for the I-DSM, no improvement can be achieved, but even worse, the maximum SQNR drops to 103 dB at a reduced MSA of around -4 dBFS, cf. Fig. 3.36 (dotted blue).

Table 3.2 Exemplary modulator scaling coefficients, derived by Schreiers' DS Toolbox [144], and by the heuristic I-DS ADC optimization from www.sigma-delta.de with a CoI3 and CoI4 reconstruction filters.

	b_1	a_1	c_1	c_2	c_3	d_{14}	d_{24}
DS Toolbox	0.320	−0.320	0.429	0.140	15.284	5.326	8.424
www.sigma-delta.de CoI3	1.364	−1.364	1.088	0.158	1.000	1.144	0.708
www.sigma-delta.de CoI4	1.417	−1.417	0.718	0.436	1.000	1.077	1.019

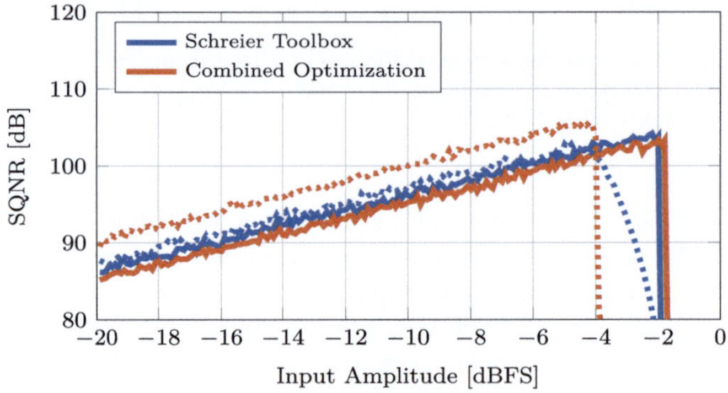

Fig. 3.36 Achievable SQNR of an exemplary third-order CIFF I-DS ADC at $M = 50$ with a 4-bit quantizer, without low-distortion path in combination with a CoI3 (solid) and CoI4 (dotted) reconstruction filter. Blue: free-running DSM scaling; coefficients are scaled for an OBG of 2.5 using the Schreier Toolbox. Red: heuristic search I-DSM scaling coefficients; using the www.sigma-delta.de website.

As an alternative, www.sigma-delta.de is used for a heuristic search based optimization of scaling coefficients for the exemplary third-order I-DSM and its corresponding CoI reconstruction filters. Thereby, we constrained all integrator outputs and the quantizer input to $\pm FS$ swing. The resulting I-DSM scaling coefficients are shown in the second row in Table 3.2. When simulating this I-DS ADC, it results in about the same SQNR = 104 dB and MSA = −2dBFS, shown in Fig. 3.36 (solid red). Advantageously, all swings are now restricted to the $\pm FS$ range.

When simulating this modulator with unchanged scaling combined with a CoI4, this actually drops the SQNR even more to approximately 102 dB at an MSA of only −4.5 dBFS. Advantageously, the combined optimization allows to reoptimize the I-DSM scaling coefficients for a combination with the CoI4 reconstruction filter. This yields slightly different scaling coefficients shown in the last row in Table 3.2. A simulation now reveals a superior SQNR = 106 dB at a slightly elevated MSA of around −4 dBFS, shown by the dotted red line in Fig. 3.36.

Concluding remarks: As a conclusion for this illustrative example, we can state that the conventional approach can deliver similar performance as the combined optimization, seen in this example for the combination of third-order I-DSM with CoI3 reconstruction filter. Still, improved performance could be achieved in this example by the combined optimization, when a CoI4 reconstruction filter is used. However, it has to be mentioned here that no low-distortion path (LDP) was used in this example. As it will be further explored in Sect. 4.1, a LDP keeps the internal swings low in the I-DSM, and therefore, not much of a difference between combined optimization and the conventional approach would be seen.

Similar results have been derived in [226], where multi-bit and single-bit I-DSMs have been combined with various reconstruction filters. Especially for low-OSR and unconventional reconstruction filters, the combined optimization always gave better SQNR and MSA. The fact that the MSA seems to reduce when a CoI3 is replaced by a CoI4 reconstruction filter, even though the underlying I-DSM remains the same, is surprising and will be further investigated in Sect. 4.1.1.2. The heuristic search optimization is finally very useful to optimize the loop filter in the presence of nonidealities, e.g. finite GBW or ELD, allows to automatically include different DAC waveforms or also employ FIR DACs.

Chapter 4
Further Properties of I-DS ADC

In the previous chapter, we have derived the theory and operation principle of I-DS ADCs. They were described in the time domain, which allowed us to derive the matched reconstruction filters. Then, the concept of the overall STF and NTF were derived to describe the incremental behavior in the frequency domain, which also allowed to accurately predict their SQNR. Finally, various reconstruction filters and their influence on SQNR performance were discussed, and the possibilities to derive loop-filter scaling by reusing free-running prototypes or by combined optimization of I-DSM and selected reconstruction filters were shown. In this chapter, we discuss further properties of I-DS ADCs. Firstly, topology choices for the I-DSM loop filter are discussed together with their influence on the incremental operation. Thereafter, we summarize loop-filter nonidealities, firstly those which show the same or similar behavior as free-running DSMs, and secondly, those which are utmost different.

4.1 I-DSM Architectural Choices and Trade-Offs

In free-running DS ADCs, almost exclusively the DSM is regarded concerning properties, performance, and nonidealities, as the digital decimation filter is assumed as an ideal brick-wall filter. It has been discussed that the choice of the reconstruction filter has a major influence on the achievable performances of I-DS ADCs throughout Sect. 3.5. Also, it has been stated that the digital reconstruction filter largely dominates the overall STF in I-DS ADCs. But still, despite their orders, OSRs, and quantizer bitwidths, further topology choices for the DSM loop-filter design can be made, which are discussed in this section.

M. Ortmanns et al., *Incremental Delta-Sigma ADCs*,
https://doi.org/10.1007/978-3-032-09468-1_4

4.1.1 Loop-Filter Topologies in I-DSMs

Section 2.3.2.1 shortly revised topologies of *free-running single-loop DSM*, where the higher-order chain-of-integrator loop filters were reviewed, in which stabilization is achieved either based on feedforward compensation (CIFF), feedback compensation (CIFB), or a mix of both (FB/FF). Also, the inclusion of local resonators to optimize NTF zeros has been reviewed in Sect. 2.3.2.1, which leads to the CRFF and CRFB loop filters. As summarized there, all these topologies can realize identical NTFs. On the contrary, the STF is merely a byproduct of the loop filter designed for a required NTF, and it is utmost different for forward or feedback compensated DSMs, as feedforward compensation paths increase the tendency for STF peaking at the band edge or OOB [2, 4, 5]. In order to alter the STF for a given architecture, distributed feed-in paths or the low-distortion architecture can be used, which add the input signal to internal loop-filter nodes, cf. Sect. 2.3.2.1.

On the complexity side, we have concluded that CIFF architectures require an adder for the forward paths in front of the quantizer, which often requires extra active building blocks, cf. Sect. 2.3.2.1. Advantageously, CIFF and mixed FB/FF architectures can omit a part or even all innermost feedback DACs, which also omits, e.g. linearization of other than the most important outermost feedback DAC in case of multi-bit designs. Also, any integrator output that is not combined with a feedback path requires reduced swing in all feedforward compensated loop filters [66], which can be a huge design advantage, especially for the first integrator. It is important to remember that design for stability is only a concern during NTF design in free-running DSMs. All ad-hoc stability measures and most SoA design approaches, cf. Sects. 2.3.4 and 2.4.2, regard solely the NTF, e.g. the Schreier toolbox regarding the $||H||_{\infty}$ and Lee's 3dB rule [129], or the design approach by Kenney-Carley [130].

Therewith, the free-running DSM becomes a BIBO stable system: the maximum inband signal, for which the quantizer is not overloaded, is called the MSA. Obviously, for nonunity OOB STF, the MSA becomes frequency-dependent, and, e.g. in the presence of STF peaking, it can be much smaller OOB than inband. When we speak of an NTF-dependent BIBO stability, we usually refer to a steady-state, free-running operation of a DSM having an inband input signal with an amplitude lower than the MSA.

The MSA becomes larger for less-aggressive NTF scaling, lower-order NTF, and higher internal quantizer resolution. For an ideal loop filter, the MSA for inband signals is then independent of the actual loop-filter topology (CIFB, CIFF, FB/FF, or with/without local resonators), and together with identically implemented NTFs, also the achievable inband SQNR is independent of the topology.

In *incremental* DSM the choice of different topologies has not been largely discussed in literature, for which the reasons might be easily understood: because the overall STF of an incrementally operating DS ADC is largely defined by the reconstruction filter, the STF peaking of CIFF loop filters known from free-running DSMs is of little concern. Moreover, as I-DS ADCs have usually been realized for very low conversion speed, the drive strength, nonideal dynamics, and power consumption of the additionally required adder are negligible. In addition, a CIFF I-DS ADC is ideally combined with a CoI reconstruction filter, as shown in the time-domain derivation throughout Sect. 3.2, as it is its matched filter and achieves superior performance. Thus, the motivation to use a CIFB topology seems low and most SoA implementations have been using CIFF I-DSMs.

Recent work has highlighted further differences of possible loop-filter topologies that are utmost different from the free-running behavior and which are derived next. Mohamed [227] and Mokhtar et al. [186] worked on using FIR feedback in CT I-DSMs and observed large initial overshoots at the beginning of every Nyquist conversion period; they derived that the DSM loop filter needs a few clock cycles to track the input signal due to the FIR-delayed feedback and that the initial overshoots of the loop-filter swings reduce the MSA. In a more general analysis of this behavior, [228] elaborated on the initial overshoots based on loop-filter topology and stated *a topology dependent MSA* due to overshoots of the internal swings derived as step response of the loop filter. Moreover, [229] showed that the MSA of I-DS ADCs appears to depend on the chosen reconstruction filters. Finally, it turns out that STF inband gain of I-DS ADC depends on the loop-filter topology, OSR, and the reconstruction filter. We will highlight some of these effects next.

4.1.1.1 Loop-Filter Topology-Dependent MSA of I-DSM

An I-DS ADC basically inherits its stability behavior and MSA from the free-running loop filter it is derived from. Aggressively scaled single-bit loop filters have the worst MSA, nonaggressively scaled multi-bit loop filters have an MSA close to fullscale. However, in free-running operation, the initial settling phase of the DSM is not included to determine its performance. We know that an I-DSM is in reset right before the start of every conversion period. Thus, the periodic initial settling is a detrimental part of the incremental operation and can further reduce the MSA. Starting a new conversion with a nonzero (and for the sake of simplicity assumed constant) input signal is equivalent to exciting an initially relaxed (free-running) DSM with a step input and observing its step response. This step response of a free-running DSM lets us predict the initial settling phase and thus potential overshoots of an I-DSM. Consequently, [228] stated that a key to low overshoot and thus large MSA in I-DSMs is to ensure that there is no overshoot in the step response of the corresponding free-running DSM. For a DT DSM, we can simply investigate the step response to its DT loop filter. In the case of a CT DSM, this can still be investigated by its *equivalent DT loop filter*, since the CT loop is excited by a stepped input to a constant DC value, and we can use the equivalent IIT loop filter for its analysis [4, 5].

Concluding, [228] claimed that the key to achieving large MSA in an I-DSM is to reduce overshoots in the step response of its (equivalent) DT loop filter, and thus to *realize a unity (equivalent) DT STF*. This can be more generally phrased: to avoid overshoots in the step response and thus achieve large MSA, the DT STF of the free-running DSM should feature no complex poles and thus be *maximally flat or be unity.*

Exemplary third-order I-DS ADC for simulations: The simulations in the following sections are performed with third-order I-DSMs with single- and multi-bit quantization. The I-DSMs are combined with CoI3 reconstruction filters unless otherwise specified and simulated for $M = 50$ and $M = 150$ for the multi-bit and single-bit operations. The loop filters have been implemented as CIFB and CIFF architectures, and the block diagram of the used DT I-DSM in CIFF architecture is shown in Fig. 4.1. It includes a direct feed-in low-distortion path (LDP). The loop-filter scaling was obtained using the DS toolbox with $OBG = 1.5$ for the single- and $OBG = 3$ for the multi-bit DSM. The scaling coefficients for the CIFF modulator are shown in Table 4.1. Note that for the sake of simplicity the individual integrator swings are not further scaled which sets the further usable coefficients in the more generally scaled CIFF architecture of Fig. 2.13 to unity. The coefficients for the CIFB architecture—as shown in Fig. 2.12—can be readily obtained from Table 4.1, too.

Simulation of loop-filter topology-dependent MSA: The behavior described above can be illustrated by the following simulations. An STF with Butterworth characteristic can be achieved using (rather nonaggressive) CIFB loop filters. A unity DT STFs can be achieved using the *low-distortion topology* [64], i.e. a DT multi-bit CIFF DSM with direct input path, cf. Sect. 2.3.2.1.

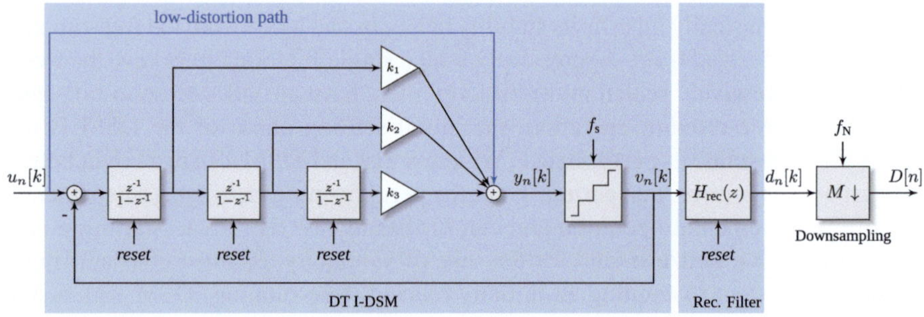

Fig. 4.1 Block Diagram of the Exemplary Simulated Third-Order DT CIFF I-DS ADC. It realizes the CIFF architecture from Fig. 2.13 but with simplified scaling ($b_1 = a_1 = 1, c_1 = 1, c_2 = 1, d_{24} = k_1, d_{14} = k_2, c_3 = k_3$), and an additional direct feed-in path. This implements the same NTF as Fig. 2.13, but does not scale the internal swings, because for the following sections less variables result in a more convenient analysis.

Table 4.1 Loop-filter coefficients for a third-order DT I-DSM derived by the Schreier toolbox. Single-bit $OBG = 1.5$, multi-bit $OBG = 3$).

	k_1	k_2	k_3
OBG = 1.5	0.800	0.288	0.044
OBG = 3	1.982	1.504	0.414

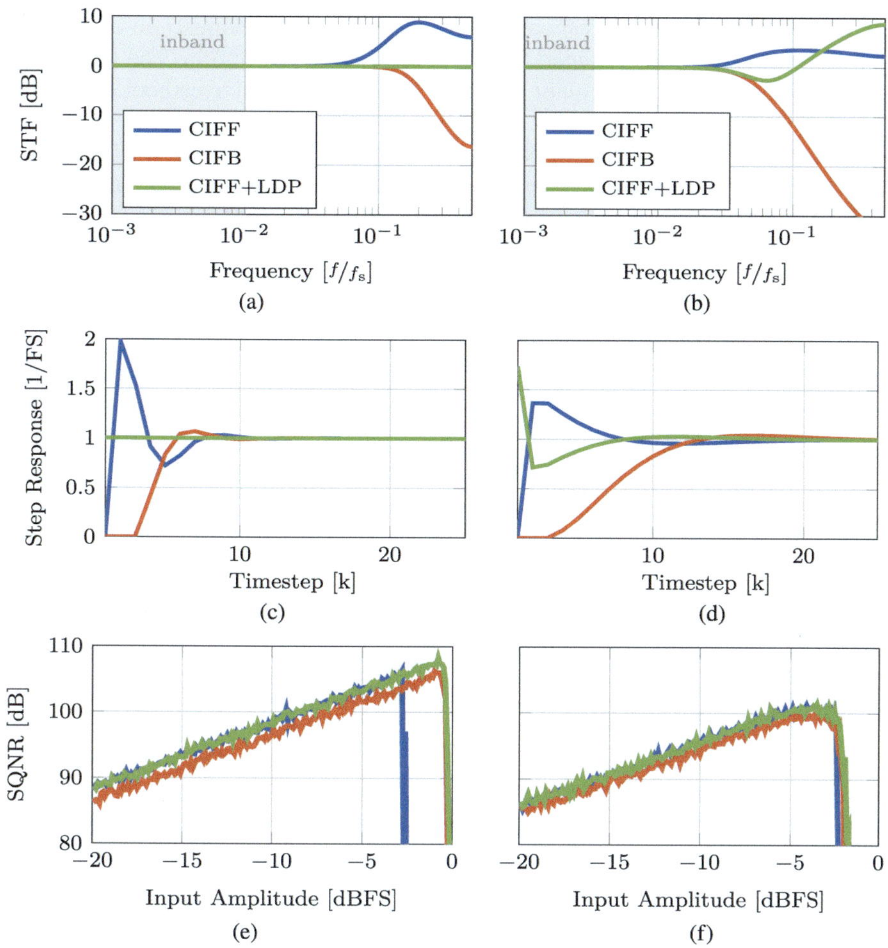

Fig. 4.2 Simulation of Loop-Filter Topology-Dependent MSA. Third-order DT DSM scaled using the DS toolbox implemented as CIFB, CIFF, and CIFF using a LDP. Left: 4-bit quantizer, $M = 50$, $OBG = 3$, right: single-bit quantizer, $M = 150$, $OBG = 1.5$. Top: STFs of the free-running DT DSM. Middle: Normalized step response of the free-running DSM STF using the linear quantizer model ($q_{\text{mb}} = 1$, $q_{\text{sb}} = 1.8$). Bottom: Simulated SQNR of the incrementally operating DSMs combined with CoI3 reconstruction filter.

Multi-bit operation: The calculated STFs of the free-running multi-bit DSM (Fig. 4.2a) show the expected behavior: a nonpeaking STF for the CIFB and a peaking STF for the CIFF architecture. The low-distortion path features the unity STF for the multi-bit DSM, as the effective quantizer gain $q = 1$ in (2.36). The calculated step responses of the free-running multi-bit DSMs (Fig. 4.2c) follow the STF shapes. The CIFF DSM shows the expected overshoot, which corresponds to the peaking STF. For the low-distortion multi-bit architecture the step response is unity, for the

CIFB architecture the step response settles smoothly with (almost) no overshoots. This is reflected in the simulated SQNR in *incremental* operation. The CIFF I-DS ADC shows an MSA reduction of about 3dB due to the initial overshoot (Fig. 4.2e). Both the multi-bit I-DS ADC with low-distortion and CIFB architectures feature the ideal MSAs of the free-running DSM. Note that the SQNR of the CIFB I-DS ADC is slightly reduced; this comes due to the triple-delaying loop filter, which prevents the reconstruction filter from processing valid data during the first sampling points (see also the step response). This leads to a slightly decreased inband STF in incremental operation and will be further shown in Sect. 4.1.1.3.

Single-bit operation: Also, the calculated STFs and step responses of the single-bit free-running DSM show the expected behavior: flat STF and no peaking in the step response of the CIFB, peaking STF and step response of the CIFF, and additionally a nonunity STF and thus also peaking step response for the single-bit DSM with low-distortion path; the latter is a result of a nonunity effective quantizer gain $q = 1.8$ which is chosen for the calculation of the single-bit STF using (2.36). Interestingly, the SQNR of all simulated I-DS ADC show the same MSA in single-bit operation. This is because the MSA is determined by the BIBO stability of the scaled higher-order DSM, and the initial overshoots do not influence the MSA in the shown single-bit examples.

Since the calculated step response of the single-bit CIFF example with low-distortion path predicts an even larger overshoot than without, this is worth a closer look. While Fig. 4.2c and d show the *calculated* step responses using $q_{\text{mb}} = 1$ and $q_{\text{sb}} = 1.8$ for the multi- and single-bit case, Fig. 4.3 shows the *simulated* maximum quantizer input of the multi- and single-bit CIFF I-DSM without and with low-distortion path. While the multi-bit results perfectly match the calculated step response from Fig. 4.2, this is different for the single-bit case. Indeed, Fig. 4.3 shows that the LDP is in fact quite helpful in reducing the initial overshoot. The difference to the calculated results comes from the fact that we assumed $q_{sb} = 1.8$ in Fig. 4.2, which is an averaged statistical value for a linearized single-bit quantizer,

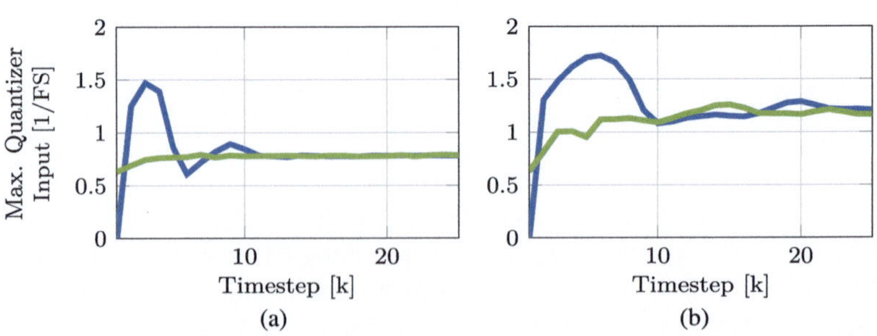

Fig. 4.3 Maximum quantizer input over the timestep k for the multi-bit (left) and single-bit (right) I-DSM examples from Fig. 4.2 simulated at -4dBFS input. Simulations are done for multiple input frequencies and averaged. Blue: CIFF without low-distortion path, green: CIFF with low-distortion path.

cf. Sect. 3.3.4; though, in the single-bit case the linearized quantizer model partly loses its validity and for the step response we would effectively need an accurate value for the quantizer gain at every timestep.

In summary, the LDP is also recommendable in the case of single-bit I-DSM as it lowers the initial overshoot and thus makes actual circuit implementation easier, even though its influence on the MSA is minor as stability is defined by the loop-filter overload due to the large amount of quantization error rather than the initial overshoot.

4.1.1.2 Reconstruction Filter Dependent "MSA" of I-DS ADCs

In Sect. 4.1.1.1 it was shown that for I-DS ADCs, the MSA depends on the architecture of the I-DSMs, even if their NTFs are identical, and the root cause was found in the initial settling operation. This is in contrast to the free-running DSM. Similarly, [218] reported that with I-DS ADCs, there also appears to be a correlation between the MSA and the reconstruction filter. However, since the reconstruction filter is an open-loop system after the modulator, it cannot possibly affect the stability of the I-DSM, which contradicts any effect on the MSA. Kaesser et al. [218] showed that it is the input amplitude giving the maximum SQNR which changes depending on the reconstruction filter, and not the input amplitude that still achieves stable operation of the I-DSM, i.e. its MSA.

A corresponding simulation is shown in Fig. 4.4. There, the SQNR of the exemplary third-order, 4-bit CIFF I-DSM is simulated over the input amplitude using four different reconstruction filters with an oversampling of $M = 60$. The I-DSM is first

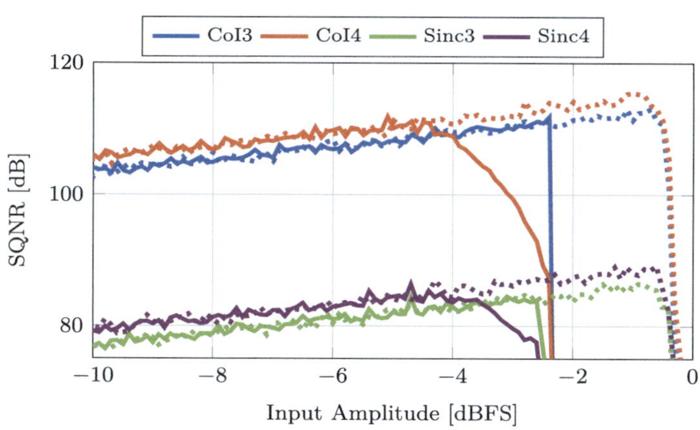

Fig. 4.4 SQNR over input amplitude of the exemplary third-order CIFF I-DS ADC with different reconstruction filters at an oversampling of $M = 60$ and a 4-bit internal quantizer. Solid: without low-distortion path. Dotted: with low-distortion path. ©2025 IEEE. Modified, with permission, from Kaesser et al. [218].

simulated without a direct input path (solid lines). Comparing the sinc3 filter with the sinc4 and the CoI3 with the CoI4, it appears that the "MSA" is drastically reduced when a higher-order filter is used, because the peak SQNR is achieved at a much lower input amplitude.

The reason for this is the overshoot that occurs in the incremental modulator during the initial settling phase. Unlike the free-running modulator, this settling occurs at the beginning of each Nyquist conversion and has a major impact on its performance, as shown in Sect. 4.1.1.1. Due to these overshoots during the settling phase, the quantizer is overloaded and the quantization noise can no longer be described as white noise, but rather shows tonal behavior. This can be seen in Fig. 4.5a. There the quantization error spectrum of the exemplary third-order CIFF I-DSM with a 4-bit quantizer is shown for an input amplitude close to the MSA, where the quantization error is not flat but increased at lower frequencies. Depending on the chosen reconstruction filter, these tones are then filtered differently. Figure 4.5b shows the overall NTFs of the third-order CIFF I-DSM combined with a CoI3 and CoI4 filter. For a white quantization noise, the CoI4 filter offers a better overall suppression, cf. Sect. 3.5.2.2, but due to the smaller suppression at lower frequencies, the SQNR degrades quickly for the non-white quantization error for input amplitudes close to the MSA.

Looking again at Fig. 4.4, the effect of a reduced "MSA" due to the reconstruction filter can be mitigated by avoiding the initial overshoots in the step response by adding the LDP in the case of the CIFF architecture. Without initial overshoots, the

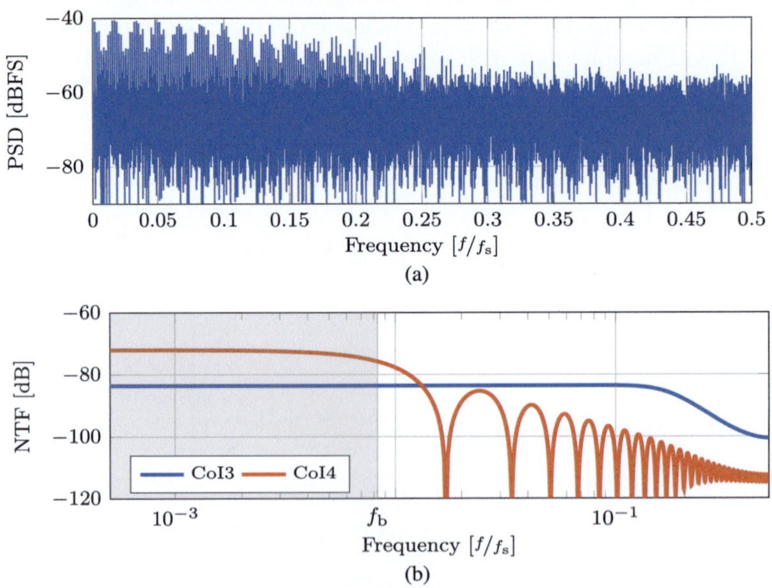

Fig. 4.5 (a) Spectrum of the quantization error close to the MSA (white at lower input amplitudes). (b) NTFs of the exemplary third-order CIFF I-DSM with a CoI3 and CoI4 filter. ©2025 IEEE. Modified, with permission, from Kaesser et al. [218].

quantizer will not produce tonal behavior due to overloading at the beginning of a Nyquist period and the quantization error has a white spectrum up to the MSA. This is shown by the simulation in Fig. 4.4 (dotted). By using the LDP, there is no difference in the input amplitude at peak SQNR depending on the reconstruction filter order, and the MSA is determined only by the I-DSM stability behavior. [218] showed that this effect is not only seen in case of a multi-bit I-DSM but also when a single-bit quantizer is applied.

4.1.1.3 STF of I-DS ADCs Dependent on Topology and Reconstruction Filter

In Sect. 3.3.3, the derivation of the overall STF of I-DS ADCs was shown. In the frequency-domain approach, we adopted the FIR model for the incremental DS ADC, and its transfer function was derived by convolving the free-running DSM and the reconstruction filter transfer functions with the rectangular window function. Thereafter, we have variously stated that the STF of incremental DS ADCs is largely dominated by the chosen reconstruction filter. Therefore, when introducing different reconstruction filters in Sect. 3.5, the STF was discussed as if it were completely determined by the digital filter. In the following, this analysis is completed by simulations.

Figure 4.6 shows the overall STFs of the same topologies used above in Fig. 4.2, i.e. a third-order I-DSM together with a CoI3 reconstruction filter. Single- and multi-bit loop filters are chosen, and the scaling coefficients are again derived using the DS toolbox with $OBG = 1.5$ for the single- and $OBG = 3$ for the multi-bit DSM. The I-DSM loop filters are realized as CIFB and CIFF architectures, the latter without and with the LDP. The overall STFs are compared to the transfer function of the CoI3 reconstruction filter. Obviously, the overall shape of all STFs follows the shape of the CoI3 filter in Fig. 4.6. In the case of the two CIFF modulators, there is no visible difference. In contrast, the CIFB-based I-DS has an inband suppression of $-1.9/-1.2$dB for the multi-bit and single-bit cases, respectively. This reduction in signal power was already visible in the SQNR simulations in Fig. 4.2, where the CIFB implementations showed a slight overall drop. The reason is that the CIFB loop filter has latency until the reconstruction filter effectively processes the input signal. This is also seen in the step response in Fig. 4.2. For the multi-bit I-DS, this reduction is worse than for the single-bit I-DS, as missing clock cycles have a relatively greater impact on $M = 50$ than on $M = 150$. The drop in inband STF gain can be alleviated by omitting the first cycles in the CoI3, i.e. by considering the latency through the I-DSM in the reconstruction filter. Figure 4.7 shows a corresponding simulation, where the multi-bit example from Fig. 4.6 is extended by a CIFB I-DSM and CoI3 reconstruction filter, which omits the first 3 samples. Another possibility to get rid of the STF inband suppression is by adding distributed feedin paths to the CIFB modulator. Thereby, not only the inband suppression problem is solved, as shown in Fig. 4.7 (dashed green), but also the SQNR is increased to have the same value as the CIFF architectures in Fig. 4.2e.

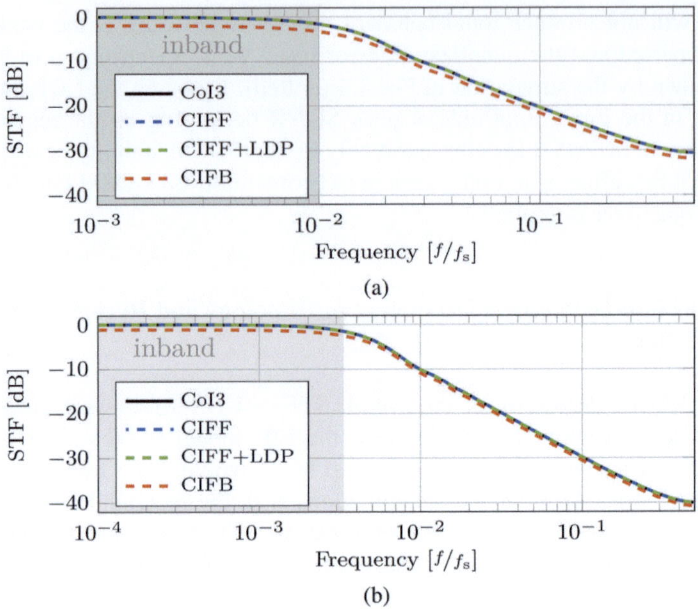

Fig. 4.6 Overall STFs of a third-order I-DSM together with a CoI3 reconstruction filter compared to the transfer function of a CoI3 only. (a) 4-bit quantizer, $OBG = 3$, $M = 50$. (b) single-bit, $OBG = 1.5$, $M = 150$.

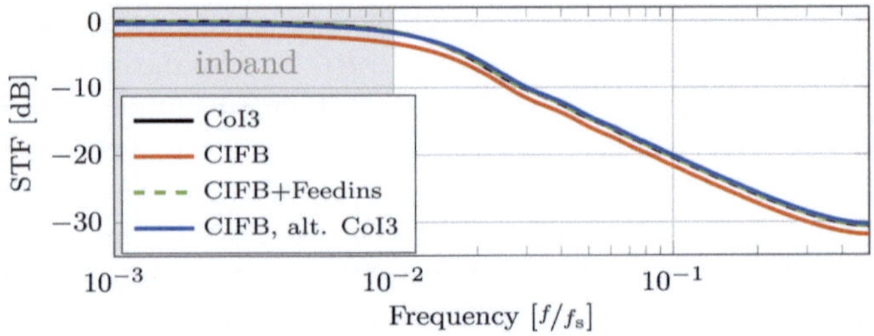

Fig. 4.7 Overall STF of a third-order CIFB I-DSM (4-bit quantizer, $OBG = 3$, $M = 50$) together with a CoI3 reconstruction filter. *CIFB, alt. CoI3* corresponds to a simulation using a CIFB loop filter combined with an altered CoI3 reconstruction filter omitting the first 3 samples from the I-DSM.

Figure 4.8 completes the discussion of loop-filter topology-dependent STF by combining the exemplary set of I-DSM with a sinc3 reconstruction filter. It can be seen that the overall course of the STF now largely follows the transfer function of the sinc3 reconstruction filter. Though, only the CIFF loop filter with LDP makes the STF identical to the reconstruction filter, which is obvious as the STF of the free-running DSM is unity in this case, and this makes $TF_{\text{free}}(z)$ in (3.34) identical to $H_{\text{rec}}(z)$ in (3.28). The CIFF without direct feed-in path and the CIFB loop filter have modified OOB shapes of the free-running STFs, cf. Fig. 4.2, which consequently

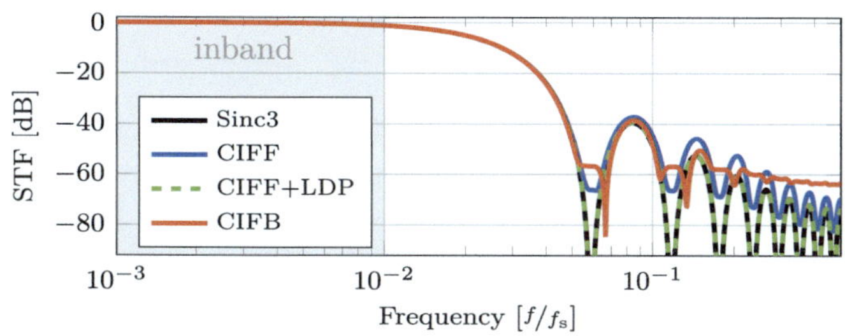

Fig. 4.8 Overall STFs of a third-order I-DSM (4-bit quantizer, $OBG = 3$, $M = 50$) combined with a sinc3 reconstruction filter compared to the transfer function of a sinc3.

influences the overall STF of the I-DS ADC. While this was almost not visible for the CoI in Fig. 4.6, it visibly alters the quality of the notches of the sinc filter in Fig. 4.8.

Another interesting fact is that there is no visible reduction of the STF inband gain when the CIFB I-DSM is combined with the sinc3 filter. This can be explained by the small weights of early samples by the sinc3, cf. Sect. 3.5.3 and Fig. 3.26, which eliminates the influence of initial latency of the signal passing through the CIFB loop filter. However, the elimination of initial latency depends highly on the oversampling used. At small oversampling ratios, the value differences in the sinc filter weights are only minor, and the inband STF suppression can still be seen.

4.1.1.4 Conclusion

In conclusion, we can summarize that the loop-filter topology of I-DSM shares some properties with their free-running counterparts, and shows some distinct differences. Most of all, we see that the MSA of I-DS ADCs not only depends on the NTF, its OOB gain and the usual BIBO stability of the original free-running modulators, but that the incremental MSA is furthermore defined by the periodically appearing initial settling of the I-DSM. This depends on the equivalent DT STF and the overshoots of its step response, wherein the loop-filter topology is a detrimental factor. Additionally, it could be shown that the choice of the reconstruction filter and the corresponding filter weights can intensify the influence of the initial settling. This appears like a further reduced MSA, but is in fact a reduction of the peak SQNR to smaller amplitudes. Finally, we have seen that the overall STF is mostly determined by the reconstruction filter, while the actual loop-filter topology only results in minor differences.

Overall, these findings show that there is no advantage to using CIFB loop filters in incremental operation. Consequently, the most commonly seen combination is a CIFF I-DSM with a direct feed-in path combined with reconstruction filters of the same order.

4.1.2 Local Resonators in I-DSM

The optimal spreading of NTF zeros over the inband frequency range is a standard technique to improve the noise shaping of higher-order free-running DSM, cf. Sect. 2.3.2.1. In practice, this is realized by combining pairs of integrators within a CIFF or CIFB loop filter by negative feedback to form local resonators, cf. Fig. 2.14. Thereby, an NTF notch is realized within the inband, cf. Fig. 2.15, which is placed to achieve overall optimized IBN suppression. This often allows about 1 or 2 bits better SQNR in SoA implementations. As the advantage of local resonators relies on the optimal inband suppression of the high-pass filtering NTF of a free-running DSM, one might wonder if a similar advantage can be achieved in incremental operation.

Disadvantageously, the periodic reset and its effect on the formation of the overall NTF of I-DS ADCs prevents this. As an example, we show the calculated overall NTF of a third-order DT single-bit I-DSM combined with a CoI3 reconstruction filter in Fig. 4.9a. In this example, reused from [201], the scaling coefficients for the I-DSM were taken from a free-running DSM derived using the Delta-Sigma Toolbox with an OOB gain of 1.5 and internal states scaled to ±0.9FS. The resulting NTFs are shown with and without optimized zeros. As derived throughout Sect. 3.3.1, the overall NTF is obtained by a convolution with the reset window in the frequency domain. Thereby, the free-running NTF notch is periodically repeated over the whole

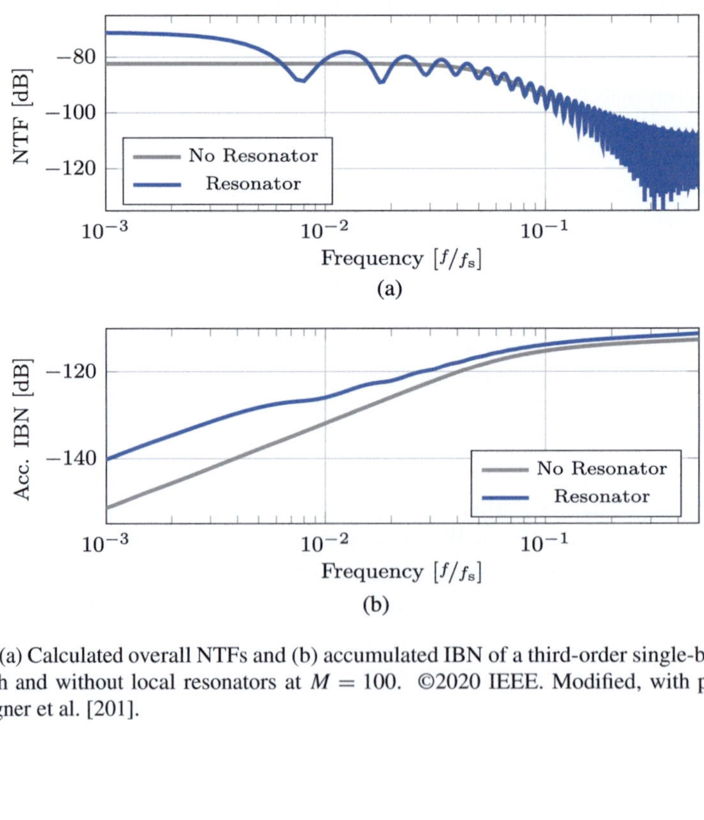

Fig. 4.9 (a) Calculated overall NTFs and (b) accumulated IBN of a third-order single-bit DT I-DS ADC with and without local resonators at $M = 100$. ©2020 IEEE. Modified, with permission, from Wagner et al. [201].

spectrum. The consequently increased main lobe of the overall NTF results in an increased IBN. This is also visible in Fig. 4.9b, where the cumulative IBNs of the I-DS ADCs are shown over the whole frequency range up to $f_s/2$, with and without local resonators. While the increased main lobe yields an increased accumulated IBN, the sinc-like notches alleviate this for higher frequencies and both IBNs converge. Still, a small loss in IBN performance remains, which supports the statement that optimized zeros using local resonators give no performance advantage in I-DS ADCs. It is interesting to note that a combined optimization of I-DSM and its reconstruction filter, e.g. using *www.sigma-delta.de* as illustrated in Fig. 3.35b, eliminates local resonators by optimizing its coefficient to be zero.

4.1.3 Continuous-Time I-DSM

In Sect. 2.3.1.2, we have reviewed the concept of CT DSM. The CT implementation of free-running DSM comes with the three distinct advantages of (1) easier drivability (resistive or Gm-based inputs), (2) reduced dynamic requirements of loop filters (lower GBW and SR), and (3) a CT STF that realizes an implicit signal filtering within the CT DSM, being known for its implicit realization of an AAF [4] and its capability to filter OOB interferers [58]. It was highlighted in Sect. 2.4.2 that CT loop filters have classically been derived from their DT counterparts by DT-CT transformation, such as IIT or modified Z-transform [4], such that the NTF of both DT and CT loop filters become identical. The SoA also proposed tools for direct synthesis, e.g. by using heuristic search [147], cf. *www.sigma-delta.de*. This gives the advantage to simultaneously achieve SQNR and MSA optimization, together with loop-filter scaling constrained for internal swings, compensation for nonidealities, and especially constraining the STF for required OOB specifications.

The synthesis of CT incremental DSM can consequently be performed using two approaches: (1) We can use DT-CT transformation of a DT I-DSM, which itself is derived using one of the two methods illustrated in Fig. 3.35. (2) A combined optimization can be performed using a heuristic search for the scaling of a CT I-DSM operating together with its reconstruction filter [157]. We have seen in Sect. 3.6 that the combined optimization yields some advantages when uncommon combinations of I-DSM and reconstruction filters are chosen, e.g. reconstruction filters with higher order than the I-DSM. When those DT prototypes are used to obtain CT I-DSM by DT-CT transformation, no impact on SQNR is expected for two reasons: (1) The overall NTF and quantization noise suppression remain identical, and thus the NTF-dependent stability and MSA are maintained. (2) The MSA is dependent on potential overshoots of the step response of the equivalent DT loop filter [228], as discussed in Sect. 4.1.1.1; consequently, this remains unaltered as well. The remaining advantage of using a combined optimization of the CT I-DSM and its reconstruction filter is that it can directly consider internal swings and implicitly compensate for nonidealities.

Example of a DT-CT transformation versus opt.-based synthesis of a CT I-DS ADC: Figure 4.10 illustrates an example. The third-order DT I-DS ADC with CIFF loop filter including a direct feed-in path from Fig. 4.1 is reused. The DT loop filter—excluding the direct feed-in path—is $H(z)$ and can be transformed to an equivalent CT loop filter $H(s)$ by IIT, whereupon both loop filters realize the same NTF. Note that while IIT was originally employed to match the DT and CT impulse responses from the DT DSM output to sampled quantizer input, the same impulse response is seen from the loop-filter (stepped signal) input to its sampled quantizer input. Thus, also the equivalent DT STFs are identical after IIT. The direct feed-in path is then added to both the DT and the CT system. The coefficients for the DT loop filter $H(z)$ (derived from the DS toolbox and reused from Table 4.1) and the IIT-derived CT coefficients for $H(s)$ are given in Table 4.2.

Similarly as for DT loop-filter scaling, the CT loop-filter coefficients can be directly derived by using the online optimization on *www.sigma-delta.de*; here, the only constraint was a limitation of the quantizer input to $\pm FS$ and an exemplary set of coefficients is shown in the last column of Table 4.2.

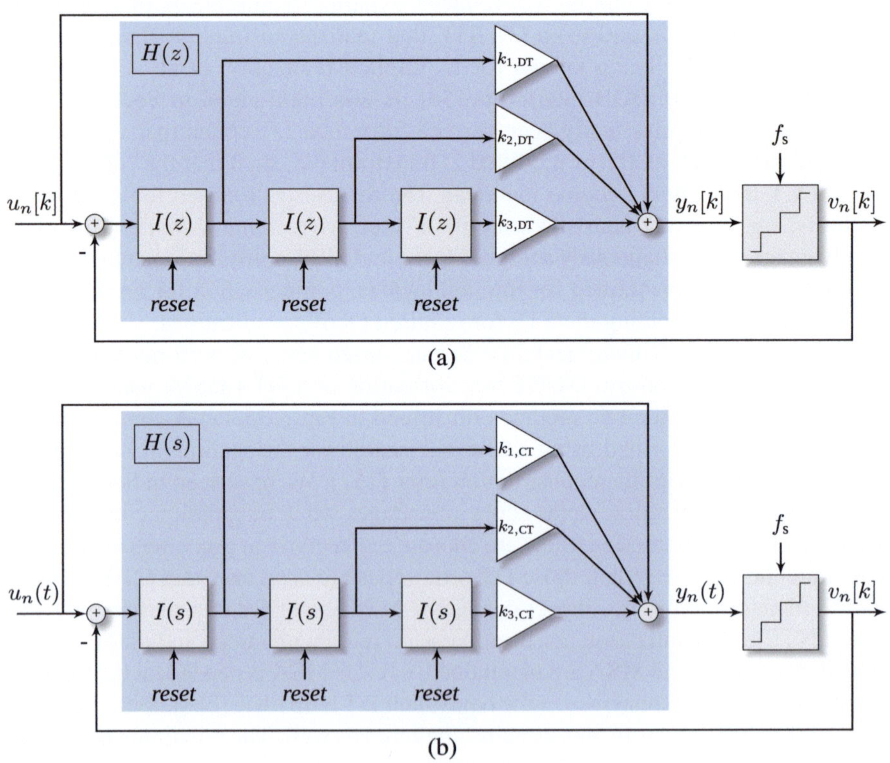

Fig. 4.10 Illustration of DT-CT conversion for a third-order I-DSM.

Table 4.2 Loop filter coefficients for a third-order I-DSM using $M = 50$ for a multi-bit I-DSM and a CoI3 filter. Left: DT derived by the Schreier toolbox (single-bit $OBG = 1.5$, multi-bit $OBG = 3$, cf. Table 4.1). Middle: CT derived by IIT from DT coefficients. Right: CT derived by online optimization, constraining the quantizer input swing below ±FS.

	$OBG = 1.5$		$OBG = 3$		OBG opt.
	DT	CT_{IIT}	DT	CT_{IIT}	CT_{opt}
k_1	0.800	0.670	1.982	1.368	1.545
k_2	0.288	0.244	1.504	1.090	1.163
k_3	0.044	0.044	0.414	0.414	0.377

The obtained SQNR for the exemplary modulator scalings are shown in Fig. 4.11. The curves of the DT and CT CIFF I-DSM are plotted in blue and red, respectively, and expectedly no difference can be seen. The I-DSM scaled with the online optimization performs slightly better, the SQNR is slightly improved while the MSA stays approximately the same. This is as the online optimizer does not constrain the NTF as chosen in the DS toolbox by the OOB gain, but only on transient simulation results.

CT versus DT STF after transformation: It is worth to remember that IIT realizes a CT DSM with the same NTF as the original DT DSM, while its CT STF is different. Also, when we add the direct low-distortion path, the CT STF is not unity. This is shown in Fig. 4.12 for the exemplary CT and DT modulators of the shown example. But as we have now stated several times, only the equivalent DT STF of the CT I-DSM is important in the context of initial overshoots and achievable MSA. As the CT modulator is derived via IIT from a DT DSM, the equivalent DT STF of the CT DSM is identical and—with the low-distortion path—becomes unity again.

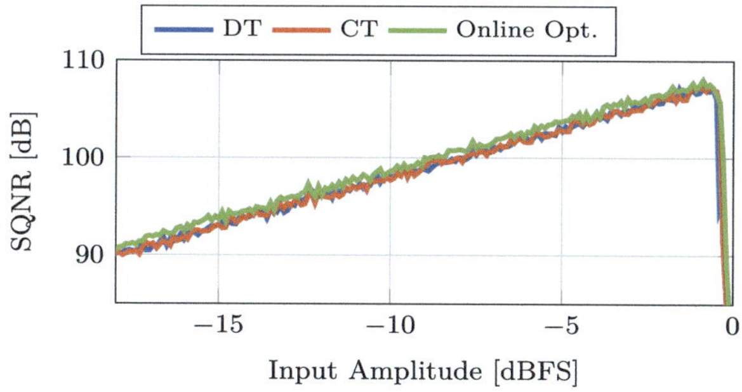

Fig. 4.11 Simulated SQNR to illustrate the DT-CT conversion for a third-order multi-bit I-DSM.

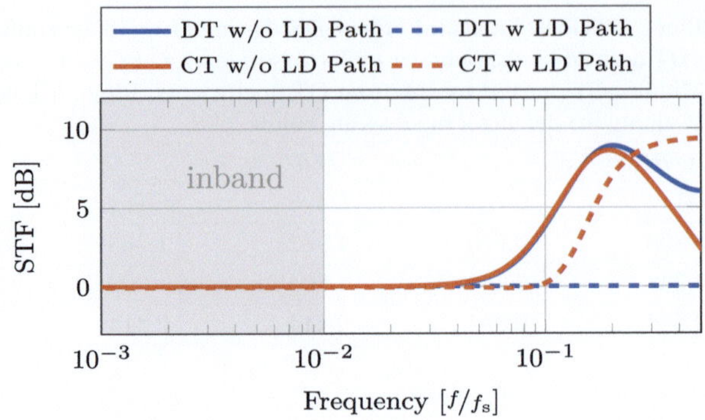

Fig. 4.12 Illustration of resulting STFs after DT-CT conversion for a third-order multi-bit I-DSM.

4.1.4 (Implicit) Anti-Aliasing Filter for (CT) I-DS ADCs

In Sect. 3.3.1, we have discussed the form of the overall STF of I-DS ADCs. In Fig. 3.8, which is repeated and extended in Fig. 4.13, we have summarized a core difference between free-running and incremental DS ADCs.

A free-running DSM is followed by an (ideal brick-wall) low-pass filter, which eliminates all parts of the DSM output spectrum outside the inband and its periodic repetition around f_s. Thus, when it comes to potential aliasing, only the small periodic inband region around multiples of f_s is regarded, as illustrated in Fig. 4.13a. For free-running DT loop filters, the input signal must be sufficiently suppressed around f_s. This translates to relaxed AAF requirements as typically seen in oversampled ADCs. For free-running CT loop filters, the free-running STF even shows a steady CT roll-off with notches around f_s and consequently features an implicit AAF [4, 5].

An incremental DSMs has no dedicated subsequent digital low-pass filter removing all OOB components before decimating. In contrast, all OOB spectral components are folded into the inband, as illustrated in Fig. 4.13b. CT loop filters in I-DSMs feature a continuous roll-off at high frequencies, e.g. shown around f_s in Fig. 4.13b. However, this is not always sufficient as AAF, as still all OOB components are folded back, i.e. not only the part around multiples of f_s, but also all parts between the inband and f_s as indicated in Fig. 4.13. In case there are potentially strong interferers outside the inband, a complex (brick-wall-shaped) AAF is required as in every other Nyquist ADC, no matter if the I-DSM is realized using a CT or DT loop filter. In case there are no OOB interferers but only OOB noise, the situation changes: a DT I-DS ADC still requires a front-end AAF, as all OOB noise, especially unattenuated around multiples of f_s, would be folded into the inband; this AAF can though be relaxed as, e.g. a first-or second-order low-pass filter already feature an acceptable effective noise bandwidth. Without strong OOB interferers, the continuous roll-off of the overall STF of a CT I-DS ADC might be a sufficient AAF and avoid the necessity

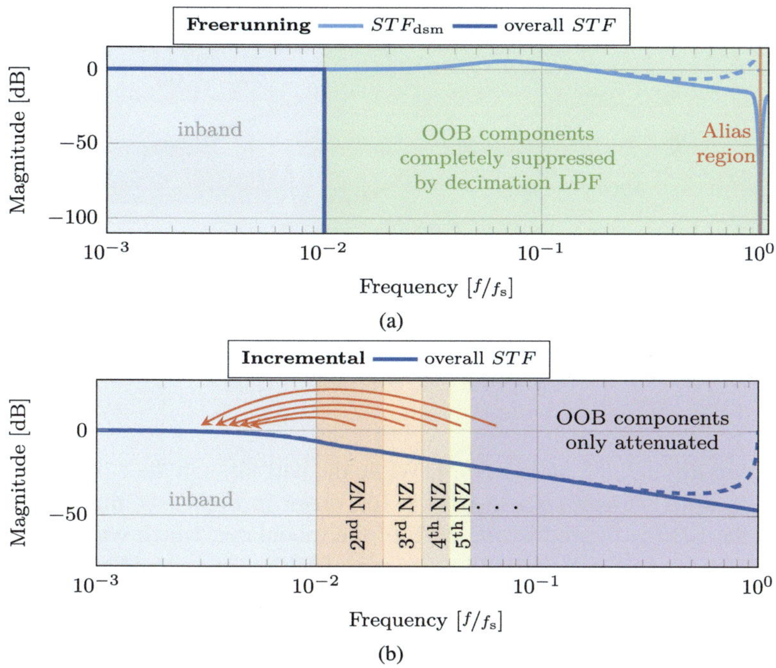

Fig. 4.13 Amended Fig. 3.8 illustrating the need for AAF in I-DS ADC. (a) STF of a CT (solid) and DT (dashed) third-order *free-running* DSM combined with an ideal brick-wall low-pass filter. Explicit AAF needed for DT DSM, implicit AAF realized by CT DSM. (b) Overall STF of a third-order CT (solid) and DT (dashed) *incremental* DS ADC without upfront S/H combined with a CoI3 having aliasing of all OOB spectral components. Both simulated at an OSR of 100 based on the same third-order DSM (CIFF without LDP) architecture and scaling (Schreier Toolbox with an OBG of 1.5).

for an explicit AAF, so we can still speak of an *implicit AAF of CT I-DSM* [204]. This situation changes if the CT I-DSM features a direct feed-in path from the overall input to the quantizer input as shown below.

CT STF and low-distortion path (LDP): If the CT I-DSM features a LDP, the high-frequency behavior of the overall STF changes and it removes the smooth roll-off toward higher frequencies. This is demonstrated in Fig. 4.14 by showing the transfer functions of an exemplary third-order CT I-DS ADC. Shown in green is the overall STF without low-distortion path. Shown in blue is the overall STF, when a LDP is added. Here, we note that the smooth roll-off of the CT STF is compromised at really high frequencies, where it flattens out. Also, we see that the disturbance in the STF happens around multiples of f_s. With this, unfiltered input noise would not be attenuated toward $f \to \infty$, and we cannot speak of an implicit AAF anymore. This effect can be explained as follows: in case an LDP is present, the overall signal input can be split into one part inserted through the LDP, and

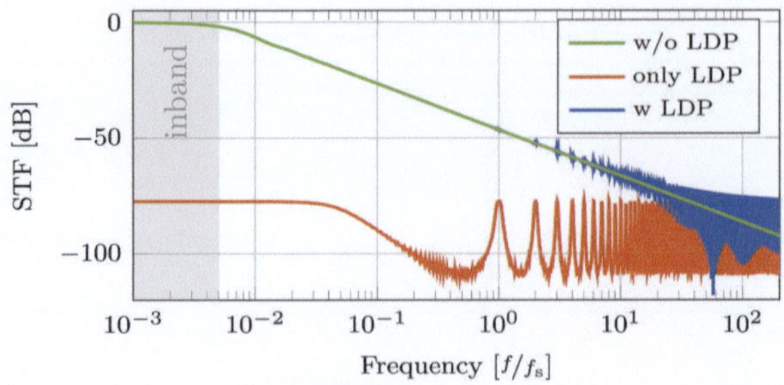

Fig. 4.14 Overall STF of an exemplary third-order CT I-DS ADC without upfront S/H combined with a CoI3. Simulated at an OSR of 100 based on the third-order CIFF DSM architecture and scaled for an OBG of 1.5. Green: without LDP. Blue: with LDP. Red: only LDP.

another part through the main I-DSM. While the part through the I-DSM is filtered by the overall STF with smooth roll-off (cf. green in Fig. 4.14), the part inserted through the LDP is filtered by the NTF of the modulator, which we plot in red in Fig. 4.14. The overall STF of the I-DSM with LDP is then an addition of the STF without LDP and the NTF. Therefore, a flattening of the STF can be seen at high frequencies. The frequency and the attenuation at which this happens depends on the modulator order, modulator scaling and the oversampling M. For small oversampling the STF starts to settle at lower frequencies and at higher magnitudes than for large oversampling. Consequently, some kind of explicit AAF is required for the LDP architecture, otherwise the input noise would be integrated to infinity. While this should be kept in mind, in practice it may play a minor role, since for usually large OSR, the flattening occurs more than one order of magnitude above f_s, and in most applications, the input noise may already be low-pass filtered by the driver or signal source.

4.1.5 FIR Feedback in I-DSMs

Even though the addition of FIR-feedback DACs is not on the forefront of DSM loop-filter design, its application has become popular in the last years, cf. Sect. 2.3.3, especially for single-bit CT loop filters. In the context of *incremental* operation, it principally inherits the same advantages as in free-running DSMs: a smoother DAC feedback signal and smaller output steps of the DAC. Consequently, the jitter sensitivity is reduced, and the dynamic requirements of the loop filter are lowered. Though, in the incremental case the regular reset and the subsequent initial settling phase that depends on the STF step response can result in severely degraded performance [186], which is shown in the following.

From its application to free-running DSMs, it is known that the addition of a FIR-feedback DAC to one or more outer loop-filter feedback paths alters the NTF of the loop filter. The ideal free-running NTF can be restored by adding a compensation

FIR DAC to one of the inner feedback paths while adapting the loop-filter scaling coefficients, cf. Fig. 2.19 [115, 117]. The compensation FIR DAC $F_C(z)$ can be derived by, e.g. IIT or using the method of moments [230]. Since the NTF of the free-running DSM is unaltered by the correctly compensated FIR DAC, the input-dependent (inband) stability and thus the (inband) MSA remain unchanged. However, we should note that the FIR compensation does not restore the STF of the DSM, which will thus be altered depending on the number and scaling of FIR taps, cf. Fig. 2.19.

Due to this change of the STF, the FIR DAC has a large influence on the MSA in I-DS ADCs. This is illustrated by the simulation in Fig. 4.15, where an exemplary third-order CT CIFF I-DSM with a 4-bit quantizer and CoI3 is simulated. The CT loop-filter coefficients are obtained by IIT from a DT DSM (Delta-Sigma toolbox using $OBG = 3$) as described in Sect. 4.1.3 and shown in Table 4.2. The DSM is enhanced by a low-distortion path. Thereafter, the CT DSM is extended by a FIR DAC in its outermost feedback path and a compensation FIR DAC is used in a direct loop around the quantizer to restore the original free-running NTF.

Even though properly compensated and in contrast to free-running operation, Fig. 4.15 shows that the MSA dramatically drops for an increasing number of FIR

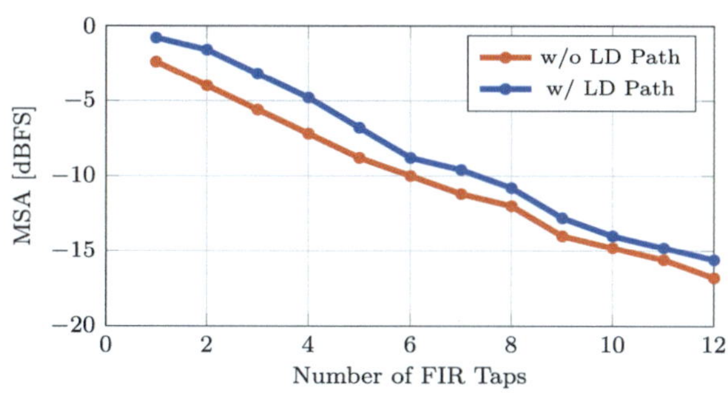

Fig. 4.15 Simulation of MSA of the exemplary third-order CT I-DSM combined with CoI3 and 4-bit internal quantizer over the number of taps in the FIR DAC.

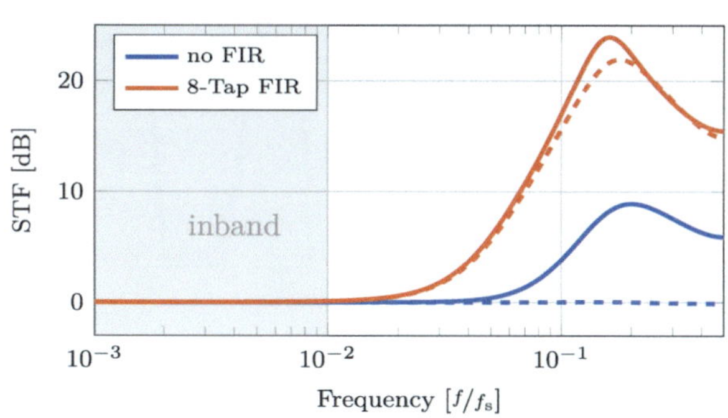

Fig. 4.16 Equivalent DT STFs of the exemplary 4-bit third-order CT modulator without FIR DAC (blue, identical to CIFF in Fig. 4.2, top-left) and with an 8-tap FIR DAC (red). Solid: w/o low-distortion path. Dashed: with low-distortion path.

taps, independent of using a low-distortion path. This behavior can be explained by investigating the equivalent DT STFs of the CT modulator, which are calculated and shown in Fig. 4.16. Without FIR DAC (blue), the equivalent DT STF of the CT DSM is obviously identical to that of the original DT DSM in Fig. 4.6 (top left, CIFF), because the CT DSM is derived from it by DT-CT transformation. Consequently, the same unity STF can be obtained with the low-distortion path (dashed). In contrast, when the CT DSM is extended by a FIR DAC (red)—even though the NTF is properly restored by compensation—the equivalent DT STF is largely modified regardless of whether or not the low-distortion path is used. Due to this peaking in the equivalent DT STF of the FIR-extended CT DSM, the MSA decreases. More FIR taps result in more peaking, and thus in more MSA drop with increasing number of FIR taps.

The problem of FIR-dependent MSA in I-DSM was already identified in [186], where it was noted that immediately after the periodic reset of the incremental loop filter, the output of the FIR DAC is not related to the input signal, and an initial settling time is required for the feedback signal to properly track the input signal. It was found that the use of FIR DAC in I-DSM leads to strong overshoots of the internal oscillations. This was illustrated in two example simulations, which are shown in Fig. 4.17. In [186] it was reported that such overshoots during the initial settling result in a reduction of MSA and DR when including FIR DACs into I-DSMs.

To counteract such initial overshoots, the SoA has contributed several solutions. [186] proposed a differential reset of the FIR DAC such that every other tap would start with positive or negative outputs. This resulted in a FIR DAC with a zeroed output instead of a full-scale positive or negative output right after reset, which helped to alleviate the problem for the employed single-bit FIR DAC. Mokhtar et al. [231] proposed that in addition to the differential resetting scheme, also unequal (symmetrical) FIR taps could further reduce the initial overshoots after reset. Alternatively, [232] proposed to preset the FIR DAC based on a coarsely quantized version of the input signal, which would then allow an initial tracking of the FIR DAC output right after reset, but the authors revealed issues with FIR coefficient matching and the required digital correction.

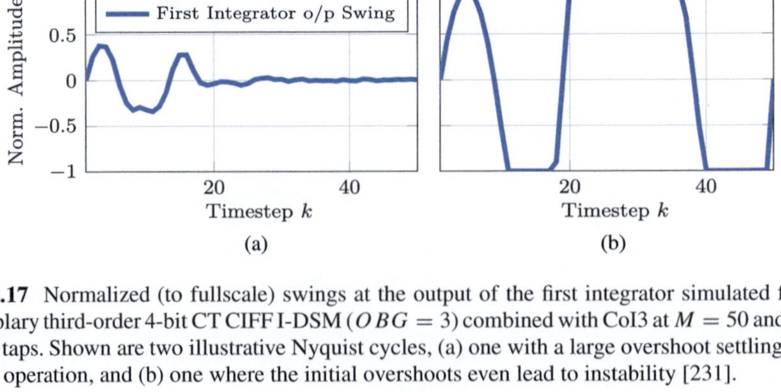

Fig. 4.17 Normalized (to fullscale) swings at the output of the first integrator simulated for the exemplary third-order 4-bit CT CIFF I-DSM ($OBG = 3$) combined with CoI3 at $M = 50$ and using 8 FIR taps. Shown are two illustrative Nyquist cycles, (a) one with a large overshoot settling into a stable operation, and (b) one where the initial overshoots even lead to instability [231].

Pavan et al. [228] analyzed the underlying problem of overshoots in the presence of a FIR DAC in a I-DSM in order to present a more general solution. The analysis follows the same idea as we discussed throughout the topology-dependent MSA in Sect. 4.1.1. There, we stated that the I-DSM output, right after reset, shows the step response of the (equivalent DT) STF weighted by the (constantly assumed) input signal. We discussed that choosing either CIFB or CIFF has a significant influence on the MSA, since—even though they can be designed with the same NTF—their (equivalent DT) STFs yield different responses. The same argument can be made for loop filters using FIR feedback, as the compensation FIR DAC $F_c(z)$ is scaled to restore the NTF, while the STF is not regarded.

Design example of a CT I-DSM employing FIR feedback: A design example is derived similar to [228], and the results of the various methods are compared afterwards. A CT I-DSM loop filter is chosen as it shows the more complex problem regarding IIT and equivalent DT STF of the otherwise CT loop filter. The ideal third-order loop filter of a CT DSM as in Fig. 4.10b can be expressed as $k_1/s + k_2/s^2 + k_3/s^3$. Employing a NRZ feedback DAC and using IIT, its DT NTF becomes

$$NTF(z) = \frac{1}{1 + \underbrace{\text{c2d}\left\{\frac{k_1}{s} + \frac{k_2}{s^2} + \frac{k_3}{s^3}\right\}}_{=H_{\text{dt}}(z)}}. \tag{4.1}$$

Now, we insert a FIR DAC $F(z)$ with N_{FIR} equally weighted taps ($1/N_{\text{FIR}}$) in the outermost feedback path. To restore the original NTF, a compensation FIR $F_c(z)$ is implemented in the innermost path and the loop-filter coefficients have to be rescaled to become $\hat{k}_i$. The resulting CT I-DSM with FIR DAC and FIR compensation DAC is shown in Fig. 4.18a.

The scaling coefficients for the FIR-enhanced DSM become [228]

Table 4.3 Compensation FIR DAC coefficients for the third-order CT DSM in Fig. 4.18a, when the original CT DSM is scaled as in Table 4.2 for single-bit or multi-bit quantization. The coefficients relate to the weighting of the ith tap of $F_c(z)$, with $i = 0...(N_{\text{FIR}} - 1)$.

N_{FIR}	Coefficients of $F_c(z)$											
	Single-Bit DSM						Multi-Bit DSM					
2	0	0.333	–	–	–	–	0	0.667	–	–	–	–
3	0	0.442	0.262	–	–	–	0	0.866	0.615	–	–	–
4	0	0.495	0.389	0.228	–	–	0	0.948	0.888	0.606	–	–
5	0	0.525	0.462	0.360	0.209	–	0	0.984	1.024	0.928	0.614	–
6	0	0.543	0.509	0.444	0.343	0.197	0	0.996	1.091	1.108	0.978	0.632

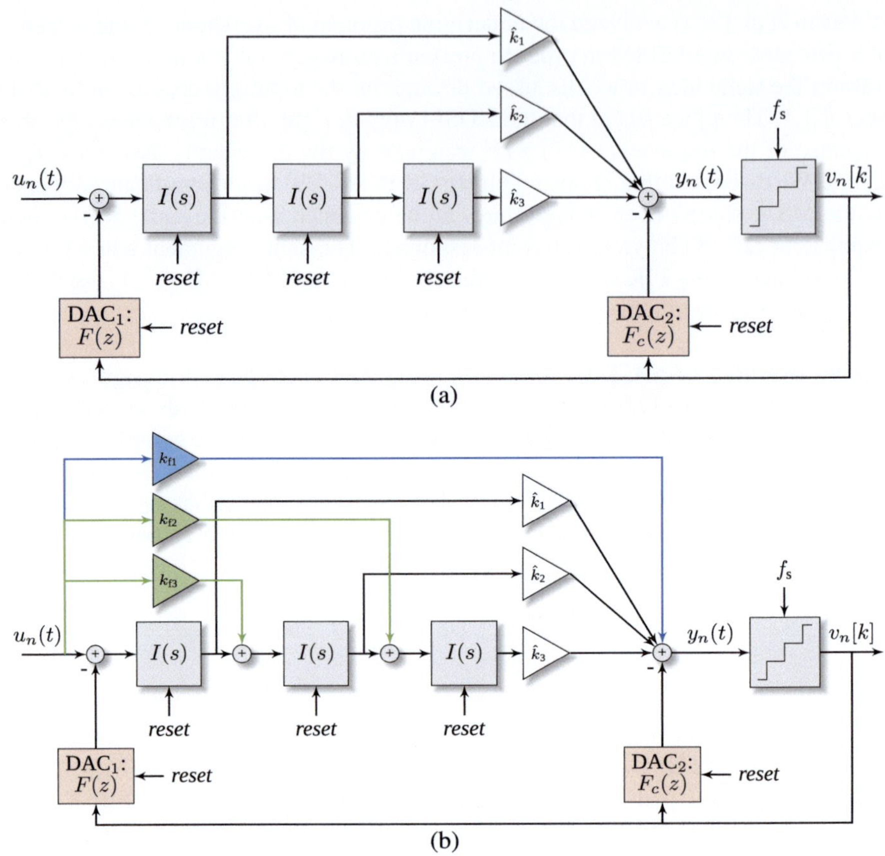

Fig. 4.18 (a) Third-order CIFF I-DSM with FIR DAC $F(z)$ and compensation FIR DAC $F_c(z)$ and (b) illustration of feed-in paths to restore a unity STF.

$$\hat{k}_3 = k_3, \hat{k}_2 = k_2 + \frac{k_3}{2}(N_{\text{FIR}} - 1), \hat{k}_1 = k_1 + \frac{k_2}{2}(N_{\text{FIR}} - 1) + \frac{k_3}{12}(N_{\text{FIR}} - 1)(N_{\text{FIR}} - 2), \tag{4.2}$$

whereas the original scaling coefficients $k_{1,2,3}$ of the CT third-order (single- and multi-bit) DSM can be chosen as in Table 4.2. Therewith, the coefficients of the compensation FIR DAC $F_c(z)$ can be derived and are exemplarily shown in Table 4.3 for up to 6 FIR taps.

Using this compensation FIR DAC $F_c(z)$ and the rescaled loop-filter coefficients $\hat{k}_{1,2,3}$, the NTF of the original CT loop filter $H(s)$ in Fig. 4.10b and the NTF of the FIR-enhanced loop filter in Fig. 4.18a become identical. But their STFs do not match. Thus, while the original CT loop filter $H(s)$ in Fig. 4.10b was derived from a DT loop filter $H(z)$, and by using an LDP, a unity (equivalent) DT STF could be achieved. The FIR-enhanced loop filter does not have this unity (equivalent) DT STF.

Consequently, the goal is to restore the unity (equivalent) DT STF of the FIR-enhanced CT loop filter. In order to do so, the loop filter must be extended such that

1. it features the unity direct input path $k_{f1} = 1$,
2. and the IIT-equivalent DT transfer function from the input $u(t)$ to the loop-filter output $y(t)$ of the FIR-enhanced CT DSM becomes equal to $H_{\mathrm{dt}}(z)$.

We know that the CT loop filter, whose DT equivalent equals $H_{\mathrm{dt}}(z)$, is the original $k_1/s + k_2/s^2 + k_3/s^3$. Thus, we extend the FIR-enhanced DSM in Fig. 4.18b by further feed-in paths ($k_{f2/3}$), such that the resulting transfer function from the signal input $u(t)$ to the loop-filter output $y(t)$ is described by

$$\frac{\hat{k}_1 + k_{f3}\hat{k}_2 + k_{f2}\hat{k}_3}{s} + \frac{\hat{k}_2 + k_{f3}\hat{k}_3}{s^2} + \frac{\hat{k}_3}{s^3} \overset{!}{=} \frac{k_1}{s} + \frac{k_2}{s^2} + \frac{k_3}{s^3}. \tag{4.3}$$

Comparing the coefficients in (4.3) and using (4.2) yields the required feed-in paths [228]:

$$k_{f2} = \frac{(N_{\mathrm{FIR}} - 1)(2N_{\mathrm{FIR}} - 1)}{12} \text{ and } k_{f3} = -\frac{(N_{\mathrm{FIR}} - 1)}{2}. \tag{4.4}$$

By using these coefficients, the DT step response of the original DT DSM, the IIT derived CT DSM, and the FIR-extended CT DSM become equal and unity, and thus no overshoots and no MSA degradation will be seen. Please note that the feed-in coefficients only depend on the length of the FIR filter $F(z)$ and become zero when the FIR is removed (N_{FIR}=1).

Simulation results and comparison of methods: The efficacy of the approach is shown by simulation in Fig. 4.19. The exemplary third-order CT I-DSM is simulated using an N_{FIR}-tap FIR DACs and a FIR compensation DAC as in Fig. 4.18a. The simulations are performed for single- and multi-bit operation, for which the original CT scaling coefficients are found in Table 4.2, the rescaled loop-filter coefficients are obtained using (4.2) depending on the number of FIR taps N_{FIR}, and the compensation FIR coefficients have been listed in Table 4.3. The simulation results in Fig. 4.19 compare the results for different methods: first, no counteraction against the loss of MSA by FIR-induced STF peaking is done. Then, the direct feed-in LDP is added and also combined with differential reset in the single-bit case as proposed in [186]. Finally, the distributed feed-in coefficients are applied as shown in Fig. 4.18b and as derived in (4.4) [228].

In the single-bit case, without countermeasures, the performance drops dramatically as the number of FIR taps increases. With differential resetting as proposed by [186] and including the direct feed-in low-distortion path, the problem is largely alleviated, and up to about $N_{FIR} \leq 6$ no significant performance degradation is seen. With the distributed input paths, the FIR-induced drop in MSA is completely solved, also for an even larger number of FIR taps. In the multi-bit case, the drop in MSA is also significant, but not as abrupt as in the single-bit case, i.e. the FIR-

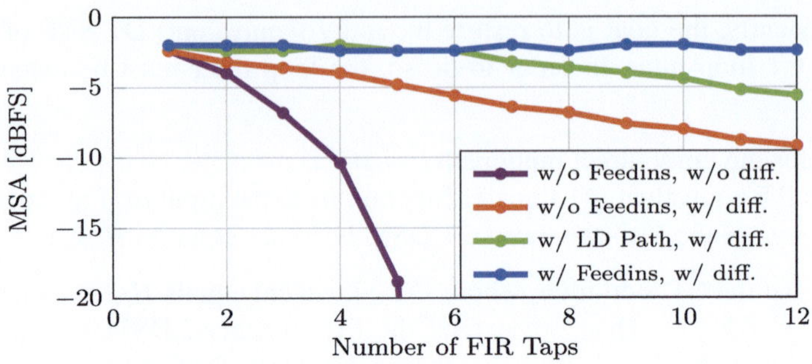

(a)

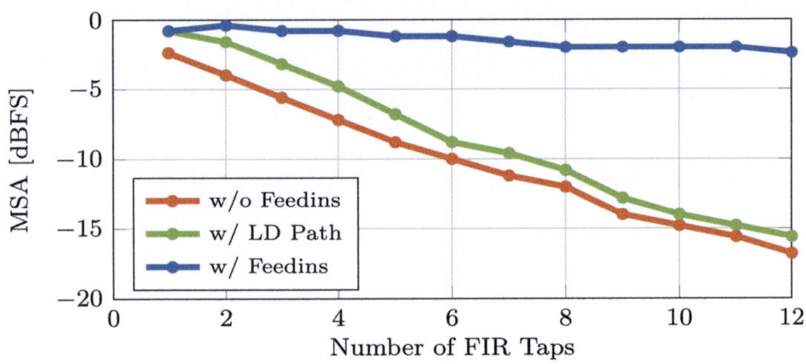

(b)

Fig. 4.19 Simulation of MSA over number of FIR taps for exemplary third-order (a) 1-bit QTZ ($OBG = 1.5$) and (b) multi-bit ($OBG = 3$) I-DSM in CIFF topology with FIR DAC and FIR compensation DAC.

enhanced I-DSM remains stable for smaller input signals. Differential reset is implicitly included in the multi-bit case, as the feedback DACs can start at mid-code. In contrast to the single-bit example, using only the direct feed-in path (with differential reset), the FIR-induced drop in MSA remains unsolved, as was also observed in Fig. 4.15. The underlying reason is the more aggressive loop-filter scaling of the multi-bit example. Only the addition of all distributed input paths—as indicated in Fig. 4.18b—removes the MSA drop.

4.2 Nonidealities Similar in Incremental and Free-running Operation

The performance of I-DS ADCs depends on the combination of I-DSMs and reconstruction filter. While the digital reconstruction filter implementation performs ideally, the I-DSM suffers from analog nonidealities. Some of these nonidealities have the same or similar effect in free-running and incremental operation, while others are distinctly different. In this section, we cover the first set of these nonidealities.

4.2.1 DAC Nonidealities

It is common knowledge that the outermost feedback DAC in DSMs—even though its resolution is as low as that of the internal quantizer—needs to be as accurate as the required resolution of the overall DS ADC. The simple reason is that all outermost DAC nonidealities are inserted at the most sensitive input node of the DSM and are thus indistinguishable from the input signal itself. This requirement applies to DAC noise and DAC linearity, both static and dynamic, and there is ostensibly no difference between free-running and incremental DSM. While there is little that can be done to counteract DAC circuit noise, DAC nonlinearity is often mitigated by the use of dynamic linearization techniques such as DEM and DWA, which have some distinct differences between free-running and incremental DSM and are discussed in Sect. 4.3.6.

Another nonideality, often called a DAC nonideality, but actually a combined nonideality of quantizer and DAC, is the finite response time of the appearance of the feedback signal in a DS loop filter, and is called ELD.

4.2.1.1 Excess Loop Delay

Unwanted loop delay, which models the time between the ideal sampling point at the quantizer to the appearance of the output signal at the feedback DAC, is commonly referred to as ELD [52], and it is known to destabilize the loop filter of a DSM. The absolute delay time $t_d = \tau_{\mathrm{ELD}} \cdot T_s$ is commonly given as a fraction τ_{ELD} of the sampling period T_s. While signal dependent latency in a comparator or quantizer is practically unavoidable due to its finite gain and resolution, its effect on the loop filter can be alleviated by relatching the quantizer output with a constant delay. In low-speed DSM, this can be done at a short instant after the quantizer sampling point [78]. In wideband DSMs, the most common solution is to use one clock edge for sampling and the other for relatching. This yields in a commonly used $\tau_{\mathrm{ELD}} = 0.5$ loop delay. If such constant delay is then called loop delay or excess loop delay has been controversially discussed in literature, but we stay with the original nomenclature of ELD [52].

In free-running DSMs, ELD leads to peaking in the OOB NTF [4]. In consequence, the MSA drops and the system is destabilized. The amount of tolerable ELD depends on the intrinsic stability of the loop filter: high-order, single-bit and aggressively scaled loop filters tolerate less, low-order, multi-bit and unaggressively scaled loop filters tolerate more ELD. It can be shown that excess-loop delay effectively adds one more order to the loop filter [4, 52]. Thus, in order to be able to restore the ideal NTF, an extra degree of freedom is needed in the loop filter. In practice, this is achieved by ELD compensation, which, e.g. implements a direct 0th-order loop around the quantizer. Various architectural and circuit-level implementations are known [145].

In Fig. 4.20, a third-order CIFF CT I-DSM with ELD is shown. It is extended by a direct 0th-order feedback path around the quantizer, which allows a compensation of

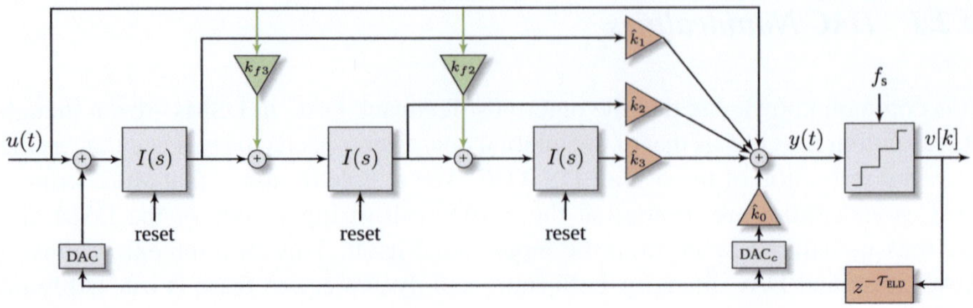

Fig. 4.20 Block diagram of a third-order CIFF CT I-DSM with low-distortion path, ELD added into the feedback and the loop filter extended by a direct feedback ($\hat{k}_0$) and feedin (k_{f2}, k_{f3}) paths to allow ELD compensation in incremental operation. ©2025 IEEE. Modified, with permission, from Kaesser et al. [229].

ELD in the NTF up to one clock cycle ($\tau_{\mathrm{ELD}} < 100\%$). In free-running DSMs, it is known that by adapting the original loop-filter coefficients k_1, k_2, k_3 to $\hat{k}_1, \hat{k}_2, \hat{k}_3$ and by allowing a direct feedback path coefficient $\hat{k}_0$, the original NTF can be restored [145]:

$$\begin{aligned}
\hat{k}_3 &= k_3, \quad \hat{k}_2 = k_2 + k_3 \cdot \tau_{\mathrm{ELD}}, \\
\hat{k}_1 &= k_1 + k_2 \cdot \tau_{\mathrm{ELD}} + \frac{k_3}{2} \cdot {\tau_{\mathrm{ELD}}}^2, \\
\hat{k}_0 &= k_1 \cdot \tau_{\mathrm{ELD}} + \frac{k_2}{2} \cdot {\tau_{\mathrm{ELD}}}^2 + \frac{k_3}{6} \cdot {\tau_{\mathrm{ELD}}}^3.
\end{aligned} \tag{4.5}$$

A loop filter in incremental operation suffers similarly from ELD as its free-running counterpart. In both cases, the ELD destabilizes the modulator. But while a free-running DSM with ELD compensation ideally restores its NTF, a DSM in incremental operation also has an MSA dependency on its (equivalent DT) STF, because the initial settling operation depends on its step response, cf. Sect. 4.1.1.3. [233] already analyzed the influence of ELD on the peaking of the CT STF of free-running DSMs. While this is disadvantageous, e.g. in wideband free-running DSM concerning OOB peaking, the MSA is influenced in incremental mode. This is illustrated next.

Figure 4.21 shows simulation results of the exemplary third-order CT CIFF I-DSM with 4-bit QTZ and direct feed-in path as in Fig. 4.20. An $ELD = 0.2 \cdot T_S$ has been added. Its scaling coefficients are derived from a DT free-running prototype with $OBG = 3$, which are given in Table 4.2. The 0th-order coefficient around the QTZ and the scaling coefficients are recalculated to compensate for ELD [4, 52], cf. (4.5). This restores the ideal NTF which therefore is identical between the original DT loop filter, the ideal CT loop filter, and the ELD-affected but compensated CT loop filter. On the contrary, the step responses and equivalent DT STF are not ideally restored which is shown in Fig. 4.21: for the uncompensated CT loop filter, the step response shows prominent ringing and the equivalent DT STF shows extreme peaking, which is in severe contrast to the DT prototype and its flat unity STF in Fig. 4.2c. Even

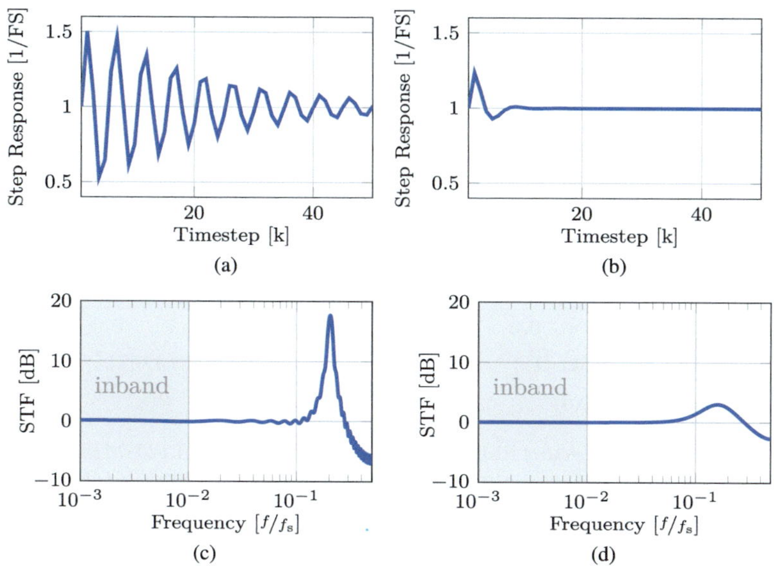

Fig. 4.21 Comparison of the step responses and the equivalent DT STFs of a third-order CIFF CT I-DSM with low-distortion path, ELD and CoI3 reconstruction filter. 4-bit quantizer, $OSR = 50$, $OBG = 3$, $ELD = 0.2 \cdot T_s$. Top: step responses. Bottom: equivalent DT STFs. Left: uncompensated. Right: compensated. ©2025 IEEE. Modified, with permission, from Kaesser et al. [229].

though the step response of the CT I-DSM rings much less after ELD compensation, the ringing is still not removed; this is also seen in the equivalent DT STF of the ELD-compensated CT loop filter which still features minor peaking. As a consequence of the remaining peaking of the equivalent DT STF and the ringing of its step response, a remaining degradation of the MSA must be expected even after ELD compensation.

This is confirmed by the simulation in Fig. 4.22, where the achievable MSA of the CT I-DSM is simulated with and without ELD compensation over increasing ELD. For completeness, the third-order I-DSM is simulated for single- and multi-bit quantization. A sharp drop of MSA is visible for increasing loop delay for the uncompensated loop filter both for the single- and multi-bit I-DSM, which is known from free-running DSM [4]. While the NTF-induced instability is solved by proper ELD compensation, the resulting nonunity equivalent DT STF and the remaining ringing of the step response still result in a gradually dropping MSA for the multi-bit I-DSM. Notably, no further influence seems to exist for the single-bit I-DSM. This higher sensitivity of the multi-bit I-DSM is again—as in the case of FIR-induced MSA drop in Fig. 4.19—attributed to the more aggressive scaling of the multi-bit loop filter.

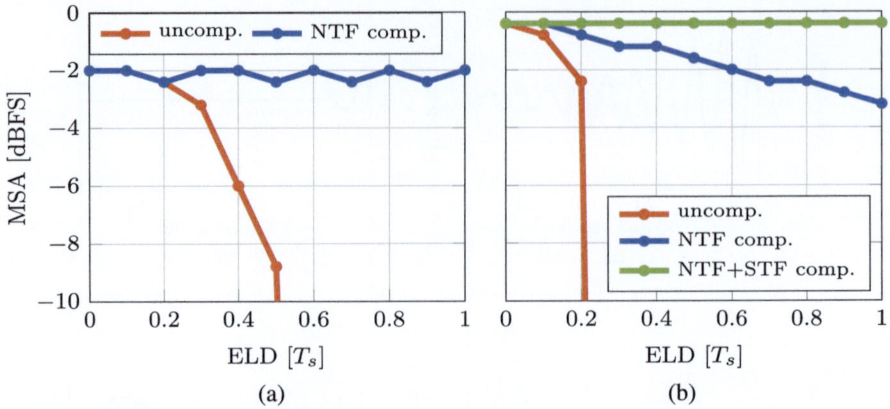

Fig. 4.22 Simulation of MSA over ELD for the exemplary third-order CT I-DSM in CIFF topology with low-distortion path, CoI3 reconstruction filter from Fig. 4.20. (a) 1-bit QTZ ($OBG = 1.5$) and (b) multi-bit ($OBG = 3$). Blue: without ELD compensation. Red: with ELD compensation by 0th order path restoring the original NTF. Green: with feed-in paths to restore original flat STF. ©2025 IEEE. Modified, with permission, from Kaesser et al. [229].

Extension from NTF- to STF-based ELD compensation: To remove the remaining influence of ELD on the equivalent DT STF of the CT I-DSM and, therefore, on its MSA, a similar compensation technique as in the case of FIR feedback (cf. Sect. 4.1.5) can be applied [229]. For this, the feed-in paths k_{f2} and k_{f3} are introduced, as indicated in Fig. 4.20. To restore the original maximally flat DT STF, which the prototype DSM had without ELD, the loop filters have to match:

$$H_{\text{ELD}}(s) = H_{\text{ct}}(s) = \frac{Y(s)}{U(s)}. \tag{4.6}$$

For the exemplary third-order I-DSM from Fig. 4.20 this leads to the condition:

$$\frac{\hat{k}_1 + k_{f3}\hat{k}_2 + k_{f2}\hat{k}_3}{s} + \frac{\hat{k}_2 + k_{f3}\hat{k}_3}{s^2} + \frac{\hat{k}_3}{s^3} = \frac{k_1}{s} + \frac{k_2}{s^2} + \frac{k_3}{s^3}. \tag{4.7}$$

Equalizing the coefficients of like powers of s and solving the equations leads to the feed-in coefficients:

$$\begin{aligned} k_{f3} &= -\tau_{\text{ELD}} \\ k_{f2} &= 0.5 \cdot \tau_{\text{ELD}}{}^2. \end{aligned} \tag{4.8}$$

This is verified by a simulation added to Fig. 4.22b in the multi-bit case (green). If both the NTF and STF are compensated for ELD, there remains no influence of the MSA on the applied ELD.

4.2.2 Integrator Nonidealities

4.2.2.1 Integrator Gain and Loop-Filter Scaling

Actual integrator gains define the scaling coefficients of a loop filter. While integrator gains in SC implementations are very precisely defined, integrator gains in CT implementations can largely vary, when taking process and temperature variations into account. If integrator gains in a free-running DSM vary, its NTF, its OOB gain, its MSA, and also all internal swings of the loop filter change. Generally, positive or negative variation of integrator gains leads to more or less aggressive loop-filter scaling with larger or smaller OOB NTF peaking. Thus, while a positive shift of scaling coefficients at first gradually improves the noise shaping and reduces the IBN, eventually it will abruptly lead to instability. On the other side, a negative shift of scaling coefficients gradually worsens the noise-shaping and increases the IBN, but stability is even improved. Obviously, the tendency to instability thereby depends on the initial aggressiveness of the loop-filter scaling. Single-bit operation shows slightly different behavior to variation of scaling coefficients: as long as the single-bit comparator operates ideally, scaling its input linearly up or down will not affect its decision, and changes in those loop-filter scaling coefficients, which directly precede a single-bit quantizer, are implicitly compensated [4].

These observations from free-running DSM do not fundamentally change when the DSM is in incremental operation. This is shown with simulations in Fig. 4.23 where the exemplary third-order CIFF I-DSM is simulated while all integrator gains are simultaneously varied. The expected behavior is seen: negative variation leads to less performance, positive variation to more aggressive scaling and eventually to instability. The gradient of the influence of parameter variation on the single-bit (blue) I-DSM is slightly less compared to the multi-bit (red) case. To explain this behavior, we include the calculation results from (3.58) in Fig. 4.23. As outlined in the derivation of the approximate expression in (3.58), the more the loop filter is scaled,

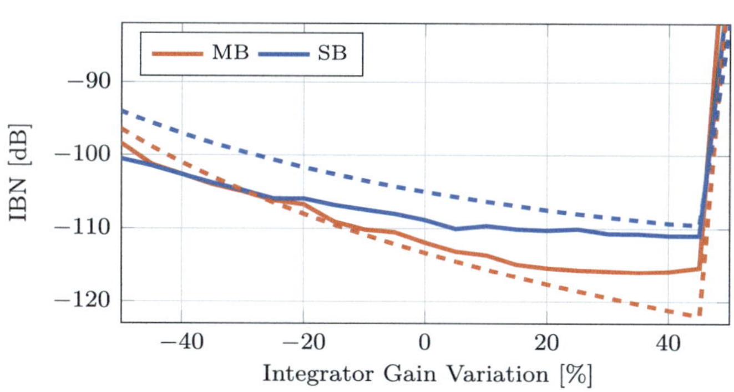

Fig. 4.23 IBN over variation of integrator gain for a third-order SB (blue, $OBG = 1.5$, $OSR = 150$) and MB (red, $OBG = 3$, $OSR = 50$) CT I-DSM at −6dBFS input amplitude.

the less accurate the prediction becomes, resulting in a slightly more noticeable mismatch between the calculated and simulated results for the single-bit I-DSM in Fig. 4.23. However, the differing slopes between the single-bit and multi-bit cases are correctly predicted by (3.58). This difference originates from the inclusion of the effective quantizer gain in (3.58), as the single-bit quantizer effectively cancels the scaling coefficient and its variation in the last integrator stage by adapting its effective quantizer gain q. This behavior was also observed in earlier work on free-running DSMs [4, Sect. 5.3.1].

4.2.2.2 Finite DC Gain and Finite GBW of Integrators

Finite DC gain of the integrators in a DSM loop filter is one of the earliest nonidealities reported in literature. It leads to leaky integration and insufficient suppression of quantization noise within the inband [2]. It is also known that low integrator DC gain cause dead-zone problems in DSM [2, 5], especially when using low-order loop filters [234]. While the resetting operation of I-DSM already breaks some of the periodicity, dither can be further used to remove the dead zone around zero inputs [235]. Concerning the leaky integration, it has been derived manifold that in single-loop free-running DSM finite DC gain A_0 of all integrators must be in the order of the oversampling ratio in order to not deteriorate the IBN performance, i.e. $A_0 \approx OSR$ [4, Sect.5.2]. In practice, also the achievable OpAmp linearity depends on its gain and this mostly determines the required DC gain to larger values than those required to avoid leaking. Finite OpAmp GBW within the integrators can be modeled as an integrator gain error and a parasitic pole in CT loop filters. Thereby, the pole can again be modeled as a loop delay [236], [4, Sect. 5.4], both of which are separately discussed in Sects. 4.2.2.1 and 4.2.1.1.

Both finite DC gain and finite GBW are LTI nonidealities, and thus they can be accurately included into the analysis of the overall transfer functions for I-DS ADCs. This is illustrated in Fig. 4.24 for the exemplary third-order CT single-bit CIFF I-DSM combined with a third-order CoI reconstruction filter. The overall NTF

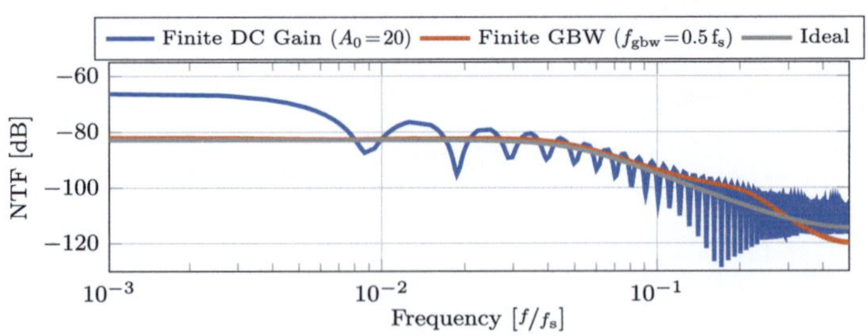

Fig. 4.24 Comparison of calculated and simulated overall NTF of a CT third-order single-bit I-DS ADC utilizing integrators with finite GBW and finite DC gain. ©2020 IEEE. Modified, with permission, from Wagner et al. [201].

is calculated for the ideal CT loop filter and for loop filters where all integrators feature finite GBW or finite DC gain. Simulation results at distinct frequencies are annotated by symbols. Firstly, it can be seen that simulations and calculations show excellent matching. Similar as in free-running DSMs, finite DC gain reduces the suppression of quantization noise within the inband. In contrast to the free-running DSM, the course of the overall NTF takes on the characteristic of a sinc function with notches around multiples of $1/M$ and an amplified main lobe caused by the circular convolution with $X_{\mathrm{rect},M}(f)$, cf. Sect. 3.3.3.1. The simulation for finite GBW in Fig. 4.24 shows that the shape of the overall NTF is mostly changed in the high-frequency part. This is again in accordance with the influence of finite GBW on free-running DSMs, which is a consequence of the additional nonideal parasitic (high-frequency) poles. In free-running DSMs, this is known to cause excessive OOB peaking of the NTF [4], which yields stability problems. In the overall NTF of I-DS ADCs, this peaking is heavily reduced by the OOB suppression of the digital reconstruction filter. In the lower frequency region, the shape of the overall NTF is slightly elevated due to the GBW-induced gain error [236]. Note that since both nonidealities can be accurately included into the analysis of the overall transfer functions for I-DS, this also allows performance predictions including nonidealities using the frequency-domain method [201] derived throughout Sect. 3.3, which is not possible when using time-domain performance predictions.

Simulations with finite DC gain: Figure 4.25 presents simulations of second-, third-, and fourth-order CT I-DSMs over finite DC gain in all loop-filter integrators. The I-DSMs operate with 4-bit internal quantization and are combined with CoI reconstruction filters of the same order. Their scaling is derived using the DS toolbox for a DT loop filter and subsequent DT-CT transformation. Interestingly, while the settling to the ideal IBN around $A_0 \approx OSR$ is seen for the second-order I-DSM, there is a tendency to slightly increased requirements on finite A_0 for higher-order I-DSM. As the difference to the expected behavior from free-running DSM is in the range of $1 \ldots 2\times$, this could be considered to be too insignificant for most practical implementations, where DC gain is usually anyhow much higher to maintain amplifier linearity.

However, the different behavior for the different I-DSM orders also indicates a possible influence of the reconstruction filters. Therefore, Fig. 4.26 shows the same simulation setup, but where the I-DSMs are combined with sinc reconstruction filters. Interestingly, the requirements on finite A_0 significantly change, and the performance settles to its ideal value for much smaller finite A_0. As we have indicated in Sect. 3.5 and Fig. 3.29, the sinc reconstruction filter features a very different overall NTF shape as the CoI filter, which suggests that also the requirements on finite A_0 changes. This effect has not been analyzed in literature so far and remains a topic for further investigation.

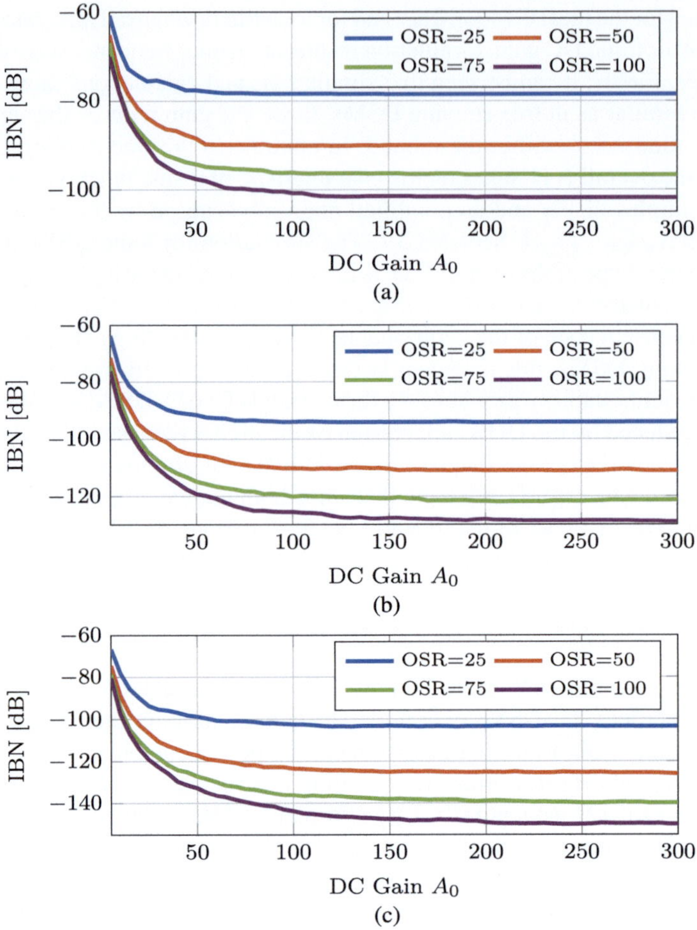

Fig. 4.25 Simulation of IBN over finite DC gain A_0 in all integrators of exemplary second-, third-, and forth-order multi-bit I-DS ADC incl. low-distortion path and *CoI reconstruction filters* over various M. Simulations were performed at −6dBFS input signal.

Simulations for finite GBW: Figure 4.27 shows simulations of second- and third-order CT I-DSMs operating with single- and 4-bit internal quantization over finite GBW in all employed integrators. Observing the single-bit I-DSM, the results match those expected for free-running DSMs [236]: at about GBW=f_s a drop of performance is seen. The more prominent sensitivity in case of a multi-bit I-DSM is also expected, even though not explicitly illustrated or mentioned in prior art [4, 236]: as finite GBW in CT integrators can be modeled as integrator gain error as well as a parasitic pole, and the latter can be modeled as a dead time or GBW-induced loop delay in case of a rectangular feedback pulse, both have a higher influence onto the performance of multi-bit I-DSMs compared to single-bit I-DSMs, cf. Sects. 4.2.1.1 and 4.2.2.1.

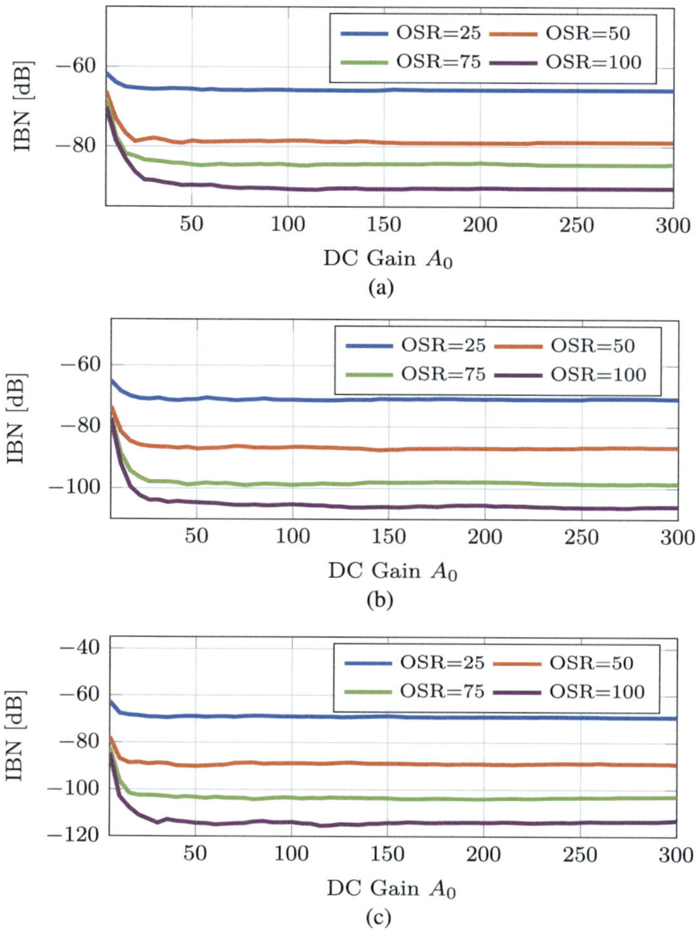

Fig. 4.26 Simulation of IBN over finite A_0 in all integrators of exemplary second-, third-, and forth-order multi-bit I-DS ADC incl. low-distortion path and *sinc reconstruction filters* over various M. Simulations were performed at −6dBFS input signal.

4.2.3 Compensation of Loop Filter Nonidealities

Since finite GBW in CT integrators can be modeled as a gain error and an additional pole, and the pole can be modeled as GBW-induced delay for a rectangular feedback DAC, it is known that finite GBW can be compensated for by rescaling the DSM to counteract the gain error and additionally compensate for the GBW-induced delay [236]. The compensated CT loop filter can be derived analytically using the method of momentums [237], impulse-response matching, or

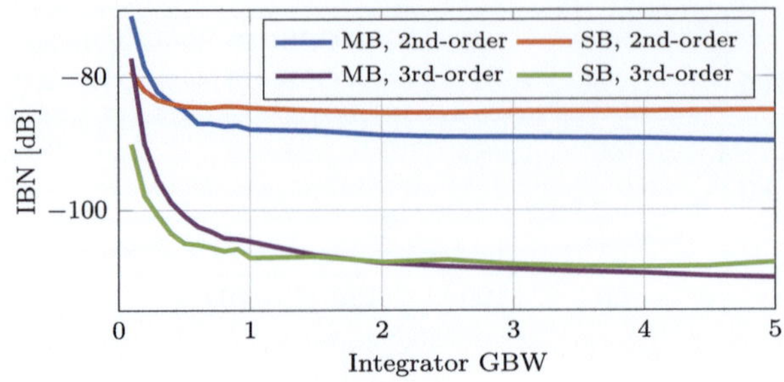

Fig. 4.27 Simulation of IBN of an exemplary second- and third-order CT single- and multi-bit I-DS ADC with low-distortion path and with M=50 and M=150, resp., over finite GBW. Simulated at an input amplitude of −6dBFS.

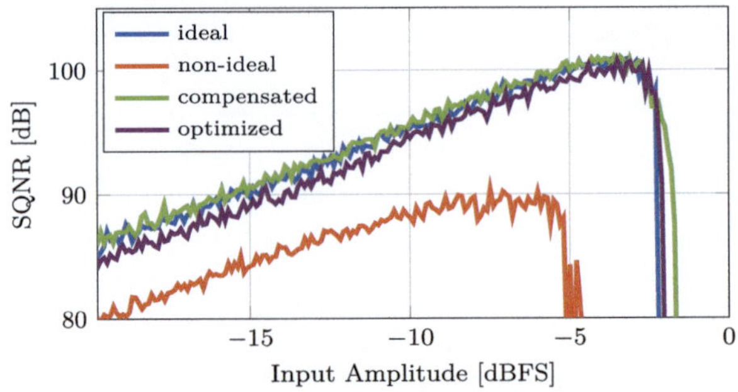

Fig. 4.28 SQNR sweep of a third-order single-bit I-DSM with CoI3 reconstruction filter and $OSR = 150$ under the influence of nonideal integrators and ELD. (a) Original scaling based on the Delta-Sigma Toolbox ($OBG = 1.5$) and DT-CT transformation without (blue) and (b) With non-idealities ($ELD = 0.5 \cdot T_s$, $GBW = 1 \cdot f_s$) (red). (c) Rescaled I-DSM coefficients compensated for finite GBW and ELD (green). (d) Optimization of nonideal I-DSM using www.sigma-delta.de (purple).

impulse-invariant transform [4, 236]. Alternatively, the scaling of the nonideal loop filter including compensation can be derived directly using a heuristic search, e.g. based on *www.sigma-delta.de*.

This is shown in the example in Fig. 4.28, where the third-order CIFF I-DSM is combined with a CoI3 reconstruction filter. The original DT scaling coefficients were derived via the DS toolbox using $OBG = 1.5$ for single-bit operation followed by a DT-CT transformation, and both were given in Table 4.2. The nonideal integrators feature $GBW = 1 \cdot f_s$ and a global $ELD = 0.5 \cdot T_s$ is added. Without compensation, the modulator significantly loses performance, both in MSA and SQNR. Both, the analytically compensated and the optimization based loop filters, restore ideal performance. The respective scaling coefficients are given in Table 4.4.

It is worth noting that the GBW-induced phase shift leads to the same STF peaking as we observed for ELD, and thus also yields loss of MSA and SQNR in I-DS ADCs. While this peaking in the equivalent DT STF can be generally counteracted with the

Table 4.4 Scaling coefficients of the I-DSMs simulated in Fig. 4.28 (A) based on the Delta-Sigma Toolbox and DT-CT transformation (SB, $OBG = 1.5$, cf. Table 4.2), (B) incl. $ELD = 0.5 \cdot T_s$ and finite $GBW = 1 \cdot f_s$ and analytical compensation, and (C) incl. ELD and finite GBW and scaling by heuristic search optimization using *www.sigma-delta.de*.

	k_0	k_1	k_2	k_3
A: CT_{ideal}	–	0.670	0.244	0.044
B: $CT_{\text{nonid.\&comp.}}$	0.460	1.008	0.391	0.066
C: $CT_{\text{nonid.\&opt.}}$	0.348	0.740	0.236	0.055

distributed feed-in paths as shown in Fig. 4.20, we have observed in Fig. 4.22 that in the case of the single-bit I-DSM, the conventional ELD compensation of the NTF was sufficient. Similarly, the NTF compensation is sufficient for the single-bit example with finite GBW in Fig. 4.28. For a multi-bit example, the feed-in paths could be employed again.

4.3 Nonidealities Distinctly Different in Incremental Operation

4.3.1 Circuit Noise

NPF Power efficient ADCs are thermal (and not quantization) noise limited and the understanding of the influence of circuit noise on (I-DS) ADCs is necessary to being able to realize highly efficient designs. In free-running DS ADCs, input-referred noise is transferred to the output of the DSM shaped by its STF, which is commonly unity within the inband. Then, the digital brick-wall filter removes all OOB components before decimation happens. This is different in I-DS ADCs.

Steensgaard et al. [192] analyzed the sampled input-referred thermal noise of DT I-DS ADCs with linear reconstruction filters. A similar approach was derived for CT I-DSM in [183, 184]. Their analysis was based on the FIR model of DT I-DSMs as introduced in Sect. 3.3.3. The overall STF determines the shaping of the input-referred noise toward the output and, thus, its integrated contribution to the final inband noise. In free-running mode, only the inband part of the input-referenced white noise is present at the output of the decimation filter. In incremental mode, however, the input-referenced noise, which is shaped by the overall incremental STF, is completely folded into the Nyquist output. Since the STF of incremental ADCs is unity within the inband, and thus the inband thermal noise contribution is identical to the free-running case, the additional STF-shaped OOB part increases the output noise, which leads to a *NPF* of I-DS ADC over their free-running counterpart. Since the STF is determined by both the I-DSM and the reconstruction filter, this *NPF* is topology dependent.

4.3.1.1 Noise Penalty Factor

NPF Next, the so-called NPF is derived following [192]. In his derivation, [192] assumes a DT I-DSM and therefore assumes that the input signal is limited to the inband. Then, he only assumes input-referred white circuit noise which is limited between $\pm f_s/2$. The input-referred thermal noise is weighted by the elements of the impulse response $h_{\text{fir,stf}}[k]$ of the overall STF, such that the noise at the output of the incremental ADC becomes proportional to

$$h_{\text{fir,stf}}[1]^2 + h_{\text{fir,stf}}[2]^2 + \cdots + h_{\text{fir,stf}}[M]^2 = \left\| h_{\text{fir,stf}}[k] \right\|_2^2, k = 1 \ldots M. \tag{4.9}$$

This becomes minimal if all elements of $h_{\text{fir,stf}}[k]={}^1/_M$ for $k = 1 \ldots M$ [192], which yields $\left\| h_{\text{fir,stf}}[k] \right\|_2^2 = \sum_{k=1}^{M} {}^1/_{M^2} = 1/M$. This is the case if a first-order CoI is chosen as reconstruction filter for a first-order DSM. An identical result is obtained in free-running DSMs, where the input-referred noise is scaled by ${}^1/_{OSR}$ toward the output of the DSM. Consequently, it is known in literature that first-order incremental DSMs with CoI reconstruction filter and free-running DS ADCs behave equivalently concerning thermal noise. For other reconstruction filters or higher-order I-DS I-DS ADCs, $\left\| h_{\text{fir,stf}}[k] \right\|_2^2$ becomes larger than minimal, and [192] defines a NPF as

$$\beta_{\text{npf}} = \frac{\textit{actual thermal noise power}}{\textit{minimum thermal noise power}}, \tag{4.10}$$

where the reference minimal thermal noise power is that of a free-running DSM or equivalently that of a first-order I-DSM with CoI reconstruction filter. (4.10) is thus a measure to compare different I-DS ADCs with regard to thermal noise. For DT I-DSMs, the NPF becomes

$$\beta_{\text{npf}} = \frac{\left\| h_{\text{fir,stf}}[k] \right\|_2^2}{\frac{1}{M}} = M \sum_{k=1}^{M} h_{\text{fir,stf}}^2[k]. \tag{4.11}$$

As we have discussed throughout Sect. 4.1, the overall STF and its impulse response $h_{\text{fir,stf}}[k]$ are dominated by the reconstruction filter impulse response $h_{\text{rec}}[k]$, which yields

$$\hat{\beta}_{\text{npf}} \approx M \sum_{k=1}^{M} h_{\text{rec}}^2[k]. \tag{4.12}$$

For illustration, this approximated noise penalty factor $\hat{\beta}_{\text{npf}}$ is plotted over M in Fig. 4.29 for different reconstruction filters (cf. Sect. 3.5) which can be used in combination with a third-order I-DSM. It can be seen that $\hat{\beta}_{\text{npf}}$ steadily increases with order. This is also the strongest disadvantage of using a reconstruction filter with a higher order than the I-DSM order, even though it gives slight improvements in SQNR, cf. Sect. 3.5.2.3. $\hat{\beta}_{\text{npf}}$ also shows dependency on M, and settles to a final value

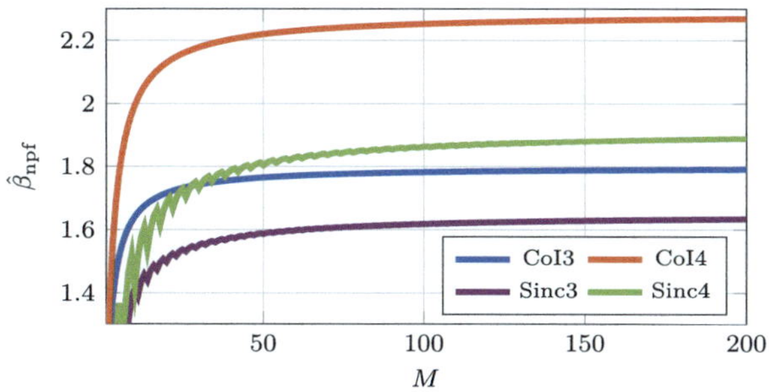

Fig. 4.29 Noise penalty factor $\hat{\beta}_{\mathrm{npf}}$ plotted over M for different reconstruction filters suitable for a third-order I-DSM.

for larger M. A small $\hat{\beta}_{\mathrm{npf}}$ is also found for the sinc reconstruction filters in Fig. 4.29, which, however, perform worse in quantization noise suppression, Sect. 3.5.3.

Note that even though [192] derived the NPF for input-referred noise of the circuit realizing an I-DSM, the same noise penalty is seen for noise coming with the input signal when it is not strictly limited to the inband, but extending to the OOB region $\pm f_s/2$.

Extension of the NPF to CT I-DSMs: In [192], only DT modulators were considered, where the analysis is performed in a sampled system assuming a proper AAF in front of the I-DSM, and where [192] analyzed the circuit noise between $\pm f_s/2$. The situation is different when we analyze CT I-DSM. By applying Parseval's theorem to (4.11), the overall STF can be used to derive a noise penalty factor of the CT incremental ADC, yielding

$$\beta_{\mathrm{npf,ct}} = \frac{\left\| STF_{\mathrm{ct,inc}}(f) \right\|_2^2}{\frac{1}{M}} = M \cdot \int_{-\infty}^{\infty} \left| STF_{\mathrm{ct,inc}}(f) \right|^2 \, \mathrm{d}f. \tag{4.13}$$

There, on the one hand, the frequency limitation between $\pm f_s/2$ is not valid. Moreover, as seen in Fig. 3.12, the overall CT STF deviates from the overall DT STF and the reconstruction filter shape as it has a continuous roll-off and no periodicity around multiples of f_s. Though this drop is only visible in the largely attenuated high-frequency range and thus it does not significantly alter the β_{npf} and the approximation by the reconstruction filter as in (4.12) also holds for CT I-DS ADCs.

In case, a direct feed-in path (LDP) is used, the explanation is slightly extended. As outlined in Sect. 4.1.4 and illustrated in Fig. 4.14, the overall STF of an I-DS ADC using an LDP in the I-DSM flattens in the high-frequency domain. In this case, let us distinguish the NPF for 3 cases:

- Noise from the I-DSM input-referred to the input summation node, i.e. the input of the first integrator: from there, the overall STF does not alter when using the LDP and thus the above statement holds, i.e. (4.12) is the NPF.

Table 4.5 Noise penalty factor $\hat{\beta}_{\text{npf}}$ of CoI and sinc reconstruction filters of different orders for large oversampling factor ($M = 150$).

Order	CoI $\hat{\beta}_{\text{npf}}$	Sinc $\hat{\beta}_{\text{npf}}$
1	1.00	1.00
2	1.33	1.32
3	1.79	1.63
4	2.26	1.88

- Noise from the LDP itself is inserted at the quantizer input. Thus it sees the NTF as transfer function and that is the reason for the flattening of the overall STF, when an LDP is used, cf. Sect. 4.1.4. Similarly as we stated for the input signal when discussing the AAF for CT I-DS DSM including the LDP, also noise through this LDP path needs some low-pass filtering, such that its integration over infinite frequency range does not yield infinite noise.
- Noise which comes together with the input and sees the overall STF including the impact of the LDP as illustrated in Fig. 4.14. Then, the statement from Sect. 4.1.4 holds that the noise from the input signal must have some low-pass filtering or an explicit AAF is required also for the CT I-DSM; either way, then the influence of LDP on the overall STF can be neglected and the above is again valid.

Exemplary NPFs of I-DS ADCs with different reconstruction filters: In Table 4.5 the noise penalty factors of commonly used reconstruction filters, i.e. CoI and sinc, are given for different orders and for a large oversampling ratio. Note that the NPF is largely independent on M, cf. Fig. 4.29, and only majorly drops for very small M. While we have shown in Sect. 3.5.2.2 that reconstruction filters with higher order than the I-DSM yield minor improvements in SQNR values, this comes at a severely increased NPF. This is the reason that the higher-order reconstruction filter is rarely used in practice, as the higher NPF would require a much smaller input-referred circuit noise for a required SNR.

4.3.2 *Jitter (Noise)*

Clock jitter has a detrimental effect on the performance of ADCs. Despite the general problem of sampling jitter, especially in CT DSMs, the continuous-time feedback DAC waveform is known to cause a severe drop in performance. The reason is that the usually employed rectangular RZ or NRZ feedback pulse is modulated by timing jitter, which translates into a feedback charge noise. This can be modeled as additional white noise at the input of the DSM [4, 5]. Compared to free-running DSMs, white noise at the input of I-DSMs impacts the resolution more by the noise penalty factor. Analysis and equations for the IBN in CT DSMs are, e.g. found in [4, Sect. 4.3] for RZ, NRZ, single- and multi-bit feedback DACs, and we can expect these to hold for incremental operation, when the NPF is regarded. To confirm this, several

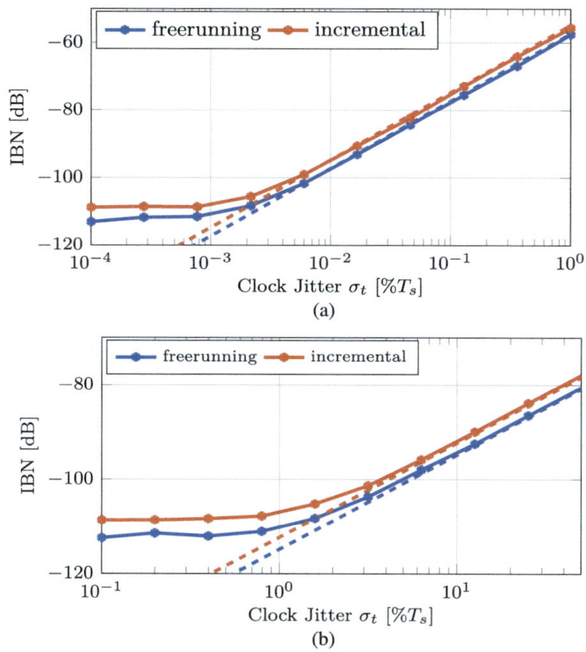

Fig. 4.30 Simulation of IBN over clock jitter σ_t in $\%T_s$ for a (a) CT and (b) DT single-bit CIFF DSM in free-running and incremental mode at an oversampling of $M = 150$ and incl. low-distortion path. For the I-DS ADC a CoI3 reconstruction filter is used. Simulations were performed at -6dBFS input signal and an input frequency of $0.3 f_b$. For the CT modulator a NRZ DAC is used. Solid: simulated. Dashed: calculated with (4.14) and (4.15).

simulations of IBN over clock jitter σ_t are shown in Fig. 4.30 for a (a) CT and (b) DT single-bit CIFF DSM incl. a low-distortion path in free-running and incremental mode. For the I-DS ADC, a CoI3 reconstruction filter is used. For the CT modulator, an NRZ DAC is used. The simulations have been performed at -6dBFS input signal and an input frequency of $0.3 f_b$. Simulation (solid) and calculation (dashed) results are shown. The calculations have been done using

$$IBN_{\mathrm{ct}} = \left(\frac{FS}{2^B - 1}\right)^2 \cdot \left(\frac{\sigma_t}{T_s}\right)^2 \cdot \frac{A_{\mathrm{NRZ}}}{OSR} \cdot \beta_{\mathrm{npf}} \tag{4.14}$$

for the jitter in CT DSMs [4, Eq. 4.25] and

$$IBN_{\mathrm{dt}} = \left(\frac{f_{\mathrm{in}}}{f_s}\right)^2 \cdot \hat{u}_{in}^2 \cdot \left(\frac{\sigma_t}{T_s}\right)^2 \cdot \frac{2\pi^2}{OSR} \cdot \beta_{\mathrm{npf}} \tag{4.15}$$

for the DT aperture jitter [238]. Thereby, $A_{\mathrm{NRZ}} \approx 0.7$ is the activity factor of a single-bit NRZ DAC [4], $\hat{u}_{in}$ and f_{in} are the amplitude and frequency of the sinusoidal input, and β_{npf} is the NPF, which is 1 for the free-running DSM and chosen from Table 4.5 for the incremental operation. The predictions match the simulations in Fig. 4.30.

4.3.3 Nonidealities at Inner I-DSM Nodes

In Sect. 3.3, the calculation of the overall noise and signal transfer functions for I-DS ADCs has been introduced. These transfer functions allow to describe the I-DS ADC in the frequency domain, which allows accurate performance predictions using the overall NTF including nonidealities and to evaluate the ADC signal filtering properties by its STF. Furthermore, as shown in Sect. 4.3.1, the STF and its impulse response can be used to calculate the NPF, as (spectrally white) input-referred noise is transferred to the output of the I-DS ADC shaped by the overall STF.

Even though the input of a DSM loop filter is the most critical node concerning nonidealities, which is why the first integrator is usually by far the most power hungry in implementations [24], in practice also circuit parts in the inner part of the loop filter contribute nonidealities. This raises the question how thermal noise, which is inserted at inner nodes of the loop filter, is filtered and attenuated before appearing at the overall output of the I-DS ADC.

These *inner transfer functions* of I-DS ADCs are analyzed next. They allow to calculate not only the attenuation of inner circuit noise but also of nonlinearity from inner loop-filter nodes toward the output of I-DS ADCs. Figure 4.31 shows a linearized L^{th}-order I-DS ADC that is used to clarify the nomenclature of the inner transfer functions. This analysis uses the full set of loop-filter scaling coefficients from Sect. 3 and, e.g. Fig. 3.4, where each integrator had its own scaling, in contrast to the simpler scaling used in Fig. 4.1 and previous sections. The quantizer is replaced by its linear model consisting of the quantizer gain q and the quantization noise source $e_\mathrm{q}[k]$. Neither forward nor feedback loop-filter scaling paths are included in the block diagram as the theory is derived independently of the architecture. The definition of the overall inner transfer functions $TF_l(f)$ and the overall NTF are illustrated in the figure. Thereby, $TF_l(f)$ describes the transfer function from the input of the lth integrator toward the overall output of the I-DS ADC.

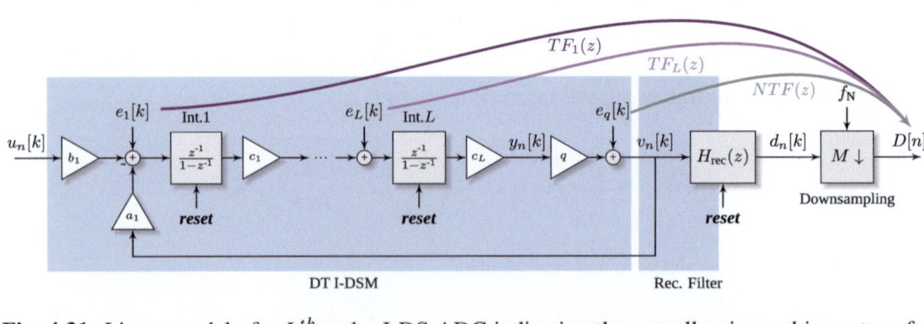

Fig. 4.31 Linear model of a Lth-order I-DS ADC indicating the overall noise and inner transfer functions. Since the theory is derived independently of the architecture, neither feedback nor feed-forward compensation paths are not shown. ©2024 IEEE. Modified, with permission, from Kaesser et al. [158].

4.3.3.1 Analysis of Inner Transfer Functions

The inner transfer function in I-DS ADCs can be similarly calculated as the overall STF introduced in Sect. 3.3.3 [158]. We start with the STF of the free-running (DT or CT) DSM

$$STF_{\text{dsm}}(f) = \frac{q \cdot H_{\text{dt/ct}}(e^{j2\pi f})}{1 + q \cdot H_{\text{dt}}(e^{j2\pi f})} \tag{4.16}$$

with the free-running loop filter $H_{\text{dt/ct}}(e^{j2\pi f})$, the quantizer gain q. The index STF_{dsm} indicates that it describes the transfer function of the DSM, but without decimation filter. Similarly, the inner transfer functions of the free-running modulator from all inner nodes to the DSM output can be calculated. Thereby, the loop-filter transfer function $H_{\text{dt/ct}}(e^{j2\pi f})$ in the nominator of (4.16) has to be replaced with the transfer function from the respective inner node to the DSM output, referred to as $H_{\text{dt/ct},l}(e^{j2\pi f})$. Then, the inner transfer function from the input of the lth integrator to the output of the (DT or CT) free-running DSM yields

$$TF_{\text{dsm},l}(f) = \frac{q \cdot H_{\text{dt/ct},l}(e^{j2\pi f})}{1 + q \cdot H_{\text{dt}}(e^{j2\pi f})}. \tag{4.17}$$

To derive the overall inner transfer function of the I-DS ADC, $TF_{\text{dsm},l}(f)$ has to be combined with the reconstruction filter and the reset, similarly as the overall incremental STF was derived in Sect. 3.3.3. This yields

$$TF_{\text{dt},l} = \left(TF_{\text{dsm},l}(f) \cdot H_{\text{rec}}(\mathrm{e}^{j2\pi f})\right) \circledast X_{\text{rect},M}(f) \tag{4.18}$$

in the DT case and

$$TF_{\text{ct},l} = \left(TF_{\text{dsm},l}(f) \cdot H_{\text{rec}}(\mathrm{e}^{j2\pi f})\right) * X_{\text{rect},T}(f) \tag{4.19}$$

in the CT case.

Example: As an example, the inner transfer functions are calculated and plotted for the exemplary DT single-bit third-order I-DS ADC with an oversampling of $M = 150$ as already introduced in Sect. 3.3.3.3. The block diagram can be found in Fig. 3.11 and the coefficients in Table 3.1. The coefficient scaling was done using an OBG of 1.5.

In combination with a CoI3 reconstruction filter the inner transfer function from the input of the second and third integrator are shown together with the overall STF and NTF of the DT I-DS ADC. As expected, the attenuation of circuit noise and distortion increases the later these nonidealities are inserted into the chain of integrators of the I-DSM. Nevertheless, the course of the inner transfer functions varies significantly from its free-running counterpart (Fig. 4.32).

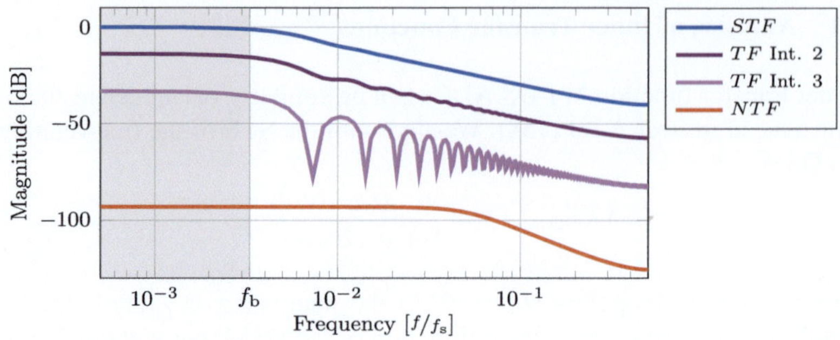

Fig. 4.32 (Inner) transfer functions of the exemplary third-order single-bit DT I-DS ADC with $M = 150$ and a third-order CoI reconstruction filter. ©2024 IEEE. Reprinted, with permission, from Kaesser et al. [158].

The employed reconstruction filter has a large influence. We have demonstrated that, in incremental DS ADCs, the reconstruction filter largely influences both the overall NTF and, consequently, the quantization noise suppression, as well as the overall STF [158]. The inner transfer functions can also be obtained for various reconstruction filters and I-DSMs using the online tool *www.sigma-delta.de* and by activating the "Calculate Inner TFs" option in the TF settings.

4.3.3.2 Circuit Noise from Inner Nodes of I-DSM

With the knowledge of the overall STF, we could analyze the influence of input-referred white noise and thus the NPF in Sect. 4.3.1. By using the inner transfer functions of I-DSMs, this analysis can be extended to noise sources inserted at inner nodes of I-DS ADCs.

Review of noise inserted within free-running DSM: The integrated inband noise $P_{\mathrm{N,free},l}$ at the output of a free-running DSM due to the noise $P_{\mathrm{N},l}$ inserted at the input of the lth integrator can be calculated using [75, 239]:

$$P_{\mathrm{N,free},l} = \frac{P_{\mathrm{N},l}}{f_\mathrm{s}} \int_{-f_\mathrm{s}/2}^{f_\mathrm{s}/2} \left|TF_{\mathrm{dsm},l}(f) \cdot H_{\mathrm{dec}}(f)\right|^2 \mathrm{d}f, \tag{4.20}$$

where $TF_{\mathrm{dsm},l}(f)$ is again the transfer function from the lth integrator input toward the output of the free-running DSM, cf. (4.17). By assuming a brick-wall decimation filter for $H_{\mathrm{dec}}(f)$, this yields

$$P_{\mathrm{N,free},l} = 2\frac{P_{\mathrm{N},l}}{f_\mathrm{s}} \int_{0}^{f_\mathrm{b}} \left|TF_{\mathrm{dsm},l}(f)\right|^2 \mathrm{d}f. \tag{4.21}$$

By assuming large a OSR, it simplifies to [239]:

$$P_{\mathrm{N,free},l} \approx P_{\mathrm{N},l} \frac{\pi^{2(l-1)}}{(2(l-1)+1)\left(a_1 \prod_{i=1}^{l-1} c_i\right)^2} \left(\frac{1}{OSR}\right)^{2(l-1)+1}. \tag{4.22}$$

Thereby, a_1 and c_i are the integrator scaling coefficients as shown in the block diagram in Fig. 4.31. For the special case of $l = 4$ for a third-order DSM, the respective inner insertion point becomes that of the quantizer, $P_{\mathrm{N},l}$ can be replaced by the quantization noise, and $P_{\mathrm{N,free},l}$ becomes the quantization IBN of the DSM. For all earlier insertion points, the attenuation of white noise obviously decreases by one order and thus mostly by $1/OSR^2$.

Extension to noise inserted within I-DSM: Also in the incremental case, the integrated inband noise $P_{\mathrm{N,inc},l}$ at the overall output of the I-DS ADC can be calculated using the derived inner transfer functions $TF_l(f)$ from (4.18) or (4.19). Then, equivalently to (4.20), integration over the whole sampling band yields

$$P_{\mathrm{N,inc},l} = \frac{P_{\mathrm{N},l}}{f_\mathrm{s}} \int_{-f_\mathrm{s}/2}^{f_\mathrm{s}/2} |TF_l(f)|^2 \,\mathrm{d}f. \tag{4.23}$$

Note that in contrast to (4.20), $TF_l(f)$ in (4.23) includes both the I-DSM and its reconstruction filter, and also that in contrast to (4.22) no further simplification concerning frequency range can be done, but the integration must be performed over the whole sampling frequency. While (4.23) thereby becomes generally valid for all kinds of I-DSMs and reconstruction filters, it does not reveal an intuitive approximation of noise attenuation within I-DSMs in contrast to what we could derive for free-running modulators in (4.22).

For the special case of CoI reconstruction filters, [158] derived an approximation of (4.23), which yielded

$$P_{\mathrm{N,inc},l} \approx P_{\mathrm{N},l} \cdot \left(\frac{L_\mathrm{d}!}{a_1 \prod_{i=1}^{l-1} c_i}\right)^2 \cdot \Phi_\mathrm{N} \cdot \left(\frac{1}{M}\right)^{2(l-1)+1}. \tag{4.24}$$

Thereby, L_d is the order of the digital CoI filter and the factor Φ_N can be determined using Table 4.6.

Table 4.6 I-DS ADC noise approximation factor Φ_N for up to a 5th order CoI, filter order L_d and loop-filter error-insertion point l [158].

$L_\mathrm{d} - l$	0	1	2	3	4
Φ_N	1	1/3	1/20	1/252	1/5184

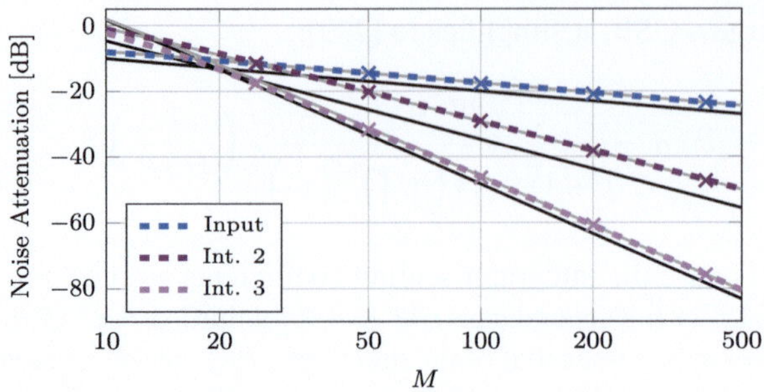

Fig. 4.33 Noise attenuation at loop-filter nodes (input, 2nd and 3rd integrators) of the exemplary third-order single-bit I-DS ADC with a CoI3 filter over M. The exactly calculated noise attenuation from (4.23) (colored dashed) strongly differ from the values of the free-running DS ADC from (4.22) (black solid). They match closely with the derived noise approximation for I-DS ADCs in (4.24) (grey). Matlab simulation points (x-marks) verify the calculated results.

The factor Φ_N depends on the order of the digital CoI filter L_d and the error insertion point l. For $l \leq L$ (order of the modulator), the approximation gives a very good fit. This is shown for an illustrative combination in Fig. 4.33, where the white noise attenuation of an exemplary third-order I-DSM using a CoI3 reconstruction filter is plotted over the oversampling M for three different insertion points: the signal input, and the inputs of the second and third integrators.

The exactly calculated noise attenuation using the inner transfer functions and (4.23) is shown in color. The calculations are verified by exemplary simulation points made in Matlab and marked with x-marks. The approximation for free-running DSMs from (4.22) is shown in solid black. The noise attenuation in the incremental case differs significantly from its free-running approximation although the course over M stays more or less the same. Looking at the solid gray line, which depicts the incremental approximation from (4.24), it can be clearly seen that the approximation matches nicely with the exact calculation for oversampling values larger than 25. Therefore, (4.24) can be used for the design of I-DS ADCs to give the designer a quick overview on the noise attenuation of the different nodes in the I-DSM if an oversampling higher than 25 and a CoI filter are used.

4.3.3.3 Noise penalty factor (NPF) of Inner Nodes of I-DSMs

Comparing the noise attenuation approximation of the free-running case from (4.22) with the incremental one from (4.24), a certain similarity between the equations becomes obvious. If an infinitely small reset is assumed leading to $M = OSR$, the only remaining difference between the two equations are the factors

$$\frac{\pi^{2(l-1)}}{2(l-1)+1} \tag{4.25}$$

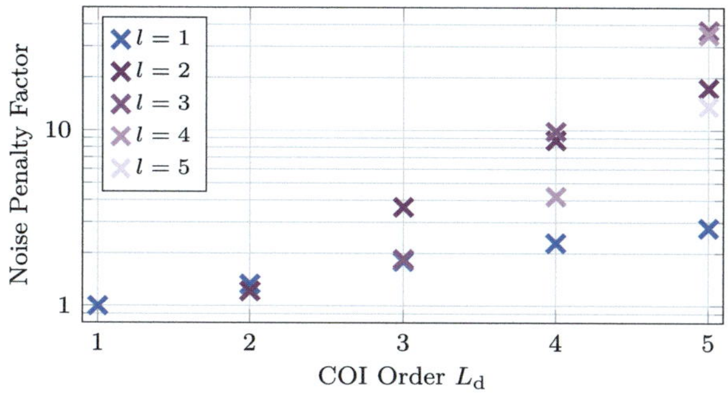

Fig. 4.34 Noise penalty factor NPF_l for different noise insertion points l and orders of the digital CoI filter L_d. For $l = 1$, this equals the known NPF at the input of I-DSMs [192]. ©2024 IEEE. Reprinted, with permission, from Kaesser et al. [158].

in (4.22) for the free-running DSM and

$$(L_\mathrm{d}!)^2 \cdot \Phi_\mathrm{N} \tag{4.26}$$

in (4.24) for the I-DS ADC. In Sect. 4.3.1, a noise penalty factor was introduced that relates the noise attenuation at the input of a I-DS ADC to its free-running counterpart. Similarly, a noise penalty factor for the inner nodes of the I-DSM can be defined by taking the ratio of (4.26) over (4.25). This yields

$$\mathrm{NPF}_l = \frac{(L_\mathrm{d}!)^2 \cdot \Phi_\mathrm{N}}{\pi^{2(l-1)}/(2(l-1)+1)} \tag{4.27}$$

for an order L_d of the CoI reconstruction filter and a certain noise insertion point l.

The derived NPFs_l are shown in Fig. 4.34. A noise insertion point of $l = 1$ (blue x-marks) refers to the input of the I-DSM, and the shown results equal the known input related NPFs, e.g. 1.3, 1.8, 2.3 for a second-, third-, and fourth-order CoI filters, cf. Sect. 4.3.1 and Table 4.5.

4.3.4 *Offset Cancellation using Fractal Chopping*

I-DSM have traditionally been used for low-frequency, near DC applications. In such scenarios, the integrator offset becomes a real bottleneck and must be eliminated. A common technique to cancel the offset of amplifiers and integrators in free-running DSM is chopping [240]. Although chopping remains effective in first-order I-DSM, there is a fundamental difference that requires careful consideration: in I-DS ADCs, the potentially nonuniform filter weighting has to be taken into account [241, 242].

In Fig. 4.35a a first-order I-DS ADC is shown, where an integrator offset is modeled as an additive input V_O. The idea of fractal sequences canceling the offset can

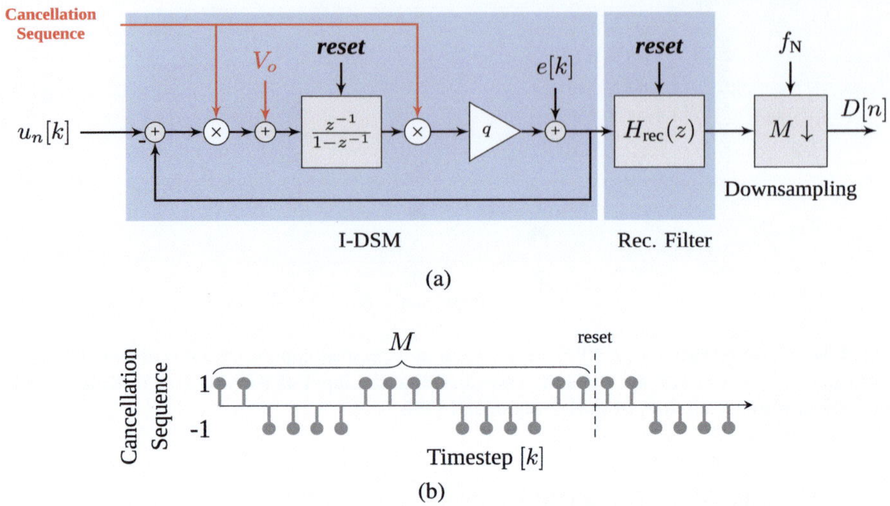

Fig. 4.35 (a) Block diagram of a first-order I-DS ADC with linearized quantizer and offset cancellation. (b) Exemplary offset cancellation sequence. ©2024 IEEE. Reprinted, with permission, from Kaesser et al. [242].

be explained in the time domain. Suppose an integrator positively integrates its input offset for $M/2$ clock cycles and negatively integrates its input offset for another $M/2$ clock cycles, while the actual input signal is positively integrated for the total M cycles. Then, the output is effectively offset-free. While this is obviously achieved by any cancellation sequence with an equal number of positive and negative offset integration cycles, e.g. the one shown in Fig. 4.35b, it is not straightforward for a chain of integrators. There, the cancellation sequence has to be determined analytically [241] depending on the DSM order and the reconstruction filter. The resulting offset cancellation using this sequence of inversions is known as *fractal chopping*.

The correct fractal chopping sequence must be chosen according to the individual sample weights of a given I-DS ADC. The weights can be determined by calculating the impulse response of the modulators' free-running transfer function

$$h_{\mathrm{DSM}}[k] = \mathcal{Z}^{-1}\left\{H_{\mathrm{DSM}}(z)\right\} \; k \in [1, M] \tag{4.28}$$

and the impulse response of the reconstruction filter

$$h_{\mathrm{rec}}[k] = \mathcal{Z}^{-1}\left\{H_{\mathrm{rec}}(z)\right\} \; k \in [1, M]. \tag{4.29}$$

The overall weights can then be derived by the convolution:

$$h_{\mathrm{overall}}[k] = h_{\mathrm{dsm}}[k] * h_{\mathrm{rec}}[k] \; k \in [1, M]. \tag{4.30}$$

It is important to note that $H_{\mathrm{DSM}}(z)$ describes the free-running transfer function from the offset insertion point to the modulator output. Usually, the most dominant offset comes from the first integrator. Then $H_{\mathrm{DSM}}(z)$ would be proportional to the free-running STF of the DSM. Since a unity STF is commonly chosen, the sample weights are dominated by the reconstruction filter, and it is a valid approximation to use only the reconstruction filter weights. The remaining offset at the overall output of the I-DS ADC can then be calculated using the offset suppression factor

$$\alpha_{\mathrm{offset}} = \left| \sum_{k=1}^{M} h_{\mathrm{overall}}[k] \cdot S[k] \right| \quad S[k] \in \{-1, +1\}. \tag{4.31}$$

The offset suppression factor α_{offset} can be minimized by choosing a suitable cancellation sequence $S[k]$. For an mth-order CoI filter, the sequence derived in [243] can be used:

$$\begin{aligned} S_{\mathrm{F0}} &= [+1 ...] \\ S_{\mathrm{F1}} &= [+1 \ \ -1 ...] \\ S_{\mathrm{F2}} &= [+1 -1 \ \ -1 +1 ...] \\ S_{\mathrm{F3}} &= [+1 -1 -1 +1 \ \ -1 +1 +1 -1 ...] \\ &\dots \\ S_{\mathrm{Fm}} &= [S_{\mathrm{Fm-1}} \ \overline{S_{\mathrm{Fm-1}}} ...]. \end{aligned} \tag{4.32}$$

In these sequences, "−1" means a sign inversion in the choppers of Fig. 4.35a. In principle, the sequence of a higher order is derived from the concatenation of the sequence of a lower order, once unchanged and once inverted. In Fig. 4.36, the

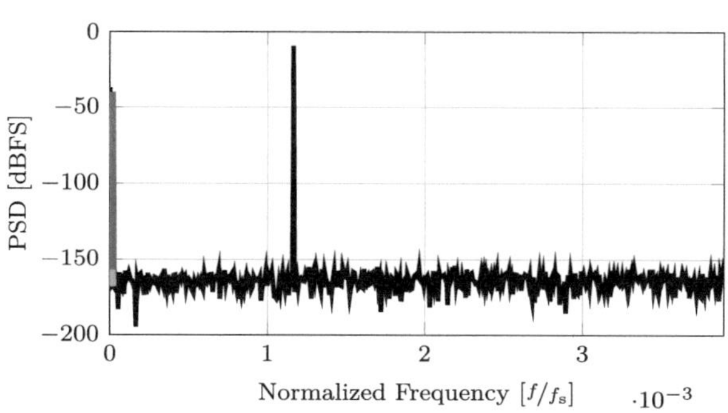

Fig. 4.36 Simulated output spectrum of the exemplary third-order DT I-DS ADC (Fig. 3.11 and Table 3.1) with a CoI3 filter, low-distortion path and a −40dBFS offset applied at the input of the first integrator. Red: without offset cancellation. Blue: cancellation with chopping at $f_s/2$. Green: cancellation with sequence S_{F3} from (4.32).

exemplary third-order DT I-DS ADC (cf. Fig. 3.11 and Table 3.1) including a direct low-distortion path and CoI3 reconstruction filter is simulated. An offset S_{off}=-40dBFS is applied at the input of the first integrator. Without any chopping, the −40dBFS offset is seen in the output spectrum. If a common alternating chopping sequence at $f_s/2$ is used, the offset is suppressed to around −75dBFS but still present. If the third-order cancellation sequence S_{F3} from (4.32) is used, the offset at the output is completely canceled.

Note that if a sinc reconstruction filter is used instead of a CoI filter the normal chopping approach is effective due to the symmetric weighting of the sinc filters [242]. Furthermore, the chopping frequency can be drastically reduced to f_{N}. Considering this in the time domain with the introduced sequences, it is enough to switch the sign of the offset one time at $M/2$ since the first half of the Nyquist conversion and the second half have equal weights due to the symmetry of the sinc filter. An example is given in [242].

4.3.5 Nonideal Reset

In I-DS ADCs, every Nyquist conversion is started without memory from prior conversion cycles, which is achieved by resetting all internal states of the I-DSM and reconstruction filter. In most theoretical works on I-DSMs, the reset is assumed to be infinitely short, implying an $OSR = M$, cf. Fig. 3.2. However, in practice one or more clock cycles M_{reset} have to be dedicated to the actual reset operation, especially for the analog memory elements, which are the capacitors of the analog integrators, cf. (3.1). The required reset then depends on the transient discharge, and finite switch resistance leads to nonideal resetting of the analog memory, retaining information from prior samples in the loop filter. An analysis of nonideal resetting in I-DS ADCs was introduced in [244].

Due to the suppression of nonidealities by the loop filter, nonidealities inserted at the first integrator are usually the dominant ones. Therefore, [244] focused on the nonideal reset of the first integrator. At the end of each Nyquist conversion, and therefore the beginning of the reset phase, the value contained in the first integrator can be statistically described with the standard deviation

$$\sigma_{\text{RSTstart}} = \frac{a_1 \cdot LSB}{\sqrt{12}}, \tag{4.33}$$

where a_1 is the scaling coefficient of the first integrator, cf. Fig. 4.31. While this is only true if the modulator is in steady-state operation at the end of its conversion cycle, in practice this is the case either by a large enough oversampling M or by using a low-distortion path. Therefore, the assumption is valid. Knowing the statistical distribution of the integrator just before the reset phase, the standard deviation at the end of the reset phase can be calculated with

$$\frac{\sigma_{\text{RSTend}}}{\sigma_{\text{RSTstart}}} = e^{-\frac{T_S \cdot M_{\text{reset}}}{\tau}}, \tag{4.34}$$

where T_s is the clock period, M_{reset} the length of the reset, and the time constant $\tau = R_{\text{on}} \cdot C_{\text{int1}}$ can be calculated using the on-resistance of the switch and the first stage's integration capacitance. With the known integrator output error distribution at the end of the reset phase, we know the integrator output error distribution at the beginning of the next conversion cycle. This error, present at the output of the first integrator in the first cycle of the next conversion cycle, is then referred to the input of the I-DSM by dividing it by the integrator scaling a_1. Moreover, this single-cycle error is modeled as an input error which is constant for all M clock cycles of the next conversion period. This can be done by multiplying σ_{RSTend} with an attenuation factor α, which equals the first weight of the reconstruction filter [244, Eq. 8]. Then, the IBN due to the nonideal reset can be calculated using

$$IBN_{\text{Reset}} = 10 \cdot \log_{10}(\alpha^2 \cdot \frac{2}{FS^2}\frac{LSB^2}{12} \cdot e^{-\frac{2 \cdot T_S \cdot M_{\text{reset}}}{\tau}}). \tag{4.35}$$

Allowing for a certain specified IBN due to the nonideal reset, it becomes possible to calculate the required R_{on} depending on the length of the reset M_{reset}. This analysis provides the designer with a quick estimate of the reset switch on-resistance sizing. A design example, simulations, and a discussion of CT I-DSMs, impact of finite GBW of opamps during reset, and various filter architectures are given in [244] confirming the results.

4.3.6 DAC Nonlinearity and DWA

As discussed through Sect. 2.3.2.2, using a multi-bit quantizer inside a DSM offers several significant advantages that enhance the performance and efficiency of the converter. It not only directly improves the SQNR by one bit per additional bit of quantizer resolution, but also enhances the DSM stability due to a more linear and uniform noise-like behavior of the quantization noise. Additionally, the enhanced stability allows for a larger MSA, and also allows a more aggressive NTF scaling. In consequence, a multi-bit quantizer allows the DSM to achieve the desired performance usually with an oversampling ratio about 3–4 times smaller as single-bit DSM, and the larger MSA allows for more input-referred thermal noise for the same SNR.

The well-known drawback of multi-bit quantization is the necessary multi-bit feedback DAC. Even though its resolution has only to be as good as that of the internal quantizer, its linearity requirements have to meet the requirement of the overall modulator resolution. Intrinsic linearity by matching above 10..12 bits is hard to achieve in practice, and various techniques for linearization have been proposed in the SoA, which were listed in Sect. 2.3.2.2.

In free-running DSMs, DEM algorithms are a popular solution to suppress the nonlinearities of the feedback DAC. However, SoA has identified inferior performance of DWA algorithms in I-DSMs [245] and attributed it to the nonuniform sample weighting in the digital reconstruction filter: in incremental operation, the error caused by DAC mismatch depends on the injection time. For example, employing a CoI reconstruction filter, the weight of a signal injected at the beginning of the conversion is larger than one injected near the end.

This can be avoided by employing a first-order I-DSM and CoI reconstruction filter with uniform sample weighting [246], though, first-order I-DS require very high OSR to achieve decent performance [246]. Alternatively, [245, 247] proposed a smart dynamic-element-matching (SDEM) technique that adapts the DEM algorithm for second and third-order I-DS ADCs such that the weighted errors are minimized. While it demonstrated efficiency in DAC mismatch error shaping, it came with a significant penalty in complexity compared to common DEM algorithms employed in free-running DSMs such as a rotational DWA, undergoing an exponential increase in complexity when going to higher-order I-DSMs [245]. The topic was further elaborated in [248], which reviewed the inferior performance of conventional DWA algorithms in I-DS ADCs. It then showed that the SDEM algorithm from [245] corresponds to higher-order DEM algorithms of the free-running case. Then, it was shown that including a regular reset of the rotation index of the DWA algorithm, synchronous with the reset of the I-DSM and its reconstruction filter, also solves the problem. In fact, this has been indicated in an earlier work [249] where a reset is employed at the DWA block in [249, Fig. 1] without further indicating how the DWA works and how it compares to other mismatch shaping techniques.

Simulation results confirm the discussion. A third-order I-DSM with CIFF loop filter, low-distortion direct input path, 4-bit internal quantizer and CoI3 reconstruction filter and scaling for an $OBG = 2.5$ is simulated in incremental mode for $M = 120$. Ideally, an SQNR of 127dB can be achieved. Then, 50 simulations for uniformly distributed mismatch in the 15 DAC unit elements are simulated and the achieved mean SNDR is plotted over various mismatch standard deviation σ_{DAC} for different DEM algorithms. Firstly, it can be seen that even though conventional DWA shows linearization of the nonideal DAC, it loses more than 40dB SNDR at a large mismatch of 1%. If the rotational DWA is reset, the linearization is superior and the loss in SNDR can be reduced to around 15dB at 1% mismatch. If the SDEM algorithm from [245] is used, this loss can be further reduced. To show that the SDEM algorithm is in fact a higher-order mismatch shaping algorithm, we compare it to a second-order

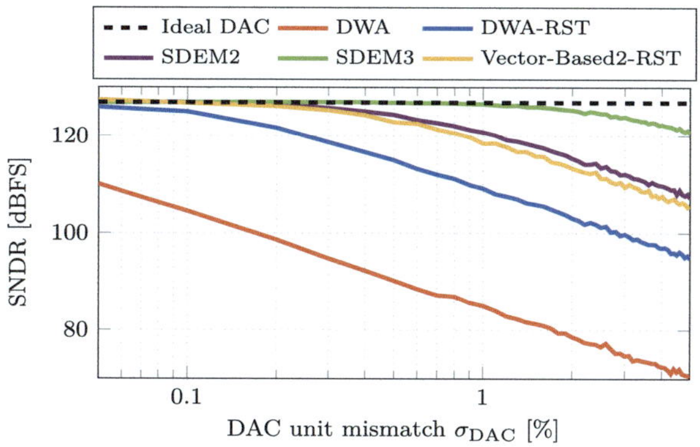

Fig. 4.37 Simulated mean SNDR (50 samples) of a third-order CIFF I-DS ADC with CoI filter over the DAC unit element mismatch σ_{DAC} for different DEM algorithms. (4-bit quantizer, $OBG = 2.5$, LD path, $M = 120$). ©2024 IEEE. Modified, with permission, from Ismail et al. [248].

mismatch shaping DWA [250]; as we also reset this DWA, it is referred to as Vector-Based2-RST. Interestingly, the second-order vector-based algorithm extended by a periodic reset performs almost identically to SDEM2. [248] also presented the results of a third-order SDEM algorithm. This algorithm demonstrated no performance loss due to its third-order mismatch shaping; however, it was very complex (Fig. 4.37).

Chapter 5
Advanced Architectures

So far, only single-loop, single-stage, and single-step I-DS ADCs have been considered throughout the basic description of I-DS ADCs in Chap. 3 and their properties in Chap. 4. These architectures have the disadvantage of reduced stability as the order of noise-shaping increases, especially when single-bit internal quantization is employed. To improve the noise-shaping properties and further reduce quantization noise beyond the capabilities of the single-loop DSM, multi-stage noise-shaping (MASH) architectures have been introduced by concatenating multiple single-loop architectures. There, later stages requantize the quantization error of earlier stages [105] as outlined throughout Sect. 2.3.2.3. Similar to the free-running case, multi-stage noise shaping can also be applied to incremental DS ADCs [13, 191]. However, due to its unique properties, which allow to extract the incremental quantization error not only from the first-stage quantizer but also from the output of the last integrator, cf. (3.18), different MASH architectures can be implemented for incremental operation. Additionally, this availability of the quantization error from within the loop filter enables multi-step I-DS ADCs [251], which are known as *extended-counting* or *extended-range* incremental ADCs. Finally, one can exploit functional reconfigurability of the I-DS ADC during the conversion of a single Nyquist sample [25, 246]. Surveys on these advanced designs can be found in [14, 179, 252], and we will derive and review them systematically in this chapter.

5.1 Incremental Multi-Stage Architectures

In Sect. 2.3.2, different loop-filter architectures for DSMs have been reviewed. Besides single-loop architectures, the architecture of multi-stage noise-shaping DSMs has also been reviewed in Sect. 2.3.2.3. Such MASH architectures have been

M. Ortmanns et al., *Incremental Delta-Sigma ADCs*,
https://doi.org/10.1007/978-3-032-09468-1_5

employed for I-DS ADCs as early as [13]. Moreover, incremental SMASH ADCs have been presented in the SoA [24, 253]. Both are described in more detail in the following section.

5.1.1 *Incremental Multi-Stage Noise-Shaping (I-MASH) ADCs*

As outlined in Sect. 2.3.2.3, the required scaling and reduced MSA of higher-order DS ADCs can be mitigated by concatenating low-order single-loop DSMs to achieve a higher noise-shaping order while maintaining large MSA and more aggressive loop-filter scaling. This is easily extended to I-DSM and an exemplary two-stage incremental MASH architecture is shown in Fig. 5.1. The commonly known MASH architecture for free-running DSMs extracts the quantization error E_1 of the first stage by subtracting the analog output of the DAC from the analog input of the quantizer [106] further referred to as MASH QE (*QE* stands for quantizer error). This is illustrated by the implementation in dashed green in Fig. 5.1. Exemplary implementations of incremental MASH QE are, e.g. found in [191]. Another option to extract the quantization error of the first stage is to exploit the unique feature of I-DSM to show the incremental ADC's quantization error at the output of the last integrator, as it was derived in (3.7) and (3.18) for a first- and third-order I-DSM.

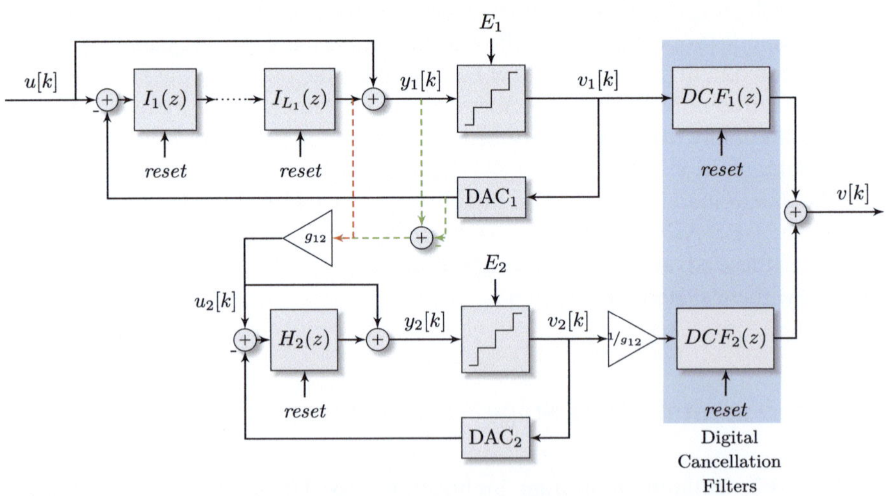

Fig. 5.1 Incremental MASH modulator passing the quantization error E_1 of a first stage to the second stage for requantization. $I(z)$ are integrators of the first stage loop filter $H_1(z)$. $H_2(z)$ is the loop filter of the second stage. (a) *MASH QE* extracts the quantization error from the difference of the quantizer input and the DAC output (dashed green). (b) *MASH LI* exploits the presence of the quantization error at the last integrator output of the first stage (dashed red).

This implementation is indicated in dashed red in Fig. 5.1 and is further referred to as MASH LI (*LI* stands for last integrator output) [13, 254–256]. It is worth noting that this variant of MASH was also analyzed for free-running operation in [257] to achieve a MASH with high robustness against circuit nonidealities.

Digital cancellation filters DCF_1 and DCF_2 are required in MASH DSM as they recombine the outputs of the individual stages such that the quantization error of the first stage E_1 is canceled and only the quantization error of the second stage E_2 remains with a noise-shaping order equivalent to that of both MASH stages at the overall output [4, 5]. The required cancellation filters depend on the employed variant of quantization error extraction and will be reviewed shortly in the next section. A simulation-based comparison of I-MASH ADCs with other incremental architectures has been presented in [256] based on extensive simulations.

The reconstruction filter required for incremental MASH DSM remains the same reconstruction filters as for single-loop incremental ADC, cf. Sect. 3.5. Thereby, the order of the reconstruction filter has to be at least the same order as the order of the overall incremental MASH, i.e. the sum of the order of all MASH stages. Consequently, incremental MASH ADC suffer from the *noise penalty factor* of higher-order reconstruction filters, where the order is that of the overall incremental MASH.

Digital cancellation filters of MASH QE: The derivation of the MASH DCFs can be performed based on the free-running transfer functions because the cancellation of the quantization noise of former stages by the preceding stages, and the digital recombination, serves the same purpose. The derivation is shown here for the two-stage example in Fig. 5.1 but can be similarly extended to more stages as illustrated in Fig. 2.17. The overall output of the two-stage MASH QE architecture, as shown in Fig. 5.1 (omitting the reset), is given by

$$\begin{aligned} V_{QE}(z) = U(z) \cdot STF_1(z) \cdot DCF_1(z) \\ + E_2(z) \cdot NTF_2(z) \cdot DCF_2(z) \\ + E_1(z)\,(DCF_1(z) \cdot NTF_1(z) - DCF_2(z) \cdot STF_2(z)) \end{aligned} \quad (5.1)$$

with $STF_i(z)$ and $NTF_i(z)$ being the (free-running) signal and noise transfer functions of the i-th stage. Canceling the quantization error E_1 of the first stage is achieved by:

$$DCF_1(z) = STF_2(z) \qquad and \qquad DCF_2(z) = NTF_1(z). \quad (5.2)$$

This is the common choice of the digital cancellation logic for MASH DSM [4, 5], which specifies that the digital cancellation filter following the first stage resembles the analog STF of the second stage, while the digital cancellation filter following the second stage resembles the analog NTF of the first stage. In this way, not only is the quantization error of the first stage canceled, but also the quantization error of the second stage E_2 is shaped by a combined analog/digital $NTF_{2|\mathrm{MASH}} = NTF_{1|\mathrm{D}}(z) \cdot NTF_{2|\mathrm{A}}(z)$, and thus the resulting noise-shaping order is the sum of the orders of the first and second stage.

It is important to remember that mismatch between digital and analog filters will lead to severe performance degradation, as the cancellation of the quantization error of the first stage will become nonideal and part of $E_1(z)$ in (5.1) will leak into the output.

Digital cancellation logic of MASH LI: The overall output of the two-stage MASH LI architecture, as shown in Fig. 5.1 (omitting the reset), is given by

$$\begin{aligned} V_{\mathrm{LI}}(z) = U(z) \cdot STF_1(z) \cdot DCF_1(z) \\ + E_2(z) \cdot NTF_2(z) \cdot DCF_2(z) \\ + E_1(z) \cdot NTF_1(z) \cdot (DCF_1(z) - DCF_2(z) \cdot STF_2(z) \cdot TF_{\mathrm{int}}(z)) \end{aligned} \tag{5.3}$$

with $TF_{\mathrm{int}}(z)$ being the transfer function from the quantizer output to the last integrator output of the first stage. Note that this is not necessarily identical to the loop filter of the first stage.

Comparing (5.3) with (5.1), it can be seen that the digital cancellation logic has to be designed differently for the MASH LI compared to the MASH QE. One possibility is to choose DCF_2 as for MASH QE and adapt DCF_1. Then

$$\begin{aligned} DCF_1(z) = NTF_1(z) \cdot STF_2(z) \cdot TF_{\mathrm{int}}(z), \\ DCF_2(z) = NTF_1(z). \end{aligned} \tag{5.4}$$

This leads not only to a cancellation of the quantization error E_1 of the first stage, but also to a higher-order filtering of E_2 by both NTFs, i.e. $NTF_{2|\mathrm{MASH}} = NTF_{1|\mathrm{D}}(z) \cdot NTF_{2|\mathrm{A}}(z)$, i.e. identical behavior as for MASH QE, but with a seemingly more complicated DCF_1.

Combination of reconstruction and digital cancellation filter: A distinct advantage of the incremental MASH is that the digital cancellation filter and the reconstruction filter can be merged [65, 191]. In [191] for example, it was shown that no dedicated digital cancellation filter is required, instead the output of the second stage of an I-MASH can be inserted at a later point into the CoI-based reconstruction filter. This obviously makes sense: with a second-order first stage in a MASH QE, DCF_2 becomes a second-order differentiator (NTF_2). Second-order differentiation can be achieved by skipping the first two integrators of a higher-order CoI reconstruction filter. Of course, this strongly depends on the used MASH architecture, loop-filter order, and scaling, as well as the employed reconstruction filter.

Similar to the single-stage case, as shown and discussed in Sect. 3.4, the IBN and ENOB of incremental MASH QE ADCs featuring unscaled, ideal higher-order differentiating NTFs ($NTF = (1 - z^{-1})^L$) can be written as

$$IBN_{\text{IMASH}} = \prod_{i=1}^{m-1} \frac{1}{g_i^2} \cdot \frac{\Delta_m^2}{12} \cdot \left(\frac{L_{\text{tot}}!}{M^{L_{\text{tot}}}}\right)^2, \tag{5.5}$$

$$ENOB_{\text{IMASH}} = \log_2 \left(\prod_{i=1}^{m-1} g_i \cdot (2_m^B - 1) \cdot \frac{M^{L_{\text{tot}}}}{L_{\text{tot}}!}\right). \tag{5.6}$$

Thereby, g_i denotes the ISGs from the ith to the $(i+1)^{\text{st}}$ stage, i.e. an attenuation or gain factor scaling the extracted quantization noise from earlier stages before feeding them to the later stages with the intention to neither over- nor underload the next stage, cf. Sect. 2.3.2.3. $L_{\text{tot}} = \sum_{i=1}^{m} L_i$ is the combined modulator order of all stages and B_m is the number of bits of the quantizer of the last (m^{th}) stage. Equation (5.6) is a direct extension of (3.51), and is only slightly different (and easier) compared to the derivation in [191]. Equation (5.5) is directly derived from (3.47).

5.1.2 Nonideal Performance of the Incremental MASH ADC

The sensitivity of MASH ADCs to mismatch between the analog loop filter and the digital cancellation filter is well known in the SoA [2, 4, 5]. The underlying reason is that the quantization error of the first stage is extracted by the higher stage, but only digitally cancelled if the conditions as in (5.2) are fulfilled. As derived in [4], the sensitivity increases the greater the amount of combined MASH resolution generated in the higher stages. While [4, Eq. 5.14] derived an expression for the IBN of a MASH DSM under mismatch, a similar expression can be found for the I-MASH as

$$IBN_{\text{IMASH|Mismatch}} \approx \frac{\Delta_2^2}{12} \cdot \left(\frac{L_{\text{tot}}!}{M^{L_{\text{tot}}}}\right)^2 \cdot \frac{1}{g_1^2} + \delta_{\text{coeff}}^2 \frac{\Delta_1^2}{12} \cdot \Phi_N \cdot \left(\frac{L_{\text{tot}}!}{M^{L_1+1/2}}\right)^2. \tag{5.7}$$

This assumes a two-stage I-MASH with $ISG = g_1$ and uses (5.5) for the ideal part (LHS) of the equation. It assumes an order L_1 for the first stage, an order L_2 for the second stage, and $L_{\text{tot}} = L_1 + L_2$ as the combined order of both stages. It assumes a quantization step width $\Delta_1 = FS/(2^{B_1}-1)$ in the first stage, and $\Delta_2 = FS/(2^{B_2}-1)$ in the second stage. δ_{coeff} stands for the coefficient mismatch between the analog and digital part. The nonideal (RHS) part of (5.7) reflects the quantization noise leakage of the first stage due to mismatch. It is derived by using [158, Eq. 15]. It is based on the quantization noise performance of the first stage, but takes into account that the leaking quantization noise is still fed into the CoI reconstruction filter of the overall I-MASH, which is higher than the order of the first stage ($L_{\text{tot}} > L_{\text{I}}$). The noise approximation factor Φ_N is given in Table 5.1 and was derived in [158, Eq. 15].

A corresponding simulation is shown in Fig. 5.2 illustrating the simulated SQNR under different mismatch conditions. An exemplary 2–1 MASH QE as in Fig. 5.1 is simulated. Its first stage is not scaled and uses a 3-bit quantizer. The oversampling

Table 5.1 Approximation factor Φ_N for MASH I-DS ADC noise leakage with the overall order L_{tot} and the order of the first stage L_1.

$L_{\text{tot}} - L_1$	1	2	3	4	5
Φ_N	1	$1/3$	$1/20$	$1/252$	$1/5184$

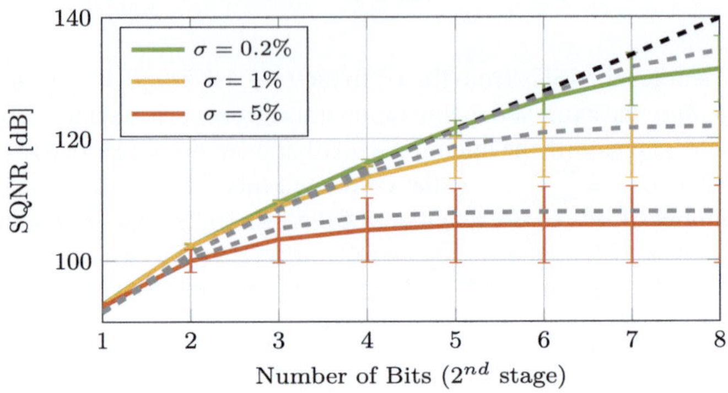

Fig. 5.2 Simulated SQNR (solid, colored) of a 2–1 incremental MASH QE over the number of quantizer bits in the second stage for different coefficient variations σ with an oversampling of $M = 50$. The first stage is not scaled and uses a 3-bit quantizer. $ISG = 3$ and a -6 dBFS input amplitude are used. Black dashed: ideal resolution using (5.6). Gray dashed: nonideal resolution using (5.7).

ratio is $M = 50$, the $ISG = 3$, and a -6 dBFS input amplitude is used. The first-order second stage uses a varying number of quantizer bits. All path coefficients in the I-DSMs and the ISG are Gaussian distributed with different standard deviations σ. Shown are the average SQNRs and the 1-sigma error bars over 20 Monte Carlo runs.

The black dashed line in Fig. 5.2 represents the ideally calculated resolution of the 2–1 I-MASH according to (5.6), while the grey dashed line represents the nonideal resolution due to coefficient mismatch according to (5.7). As is known from the free-running DSM, the I-MASH is also very sensitive to analog/digital mismatch. The influence of the mismatch becomes stronger as the ideal resolution of the combined MASH I-DS ADC differs from the ideal resolution of the first stage alone, as shown by an increasing number of bits in the 2nd stage. While the general trend of the simulation results matches with (5.7), the prediction tends to be at the upper end of what is simulated using the Gaussian variation of all coefficients. This makes sense because (5.7) assumes only the deviation of a single coefficient, whereas in the Monte Carlo simulation, all coefficients are subject to statistical variation.

Note, even though the coefficient sensitivity of the MASH-LI architecture is not explicitly discussed in this context, it shows very similar behavior to coefficient variation.

5.1.3 *Incremental Sturdy MASH (I-SMASH) ADC*

As outlined in Sect. 2.3.2.3 and in comparing Figs. 2.17 and 2.18, the SMASH architecture was proposed by [112] to overcome the required matching of the digital cancellation logic with the analog loop-filter transfer functions, which can lead to severe noise leakage from earlier stages to the overall output in a MASH DSM. Incremental SMASH ADCs employ any ordinary SMASH architecture including a reset and a reconstruction filter of at least equal order to the overall SMASH noise-shaping order and were first introduced in [253]. As already discussed in Sect. 2.3.2.3, unlike MASH, no digital cancellation logic is required. The basic idea is to employ a 0-X MASH quantizer within a single-loop DSM as indicated in Fig. 2.18; if the X-th order second stage features a unity STF, and since the zeroth-order first stage has no noise shaping and thus its NTF is also unity, the digital cancellation filters of the 0-X MASH are unity, i.e. they can be omitted, and this omits the analog-digital mismatch problem. As the digital summer within the loop in Fig. 2.18 would add significant latency within the DSM loop, SoA implementations of SMASH DSM split the adder into an out-of-the-loop digital adder, as well as an additional analog adder from the output of the second stage to the overall input, realized by DAC_{21} in Fig. 2.18. A corresponding block diagram of such an incremental SMASH modulator is shown in Fig. 5.3. The overall output of this two-stage SMASH DSM is given by

$$V(z) = U(z) \cdot STF_1(z) + E_1(z) \cdot NTF_1(z) \cdot (1 - STF_2(z)) \\ - E_2(z) \cdot NTF_1(z) \cdot NTF_2(z). \quad (5.8)$$

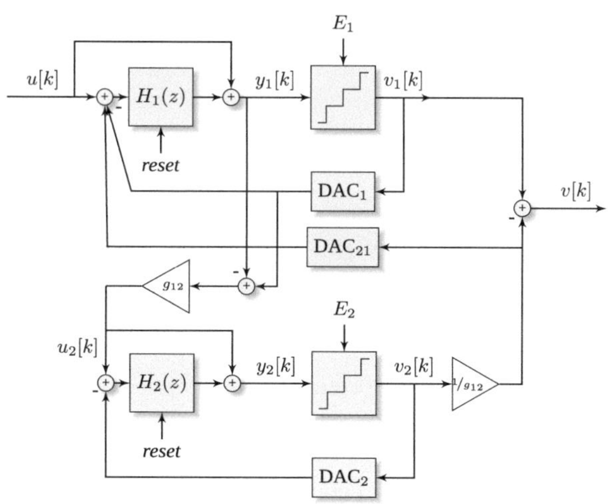

Fig. 5.3 Block diagram of a SMASH I-DSM.

In order to fully cancel the quantization error E_1 of the first stage, $STF_2 = 1$ must be fulfilled. However, due to imperfect matching, quantization noise leakage can occur. A further disadvantage of the SMASH architecture is that the second feedback DAC_{21} is required at the overall input of the modulator. Any nonideality of this additional DAC_{21} directly influences the DSM performance as it is passed to the digital output with the overall STF of the ADC. If the second stage only processes quantization noise, this leads to an increased noise floor. However, if the second stage processes also signal content, as would be the case if, e.g. a single-bit quantizer is used in the first stage, the overall linearity of the modulator is heavily decreased. This is in severe contrast to MASH DSM, where multi-bit quantization in the higher stages comes without significant degradation by a nonlinear DAC, as the outputs of the higher stages are fed through high-pass filtering digital cancellation logic. More recently, advanced architectures have been presented in SoA that use SMASH to achieve higher-order noise shaping while omitting the necessary digital reconstruction filters, but at the same time avoid the disadvantage of the high linearity required in the additional feedback DAC [24, 258].

5.1.4 Nonideal Performance of I-SMASH ADCs

It has been reported in the SoA that in SMASH ADCs—even though they were invented to reduce the impact of mismatch between the analog loop filter and the digital cancellation filters—mismatch sensitivity still exists. This is not only due to the required STF_2=1 of the second stage, cf. (5.8), but more importantly due to the matching of DAC_1 and DAC_{21} in Fig. 2.18. For example [259, Sect. III-D] stated that a 10-bit matching is required between these two outermost DACs in order to achieve SQNR≈90dB; similarly, [114] reported a required matching between the two outermost DACs in the range of $\ll 0.1\%$. The reason is found by deriving the noise cancellation condition in (5.8) but assuming an erroneous nonunity gain $GE_{DAC21} = 1 + \delta_{DAC21}$ in series with DAC_{21} in Fig. 5.4, which yields

$$\begin{aligned} V(z) = U(z) \cdot STF_1(z) + E_1(z) \cdot NTF_1(z) \cdot (1 - STF_2(z)) - E_2(z) \cdot NTF_1(z) \cdot NTF_2(z) \\ + \underbrace{E_1(z) \cdot \delta_{\mathrm{DAC21}} \cdot STF_1(z) \cdot STF_2(z)}_{\text{Not noise-shaped } E_1 \propto \delta_{\mathrm{DAC21}}} + \underbrace{E_2(z) \cdot \delta_{\mathrm{DAC21}} \cdot STF_1(z) \cdot NTF_2(z)}_{2^{\text{nd}}\text{-stage noise-shaped } E_2 \propto \delta_{\mathrm{DAC21}}}. \end{aligned} \tag{5.9}$$

We observe that together with the original cancellation condition in (5.8), where we required $STF_2 = 1$ to cancel the quantization noise E_1 of the first stage, two additional terms appear which are dependent of the DAC_{21} mismatch δ_{DAC21}. The critical term is the non-noise-shaped part of E_1, which is directly proportional to the mismatch of δ_{DAC21}. This explains the tremendously high matching requirements of the additional feedback DAC in the SoA.

With this, we can derive an approximate expression to predict the IBN of the incremental SMASH ADC:

$$IBN_{\mathrm{ISMASH|Mismatch}} \approx \frac{\Delta_2^2}{12} \cdot \left(\frac{L_{\mathrm{tot}}!}{M^{L_{\mathrm{tot}}}}\right)^2 \cdot \frac{1}{g_1^2} + \delta_{\mathrm{DAC21}}^2 \frac{\Delta_1^2}{12} \cdot \frac{\beta_{\mathrm{npf}}}{M}. \tag{5.10}$$

The LHS term is the ideally expected performance of the I-SMASH ADC depending on the combined noise-shaping order of both loops (L_{tot}), the oversampling ratio (M), the resolution of the quantizer of the second stage (Δ_2) and the ISG (g_1); the latter was shown in Fig. 5.3 but not included in the derivation of (5.8) and (5.9) for the sake of simplicity. The RHS term of (5.10) is proportional to the DAC_{21} mismatch $\delta_{\mathrm{coeff}}^2$ and the unshaped quantization error E_1, i.e. the resolution of the quantizer of the first stage (Δ_1). As we can regard the unshaped E_1 as white (quantization) noise, the subsequent CoI reconstruction filter acts like a brick-wall filter and corresponds to a gain $1/M$, but including the NPF of the higher-order CoI filter as derived in Sect. 4.3.1. Corresponding simulation results are shown in Fig. 5.4.

Similar as for the simulation of incremental MASH under coefficient mismatch in Fig. 5.2, the simulated SQNR under different mismatch conditions is shown. An exemplary 2–1 incremental SMASH is simulated, where the first stage is an unscaled second-order loop filter with a 3-bit quantizer. The oversampling ratio is $M = 50$, the $ISG = 3$, and a −6 dBFS input amplitude is used. The first-order second stage uses a varying number of quantizer bits. Shown in Fig. 5.4 are the average SQNR obtained over 20 Monte Carlo runs for different coefficient standard deviations σ. For comparison, the black dashed line represents the ideal resolution of a corresponding incremental MASH ADC calculated using (5.6), which was also provided in Fig. 5.2;

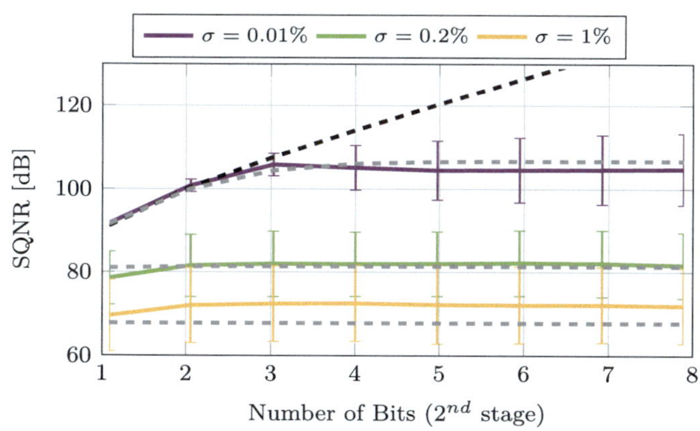

Fig. 5.4 Simulated (solid, color) SQNR of a 2–1 I-SMASH over the number of bits of the second stage quantizer for different coefficient variations σ (20 Monte Carlo runs) with an oversampling of $M = 50$. The first stage is not scaled and uses a 3-bit quantizer. $ISG = 3$ and a −6 dBFS input amplitude is used. Black dashed: ideal resolution using (5.6). Gray dashed: nonideal resolution using (5.10).

even though it is derived for a MASH ADC, under ideal conditions, with the same orders and quantizer resolution in both stages, both the 2–1 SMASH and the 2–1 MASH achieve almost identical ideal performance.

On the contrary, Fig. 5.4 reveals a tremendous performance drop even for tiny coefficient variations. We have derived in (5.10) that the performance drop is largely dominated by the mismatch between DAC_1 and DAC_{21}. For very large mismatches, the other nonideal terms in (5.9) also influence the performance, and the matching to (5.10) becomes inferior. Thus, while in a SMASH the general (absolute) coefficient variation is of lesser importance, the relative matching of the two outermost feedback DACs is extremely critical and must be intrinsically excellent, as it was found experimentally in [114, 259].

5.2 Multi-Step Incremental ADCs: Extended Counting and Range

While MASH and SMASH incremental ADCs operate on the same principles and suffer from the same nonidealities (such as matching and noise leakage) as their free-running counterparts, the idea of *extended counting (EC)* or *extended range (ER)* is unique to incremental DS ADCs. In Sect. 3.2, the time-domain description of I-DSMs was derived. For a first-order I-DSM, as shown in Fig. 3.3, we found for (a constant assumed) input voltage of $\hat{u}_{\text{const}}$ the following representation in (3.7):

$$\hat{u}_{\text{const}} = \underbrace{\frac{V_{\text{ref}}}{M}\sum_{l=0}^{M-1} v_n[l]}_{\substack{\text{Summation of IDSM output samples}\\ \text{=digital representation of input.}}} + \underbrace{\frac{V_{n,i1}[M]}{b_1 \cdot M}}_{\substack{\text{Residual quantization error}\\ \text{after M cycles.}}} . \tag{5.11}$$

Here, we see that the analog input signal is represented by a first rhs part, which is a summation of the M coarsely quantized I-DSM output samples; this motivated us to use a first-order CoI (i.e. a counter) as reconstruction filter for the first-order I-DSM. The second rhs part is then interpreted as the residue of the quantization by the I-DS ADC. Thus, if we take the digital output of the first-order incremental ADC—consisting of a first-order I-DSM and a counter—as digital representation of the analog input signal, the output $V_{n,i1}[k]$ of the analog integrator of the first-order I-DSM is a scaled representation of the remaining quantization error.

The interpretation of (5.11) leads to the idea of a multi-step operation, where the I-DS ADC operates for M clock cycles to build a digital output, and the residual (scaled) quantization error is further quantized and digitally removed. This technique was introduced as *Extended Counting* in [260]. There, it was proposed to operate a first-order incremental ADC in two distinct modes during every single A/D conversion: in the first mode, the system functions as a resettable first-order I-DSM to process the most significant bits. In this phase, the first-order IDSM with its CoI reconstruction filter (i.e. the counter) operates as counting converter [11, 174]. For the second part of the operation, [260, Sect. II-A] stated: "The rest value could be sampled and converted by some separate suitable ADC." Though, instead of realizing an explicit ADC for the quantization of the residue, it was proposed that hardware can be saved by using components already existing in the first-order I-DSM, more precisely, that the I-DSM can be reconfigured into an algorithmic ADC to convert the least significant bits from the analog residuum. As this second mode extends the counting conversion of the first mode, [260] introduced the term *extended counting*.

In fact, [260, 261] as the first source for extended counting. But a closer look at the original 1983 book shows that [261] had a completely different architecture in mind: [261, Sect. 3.4.2] describes a variant of a ramp ADC, i.e. a digital counter feeds a DAC to successively generate all possible quantization levels, while a comparator stops the counter when the DAC output becomes 1 LSB greater than the analog input voltage. Since such a ramp ADC is a one-level-at-a-time converter, it is quite slow. Therefore [261] refers to an *extended-counting* method (without further reference to previous work), which steps the ramp ADC by more than 1 LSB, e.g. 2 LSB, until the DAC output crosses the analog input, and then steps back by 1 LSB. By this the conversion time is almost halved while the final resolution is unchanged. Seitzer et al. [261] goes on to discuss larger initial steps and derives the function of a SAR ADC where the first step is 1 MSB and the last step is only 1 LSB. But obviously this whole method has nothing in common with the *extended-counting incremental ADC* described by [260] and in this section, although it was referred to with the same name in [261].

Thus [260] remains the first to describe an *extended-counting incremental ADC* and it is important to recognize that almost all of the architectures published subsequently, where separate ADCs or other ways of reconfiguring the I-DSM for requantizing the residuum are proposed, were originally claimed by [260]. For example, [262] essentially describes the same technique once again, referring to it as feed-forward residue compensation (FRC). Unlike [260], who did not extend his circuit diagram in [260, Fig. 1] to include the additional ADC quantizing the residue (which the authors of [260] called the "rest value"), this aspect was explicitly illustrated in [262, Fig. 2] and both block diagrams are reproduced in Fig. 5.5 for illustration.

Higher-order I-DSM with extended counting: The extended-range ADC: A unique contribution of [262] was that he was the first to (briefly) elaborate on the extension of the basic idea to higher-order I-DSM. Similar to the time-domain derivation of the first-order I-DSM resulting in (5.11) and motivating the extended-counting incremental ADC in the first place, we derived the same time-domain expression in (3.18) for a third-order I-DSM with CIFF loop filter combined with a CoI3 filter, and also for an I-DSM with CIFB loop filter, where we found that the required reconstruction filter is not a CoI but a digital filter matched to the analog loop filter, cf. Sect. 3.2.3 and (3.25). In fact, [262] used a second-order CIFB in [262, Fig. 16] and derived an expression similar to that in (3.25) in [262, Eq. 16].

More commonly known for the first implementation of a high-order I-DSM employing the idea of *extended counting* is [263, 264]: there, a second-order CIFF

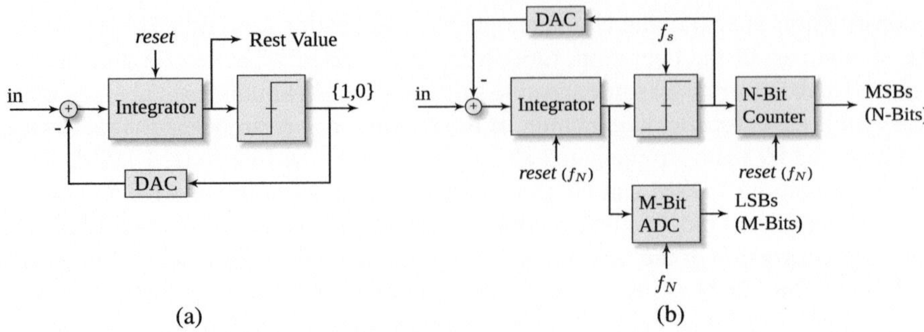

Fig. 5.5 (a) [260, Fig. 1] and (b) [262, Fig. 2] with their illustration of an *Extended-counting I-DSMs*, where [260] calls the (last) integrator output the *rest value* and does not further specify how to requantize it, while [262] explicitly draws an M-bit ADC for requantization.

I-DSM is used in combination with a CoI2 reconstruction filter. The residue is then converted by a SAR ADC. Since the reconstruction filter in this case is not a counter, [263] refrains from calling the resulting architecture *extended counting*, since there is effectively no counting phase anymore. Thus, without further ado the higher-order I-DSM with a requantization of the analog residue was called to feature an *extended range*, as the underlying two-step operation extends the dynamic range of the I-DSM. Later work has referred to *extended-counting* or *extended-range* incremental ADCs, and unfortunately, not always consistently. But the underlying principle remains

- employ a reconstruction filter that matches the filter function from the I-DSM quantizer output to the output of the last integrator. This yields a higher-order CoI for CIFF I-DSM or other digital filters for other I-DSM loop filters,
- use the fact that the output of the last integrator at the end of each Nyquist conversion is then an analog representation of the residue quantization error,
- requantize the analog residue by an explicit ADC or by a reconfiguration of the I-DSM,
- add the requantized residue to the I-DS ADC output to obtain an overall smaller quantization error.

Implementation variants of EC/ER I-DS ADCs: Even though [260] claimed almost all variants of an extended-counting I-DS ADC by saying that the analog residue can be converted by some suitable ADC, the SoA has still suggested many creative ways of actually quantizing the output of the last integrator and realizing the overall extended-range operation. In [265–269], the I-DSM was reconfigured at the end of each Nyquist conversion to quantize the integrator output in the second step. In [184], a CT implementation was presented as an ER I-DS ADC. In [251], a second-order I-DSM is employed in the first phase, which is then reconfigured into a hold amplifier for the residue and a first-order I-DSM in the second phase; even though it uses a second-order I-DSM, the structure was called extended-counting incremental ADC. In [270], the approach was further extended toward a multi-step architecture

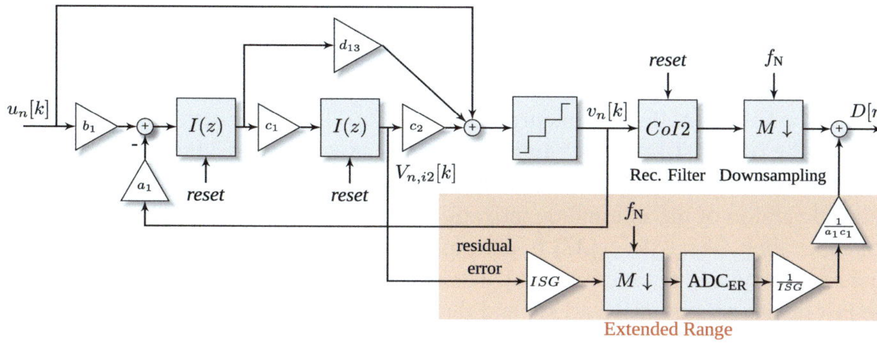

Fig. 5.6 Exemplary second-order I-DS ADC using extended range in a second conversion step.

where the I-DSM was reconfigured multiple times during each Nyquist conversion. In [271], the operational amplifier was reused in a second step to be used in a SAR ADC.

Practical considerations for EC/ER I-DS ADCs: For the discussion of a practical implementation, we show the second-order incremental ADC in Fig. 5.6. Since we employ a higher-order I-DSM we use the common naming extended range (ER) [264]. The CIFF loop filter is combined with a CoI2 reconstruction filter, which generates the residue of the incremental ADC at the output of the second integrator. The loop filter includes all practical scaling coefficients, which are individually used to (1) achieve stable operation and (2) limit the swings of the integrators. The extended-range operation is w.l.o.g. drawn with an extra ADC_{ER} having B_{ER}-bit resolution.

If we conduct the same derivation as in Sect. 3.2.2 to obtain (3.18), but for the second-order I-DSM as in Fig. 5.6, we obtain

$$\hat{u}_{\text{const}} = \frac{V_{\text{ref}}}{M(M-1)/2} \underbrace{\sum_{l=0}^{M-1} \sum_{k=0}^{l-1} v_n[k]}_{CoI2 \text{ Output}} + \underbrace{\frac{1}{M(M-1)/2} \cdot \frac{V_{n,i2}[M]}{a_1 c_1}}_{\text{Residue}} . \tag{5.12}$$

When taking a CoI2 reconstruction filter, we see that the residual quantization error, represented by the output of the second integrator in the CIFF loop filter ($V_{n,i2}[M]$), requires requantization w.r.t. the same reference V_{ref}. Further, the requantized V_{ref} needs scaling by $1/a_1c_1$ to match the residue which is added to the CoI2 output in (5.12). Looking more closely at the exemplary ER I-DS ADC in Fig. 5.6, it becomes clear that requantizing the analog quantization error ($V_{n,i2}[M]$) with a B_{EC}-bit resolution can only significantly improve the overall resolution, if ADC_{ER} operates on a full-scale input: e.g. if the output of the last integrator $V_{n,i2}[M]$ only changes in the LSB range of ADC_{ER}, its high resolution is meaningless; if, e.g. the ADC_{ER} is largely overdriven, it may even worsen the quantiziation error in the overall output. For this reason, a proper inter-stage gain (ISG) is needed, which scales $V_{n,i2}[M]$ to fit the full-scale reference voltage of ADC_{ER}.

More generally, we can note that for an Lth-order I-DSM operating for M cycles, the residue visible at the last integrator output ($V_{n,iL}[M]$) is the scaled quantization error of the overall I-DS ADC. In the special case when the STF of the equivalent DT DSM is unity and its NTF is an unscaled high-pass filter $(1-z^{-1})^L$, $V_{n,iL}[k]$ is a delayed version of the internal quantization error $e[k]$, with $V_{n,iL}[M] = e[M-L]$. For other loop filters, $V_{n,iL}[M]$ is determined by a linear superposition of several past internal quantization errors $e[k]$. Therefore, the distribution of the amplitude of the last integrator output $V_{n,iL}[M]$ and thus the optimal choice for the ISG depends on QTZ resolution (B), order (L), and loop-filter scaling. All these observations lead to some take-away messages for extended-counting/range incremental ADCs:

- The reconstruction filter needs to match the transfer function from QTZ output to last integrator output. This matched filter is a CoI filter of the same order as that of a DT CIFF I-DSM, and is a different filter for other (e.g. CIFB) loop-filter architectures.
- In case of a CT modulator, the digital reconstruction filter has to be matched to the DT equivalent of the CT transfer function from QTZ output to last integrator output.
- Employing the direct feed-in, low-distortion path in the CIFF I-DSM is highly recommended to remove signal content in the integrator output swing as much as possible. Other techniques to reduce the signal content before the EC/ER phase are possible and explained in Sect. 5.4.4.
- Reducing the signal content in the residue allows for choosing a maximally large ISG before entering the $\text{ADC}_{\text{EC/ER}}$.
- ISG should be selected such that the integrator output (analog residue) is scaled to the full scale of the $\text{ADC}_{\text{EC/ER}}$ in order to exploit its full resolution. It is best derived by simulation.
- Depending on the modulator architecture, the quantization error and its analog representation at the last integrator output may be signal dependent, which leads to less extended resolution.
- The location of the rescaling $1/a_1c_1$ in the exemplary structure of Fig. 5.6 can be moved and combined with the required ISG. Moreover, the digital scaling is efficiently implemented using bit-shifts and, thus, powers of two.
- It is important to note that the analog scaling coefficients (like a_1,c_1, and ISG) are prone to mismatch, while their digital counterparts $1/a_1c_1$ and $1/\text{ISG}$ are not. This mismatch will lead to noise leakage similar to that known from MASH architectures.

Derivation of ENOB and IBN of EC/ER I-DS ADCs: When those conditions are fulfilled, the dynamic range of extended-counting/range architectures is effectively enhanced by the resolution of the $\text{ADC}_{EC/ER}$ used to convert the analog residue [264]. In case the modulator is unscaled and the EC/ER ADC has a full-scale input, the overall resolution of the EC/ER I-DS ADC can be obtained by summing the resolution of the I-DS ADC and the resolution of the EC/ER operation, which we refer to as $B_{ER/EC}$. Using the expression (3.46) or (3.58) from Sect. 3.4 to calculate the resolution of the I-DS ADC, the combined ENOB becomes

$$ENOB_{\mathrm{ER/EC}} \approx \log_2\left((2^B - 1)\frac{M^L}{L!}\right) + B_{\mathrm{ER/EC}}, \tag{5.13}$$

which we can also write for the effective IBN of ER/EC I-DS ADC as

$$IBN_{\mathrm{EC/ER}} = \frac{\Delta^2}{12} \cdot \left(\frac{L!}{M^L}\right)^2 \cdot \left(\frac{1}{2^{B_{\mathrm{EC/ER}}} - 1}\right)^2, \tag{5.14}$$

with $\Delta = V_{FS}/(2^B - 1)$ being the resolution of the B-bit internal quantizer of the I-DSM, L is the order and M is the oversampling ratio of the I-DSM. $B_{ER/EC}$ is the resolution of the $ADC_{ER/EC}$.

Note that (5.13) and (5.14) assume an optimally chosen ISG and an unscaled I-DSM, for which—as we discussed above—the residual analog quantization error at the last integrator output ($V_{n,iL}[M]$) follows the statistics of the quantization error of the internal quantizer. Then, the ideal $ISG = 2^B - 1$ is chosen such that the residual quantization error $V_{n,iL}[M]$ is amplified to the FS of ADC_{ER}. If ISG is reduced, this acts like having less resolution in the ADC_{ER}; e.g., losing x2 in ISG effectively reduces the additional resolution of the EC/ER operation by 1 bit. To include this, we can extent (5.13) by an ISG-induced penalty as

$$ENOB_{\mathrm{ER/EC}} = \log_2\left((2^B - 1)\frac{M^L}{L!}\right) + B_{\mathrm{ER/EC}} + log_2\left(\frac{ISG}{2^B - 1}\right). \tag{5.15}$$

The same can be included into the IBN calculation from (5.14):

$$IBN_{\mathrm{EC/ER}} = \frac{\Delta^2}{12} \cdot \left(\frac{L!}{M^L}\right)^2 \cdot \left(\frac{1}{2^{B_{\mathrm{EC/ER}}} - 1}\right)^2 \cdot \left(\frac{2^B - 1}{ISG}\right)^2. \tag{5.16}$$

Note that this becomes different when a scaled I-DSM is chosen, as not only the scaling must be considered in the ENOB, cf. Sect. 3.4.3, but also the statistics of the residual quantization error ($V_{n,iL}[M]$) vary.

Ideal simulations of extended-range I-DS ADCs: An illustrative simulation using the example of the second-order I-DS ADC from Fig. 5.6 is given in Fig. 5.7.
Multi-bit I-DS ADC: The I-DS ADC with 3-bit quantizer, $M = 50$, and unscaled $OBG = 4$, is simulated for a large, but not close to MSA, input power of -6 dBFS. The maximum swing of the last integrator output ($V_{n,i2}[M]$) is limited around $\pm\Delta/2$, which is the quantization noise of the internal quantizer. It only increases above that for large input signals. Thus, a large ISG$=2^3 - 1 = 7$ can be chosen. We see that by adopting the digital rescaling by $1/a_1c_1 = 1$ and $1/ISG$ as shown in Fig. 5.6, an almost ideal increase in resolution by the extended-range ADC_{ER} can be achieved. Only for

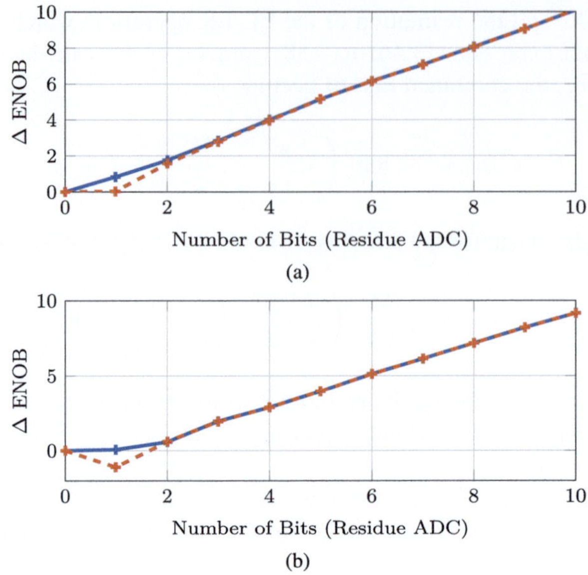

Fig. 5.7 Simulated difference of the achieved ENOB over the bits $B_{\mathrm{ER/EC}}$ used in the residue ADC. The exemplary second-order CIFF I-DSM is used at -6 dBFS input amplitude. (Red) without and (blue): With regarding effective quantizer gain $q_{\mathrm{eff}} \neq 1$ for low number of bits. (a) 3-bit quantizer, $M = 50$, $ISG = 7$ and $OBG = 4$. (b) Single-bit quantizer, $M = 100$, $ISG = 2.25$, $OBG = 1.5$.

a (practically rather irrelevant) single-bit $\mathrm{ADC_{ER}}$, the ideal increase is not achieved. The reason is that a single-bit QTZ comes with a nonunity effective QTZ gain q, which needs to be adressed by dividing the requantized residue by q before adding it to the I-DS ADC output. When this is done, the exemplary simulation in Fig. 5.7a shows a perfect increase in overall resolution by the number of bits of the $\mathrm{ADC_{ER}}$.
Single-bit I-DS ADC: The I-DS ADC with single-bit quantizer, M=100, is scaled for an OBG=1.5, and simulated for a large input power of -6 dBFS. As stated above, the maximum swing of the last integrator output ($V_{n,i2}[M]$) is now not limited around $\pm\Delta/2$, which would be the large quantization noise of the single-bit quantizer. Due to loop-filter scaling, it is found slightly below $\pm 0.5FS$ for small and slightly above $\pm 0.5FS$ for large input signals. Thus, an optimum $ISG = 2.25$ was found by simulation for the given example. Figure 5.7b reveals that even in this case, an impressive increase in resolution is achieved by an increasing resolution of the $\mathrm{ADC_{ER}}$. Only for a low number of bits does the behavior differ again. Applying the same measures as for a single-bit $\mathrm{ADC_{ER}}$ in the multi-bit case of the I-DSM, alleviates but does not fully resolve the deviation. In the given example, about 1 bit of the ADC_{ER} resolution is lost in the performance increase of the extended-range operation.

5.2.1 *Nonideal Performance of the Extended-Range I-DS ADCs*

The performance of extended-range I-DS ADCs is also evaluated under nonideal conditions by analyzing the impact of coefficient variations on the system performance. As in the case of MASH ADC, the origin of mismatch is the analog versus digital matching within Fig. 5.6. Similarly to how we derived an approximation for the nonideal IBN for the incremental MASH in (5.7), we can extend the ideal IBN from (5.14) to be

$$IBN_{\mathrm{EC/ER|mismatch}} = \frac{\Delta^2}{12} \cdot \left(\frac{L!}{M^L}\right)^2 \cdot \left(\left(\frac{1}{2^{B_{\mathrm{EC/ER}}}-1}\right)^2 \cdot \left(\frac{2^B-1}{ISG}\right)^2 + \delta_{\mathrm{coeff}}^2\right). \quad (5.17)$$

It includes the ideal IBN of the I-DS ADC enhanced by the resolution of the $\mathrm{ADC}_{\mathrm{ER/EC}}$, penalized by too small ISG, and in addition a leakage part that stems only from the I-DS ADC without extended range. In the following, we illustrate the achievable SQNR behavior under different mismatch scenarios.

Figure 5.8 presents the simulated SQNR of a second-order I-DSM with ER as a function of the number of bits in the residue ADC. The results are obtained for different levels of coefficient variation, denoted by σ, using 20 Monte Carlo simulation runs. The oversampling ratio is set to $M = 50$, with the first stage employing a 3-bit quantizer and an unscaled loop filter. The $ISG = 3$, and the input signal is maintained at a -6 dBFS amplitude. The simulation demonstrates the degradation of achievable SQNR with increasing coefficient mismatch. For a certain coefficient variation, the SQNR boost achieved by the EC/ER operation is limited and the SQNR can only be further increased by enhancing the resolution of the I-DSM operation. Both the ideally achievable resolution as well as the mismatch affected resolution

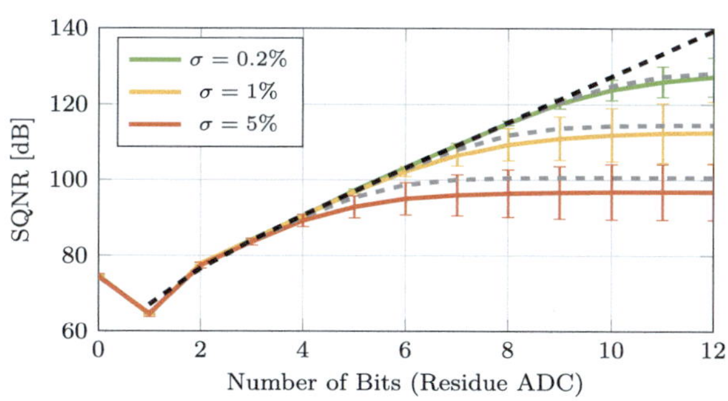

Fig. 5.8 Simulated SQNR (average and 1-sigma error bars) of a second-order I-DSM with ER over the number of bits of the residue ADC for different coefficient variations σ (20 Monte Carlo runs) with an oversampling of $M = 50$. The first stage is not scaled and uses a 3-bit quantizer. $ISG = 3$ and a -6 dBFS input amplitude is used. The ideally achievable resolution is obtained with (5.16) (black dotted), while the resolution limited by mismatch is obtained with (5.17) (grey dotted).

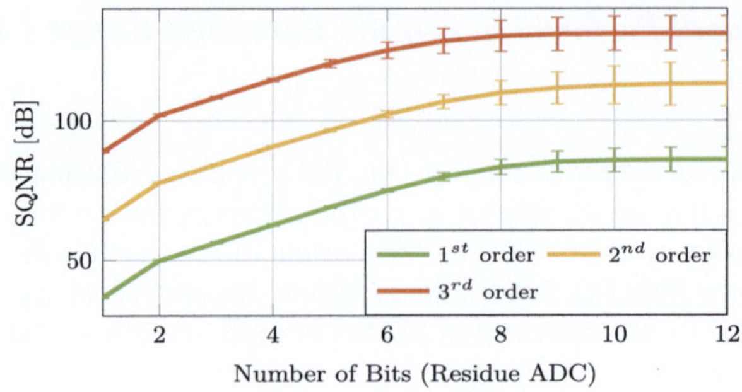

Fig. 5.9 Simulated SQNR (average and 1-sigma error bars) of a first- to third-order I-DSM with ER over the number of bits of the residue ADC for a coefficient variation of $\sigma = 1\%$ (20 Monte Carlo runs) with an oversampling of$M = 50$. The first stage is not scaled and uses a 3-bit quantizer. ISG=3 and a -6 dBFS input amplitude is used.

predicted by (5.16) and (5.17) track the simulation results well. For mismatch as large as $\sigma = 5\%$, the swing of the residual quantization error entering the $\text{ADC}_{\text{EC/ER}}$ changes significantly, which is why the ISG was limited to 3 in the first place to avoid overloading the $\text{ADC}_{\text{EC/ER}}$ in the shown simulations.

Figure 5.9 extends the analysis of the mismatch affected EC/ER I-DS ADC by considering the impact of I-DSM order on the achievable performance. The SQNR is plotted for first- to third-order I-DSM implementations with ER, again over the number of bits of the residue ADC. The oversampling and number of quantizer bits in the I-DSM remain at $M = 50$ and $B = 3$. The achievable SQNR is simulated for a coefficient variation of $\sigma = 1\%$ for 20 Monte Carlo simulation runs. While the second-order I-DS ADC is identical to Fig. 5.8, we can observe in Fig. 5.9 that for the same mismatch, an increased SQNR of the I-DS ADC will result in the same increase in achievable resolution when using EC/ER. That is, it is the relative performance boost which is defined by the number of bits of the residue $\text{ADC}_{\text{ER/EC}}$ and the coefficient mismatch, while the absolute SQNR is defined by what the I-DS ADC achieves without EC/ER (cf. (5.17)).

5.3 Linear-Exponential Incremental ADC

5.3.1 Ideal Behavior of the Linear-Exponential I-DS ADC

Another two-phase I-DS ADC was proposed by [272] namely the linear-exponential (LE) operation. The block diagram of a first-order LE I-DS ADC is shown in Fig. 5.10a. In a first (linear) phase, the first-order multi-bit I-DSM operates for M_{LP} clock cycles. Afterwards, the I-DSM is reconfigured to a second (exponential) operating phase for another M_{EP} clock cycles. In this second phase, the quantization error is exponentially increased by the factor $(1 + k_{\text{E}})$ and fed back into the loop.

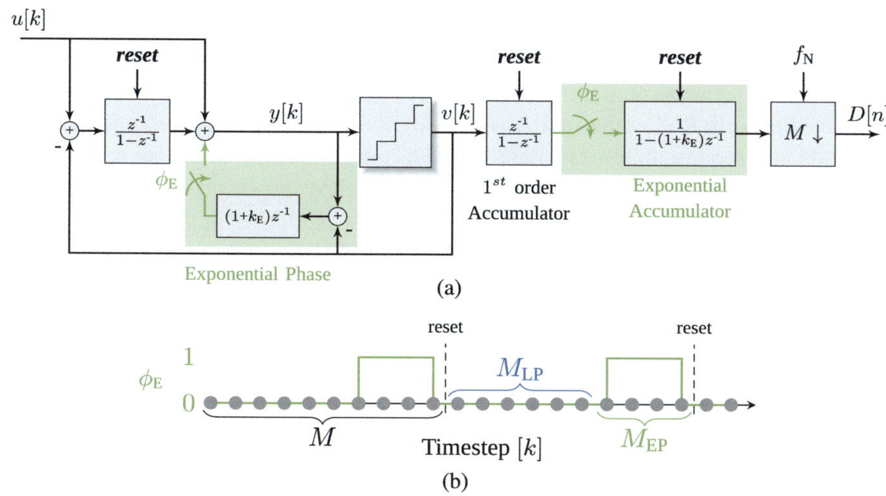

Fig. 5.10 (a) Block diagram of the linear-exponential I-DS ADC and (b) its timing diagram showing the two phases and the reset [246]. ©2023 IEEE. Modified, with permission, from Kaesser et al. [273].

The timing diagram of the LE operation of a I-DS ADC is shown in Fig. 5.10. After M=M_{LP}+M_{EP} clock cycles, a single Nyquist conversion is performed.

The exponential phase allows a large increase in the resolution of the I-DS ADC, exponentially enhancing the achievable SQNR for more clock cycles M_{EP}. However, the modification of the I-DSM loop filter requires the reconstruction filter to be adapted accordingly, as shown in Fig. 5.10a. There, a first-order CoI is used during the linear phase and concatenated with an exponential accumulator during the final exponential phase. Obviously, this affects the sample weights of the I-DS ADC, and the sample weights undergo an exponential decrease during the exponential phase. As we derived in Sect. 4.3.1, unequal and decreasing weights as in the case of higher-order CoI result in a detrimental effect on thermal noise, leading to a disadvantageous noise penalty factor (NPF), cf. Table 4.5. This is also the case for the LE I-DS ADC.

This was one of the reasons why the authors of the original work in [272] did not propose to employ an I-DS ADC solely operating in the exponential phase (M_{LP}=0, M_{EP}=M), as this would result in a large NPF. Consequently, [272] proposed a LE I-DS ADC based on a first-order I-DSM using M_{LP}=246 samples and M_{EP}=10 samples, as this would sufficiently boost the SQNR performance, but only increase the NPF slightly above the ideal value of one.

In addition, the LE I-DS ADC only works when a multi-bit quantizer is used. If a single-bit quantizer is used, the modulator would be unstable because the quantization error is multiplied by $(1 + k_{\mathrm{E}}) > 1$ and then fed back into the loop. In the single-bit case, the quantization error becomes as large as $\pm^{FS}/2$, and feeding such a large error back into the quantizer input along with the signal and quantization error from the loop filter leads to overload and instability.

Extension to higher-order linear-exponential I-DS ADCs: Kaesser et al. [273] have extended the original first-order LE I-DS ADC to higher orders. The example of a second-order version is shown in Fig. 5.11, where the reconstruction filter has to be adapted as well. Rather than a first-order CoI, it is a second-order CoI together with the exponential accumulator. The LE I-DS ADC only works as long as the reconstruction filter matches the modulator. As long as the I-DSM loop filter remains unscaled (i.e. the free-running NTF during the linear phase is $(1 - z^{-1})^L$), the reconstruction filter can be easily adapted by using a Lth-order CoI filter together with the exponential accumulator. However, in case the modulator is scaled, e.g. necessary for even higher-order loop filters, the required reconstruction filter becomes more complex. Its implementation was introduced in [274] and is shortly revised in the following.

In Figs. 5.10a and 5.11, the reconstruction filter was implemented as a time-variant implementation. However, since there is no feedback path in the reconstruction filter, it is also possible to implement it as a time-invariant FIR filter. This can be generally done for any modulator order and scaling. The required matched filter for the LE I-DSM is different from the required matched filter for the ER I-DSM. Here, it has to be of the form $1/NTF_{\text{dsm}}$, where NTF_{dsm} is the free-running NTF of the analog loop filter. Since the exponential phase only works with a multi-bit quantizer, an effective quantizer gain $q = 1$ can be assumed. The free-running $NTF_{\text{dsm},E}$ of the DSM during the exponential phase can be calculated with

$$NTF_{\text{dsm,E}}(z) = NTF_{\text{dsm,L}}(z) \cdot (1 - (1 + k_{\text{E}})z^{-1}) \tag{5.18}$$

with $NTF_{\text{dsm,L}}$ being the free-running transfer function of the DSM during the linear phase. Following this, the matched reconstruction filters during both phases are

$$H_{\text{rec,L}}(z) = \frac{1}{NTF_{\text{dsm,L}}} \tag{5.19}$$

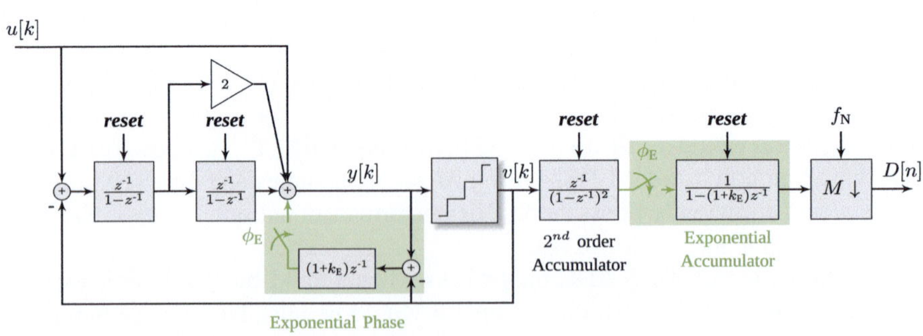

Fig. 5.11 Block diagram of a second-order DT LE I-DS ADC. ©2023 IEEE. Modified, with permission, from Kaesser et al. [273].

and

$$H_{\text{rec,E}}(z) = H_{\text{rec,L}}(z) \cdot \frac{1}{1-(1+k_{\text{E}})z^{-1}}. \tag{5.20}$$

As we intend to realize the reconstruction filters as FIR filters, we calculate the M FIR-filter weights. Therefore, the impulse response of the linear phase

$$h_{\text{rec,L}}[k] = \mathcal{Z}^{-1}\{H_{\text{rec,L}}(z)\},\ k \in [1, M] \tag{5.21}$$

is taken for M samples, as well as the impulse response of the exponential accumulator

$$h_{\text{rec,E}}[k] = \mathcal{Z}^{-1}\left\{\frac{1}{1-(1+k_{\text{E}})z^{-1}}\right\},\ k \in [1, M_{\text{EP}}+1] \tag{5.22}$$

for $M_{\text{EP}}+1$ samples. Convolving the two impulse responses leads to the weights of the LE reconstruction filter

$$w[k] = h_{\text{rec,L}}[k] * h_{\text{rec,E}}[k],\ k \in [1, M]. \tag{5.23}$$

Normalizing the FIR filter leads to the required reconstruction filter

$$H_{\text{rec}} = \frac{\sum_{k=1}^{M} w[k] z^{(k-1)}}{\sum_{k=1}^{M} w[k]}. \tag{5.24}$$

Ideal performance of the LE I-DS ADC: The linear-exponential I-DS ADC performance depends on the order L of the I-DS, the exponential-phase coefficient k_{E}, and the exponential-phase duration M_{EP}. The ENOB of the first-order purely exponential I-DS ADC can be estimated as [272]

$$ENOB_{\text{E}} \approx \log_2\left((2^B-1)\cdot\frac{(1+k_{\text{E}})^{M_{\text{EP}}}}{k_{\text{E}}}\right). \tag{5.25}$$

This expression highlights that the quantization error is exponentially suppressed, increasing by a factor of $(1+k_{\text{E}})$ per exponential-phase timestep. Consequently, even for small oversampling ratios M_{EP}, exceptionally high SQNR values can be achieved in an ideal scenario. In practice, however, this approach imposes stringent matching requirements and introduces a significant NPF, due to the exponentially non-uniform sample weighting [272]. As a result, purely exponential higher-order I-DS designs offer no practical advantage, as they suffer excessively from these drawbacks. To address this limitation, the hybrid approach proposed by [272], combining an exponential phase with a linear phase, proves beneficial. The performance of a LE I-DS ADC can be estimated as

$$ENOB_{\text{LE}} \approx \log_2\left(\underbrace{(2^B-1)}_{QTZ}\cdot\underbrace{\frac{M_{\text{LP}}^L}{L!}}_{\text{Linear Phase}}\cdot\underbrace{\frac{(1+k_{\text{E}})^{M_{\text{EP}}+1}}{k_{\text{E}}}}_{\text{Exponential Phase}}\right), \tag{5.26}$$

which we can also write for the effective IBN of LE I-DS ADCs as

$$IBN_{\text{LE}} = \frac{\Delta^2}{12}\cdot\left(\frac{L!}{M_{\text{LP}}^L}\right)^2\cdot\left(\frac{k_{\text{E}}}{(1+k_{\text{E}})^{M_{\text{EP}}+1}}\right)^2. \tag{5.27}$$

As in the case of the EC/ER I-DS ADC, the initial part of the resolution is contributed by the linear-phase operation of the I-DSM with a B-bit quantizer, while the exponential phase only contributes a boost in resolution. Nonidealities such as coefficient matching and the increased NPF are constrained by the intended boost, not the resolution of the initial linear phase. This makes the LE I-DS ADC a practical and realizable approach.

Noise-penalty factor increase by the exponential phase: As shown in Sect. 4.3.1, the I-DS ADC suffers a noise penalty factor (NPF) depending on the weight distribution of the reconstruction filter. The NPF defines how much the input referred noise is amplified toward the output of the I-DS ADC compared to a free-running DS ADC. The larger the deviation between the individual weights of the reconstruction filter, the higher the NPF. While a first-order I-DSM combined with a first-order CoI has NPF=1, higher-order I-DSM combined with a higher-order CoI have an increasingly larger NPF. Thus, it is not surprising that the exponentially decreasing weights in the exponential-phase reconstruction filter make the NPF much worse. Looking at the definition of the NPF in (4.12), we see that adding and additional M_{EP} cycles yields

$$\hat{\beta}_{\text{npf}}|_{LE} = (M_{\text{LP}}+M_{\text{EP}})\cdot\left(\sum_{k=1}^{M_{\text{LP}}} h_{\text{rec/LP}}^2[k] + \sum_{k=M_{\text{LP}}+1}^{M_{\text{LP}}+M_{\text{EP}}} h_{\text{rec/EP}}^2[k]\right). \tag{5.28}$$

Since the reconstruction filter weights of the exponential phase ($h_{\text{rec/EP}}$) are very small, they contribute little to the sum, while they linearly increase M=M_{LP}+M_{EP}. Consequently, the NPF of the linear-exponential I-DS ADC can be approximated by

$$\hat{\beta}_{\text{npf}}|_{\text{LE}} \approx \hat{\beta}_{\text{npf}}\cdot {}^{M}\!/\!_{M_{\text{LP}}} \tag{5.29}$$

i.e. the linear-exponential I-DS ADC experiences an increase of the NPF of ${}^{M}\!/\!_{M_{\text{LP}}}$ over the linearly operating I-DS ADC.

Frequency-domain description of LE I-DS ADCs: An analysis of the linear-exponential I-DS ADC in the frequency domain was given in [274], where a general methodology was shown for how the transfer functions of periodically reconfigured I-DS ADCs can be calculated as an LTV system in general, with the linear-exponential reconfiguration serving as an example. While a review of the method is beyond the scope of this chapter, the interested reader is referred to the published work, as it bridges the gap between time-domain simulations and frequency-domain performance predictions. Since the frequency-domain approach also allows for the inclusion of nonidealities, it will be used to analyze the nonideal operation of the linear-exponential I-DS ADC in Sect. 5.3.2.

Simulated performance and NPF increase for first-order LE I-DS ADC: After having discussed the theory of the LE I-DS ADC, it will next be investigated through simulations. Figure 5.12 presents the simulated (solid) and calculated (dashed, using (5.27) and (5.29)) SQNR and NPF of a first-order LE I-DS ADC as function of the exponential-phase length M_{EP} and the feedback factor k_{E}. The simulations use a 4-bit quantizer with an input amplitude of -2 dBFS at a total oversampling ratio of $M{=}M_{\mathrm{LP}}{+}M_{\mathrm{EP}}{=}256$. For these simulations, only one parameter is swept while the other remains fixed: M_{EP} is set to 10 when varying k_{E}, and k_{E} is set to 1 when varying M_{EP}.

We can observe that both increasing M_{EP} and k_{E} contribute to an exponential increase in SQNR. However, increasing k_{E} eventually leads to instability. This is expected, as stronger feedback of the quantization noise into the loop causes the quantizer to overload. Regarding the NPF, we see that k_{E} has little effect, while increasing M_{EP} increases it linearly. The simulations match the predicted SQNR in (5.27). The simulation also agrees well with the above derived approximation of the NPF, which is that of the underlying I-DS ADC—in the shown simulation NPF=1 due to the first-order CoI—increased by M/M_{LP}. In case of a variation of k_{E}, there is a minor dependency, and the simulated NPF settles toward the approximation of (5.29).

Simulated performance and NPF increase for higher-order LE I-DS ADC: Figure 5.13 illustrates the SQNR of first- to third-order linear-exponential I-DS ADCs

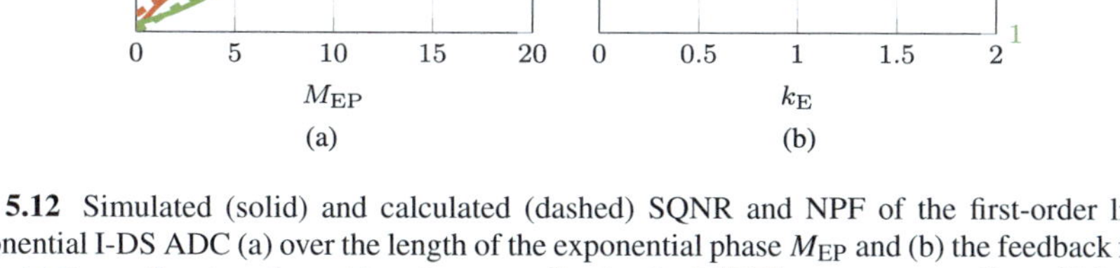

Fig. 5.12 Simulated (solid) and calculated (dashed) SQNR and NPF of the first-order linear-exponential I-DS ADC (a) over the length of the exponential phase M_{EP} and (b) the feedback factor k_{E}. A 4-bit quantizer together with an input amplitude of -2 dBFS at an oversampling of $M = 256$ is used. For the calculations (5.27) (SQNR) and (5.29) (NPF) are used. ©2023 IEEE. Modified, with permission, from Kaesser et al. [273].

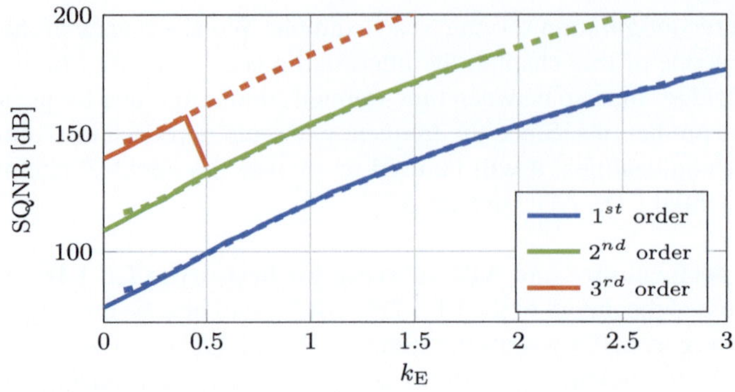

Fig. 5.13 Simulated SQNR of different-order linear-exponential I-DS ADCs over the factor k_E with calculations (dashed) from (5.27). Simulations are done with a 4-bit quantizer at a -10 dBFS input amplitude, $M = 100$, and $M_{EP} = 10$. ©2023 IEEE. Modified, with permission, from Kaesser et al. [273].

as a function of the feedback factor k_E. The simulations use a 4-bit quantizer, an oversampling ratio of $M = M_{LP} + M_{EP} = 100$, with $M_{EP} = 10$. All LE I-DSMs, even the third-order one, employ an unscaled loop filter, such that the reconstruction filters remain a concatenation of a (higher-order) CoI and the exponential accumulator, as shown for the second-order LE I-DS ADC in Fig. 5.11. Therefore, the simulation is performed with a small input amplitude of -10 dBFS, since the unscaled third-order I-DSM encounters severe stability issues even at small k_E values. In a practical implementation, a scaled loop filter would be required to achieve an acceptable MSA, and the reconstruction filter would become more complex as outlined in (5.18)–(5.24).

As discussed above, it can be seen that the LE I-DS ADC starts off from the SQNR of the linear phase improving for higher order L (shown in Fig. 5.13), but also for longer linear-phase duration M_{LP}, or higher internal bitwidth. The exponential phase then ideally boosts the performance by an equal ΔSQNR for a given exponential-phase factor k_E and duration M_{EP}. The theoretically predicted SQNR values, calculated using (5.27) and plotted as dashed lines in Fig. 5.13, closely match the simulation results. We can also observe that the third-order LE I-DSM becomes unstable for very small $k_E < 0.5$, while the second-order LE I-DSM operates up to $k_E \leq 2$ for the used unscaled loop filters. To calculate the NPF of higher-order LE I-DS ADCs, the approximation given in (5.29) can be used. It closely matches the exact calculation—particularly for $k_E \geq 1$ (cf. Fig. 5.12b). Assuming ideal operation and considering at the same time the worse NPF, it seems favorable to stay with low-order LE I-DS ADC as [272] proposed.

5.3.2 Nonideal Performance of the Linear-Exponential I-DS ADC

In [273, 274], it was shown that the significant resolution improvement provided by the exponential phase of the linear-exponential I-DS ADC is highly dependent on precise matching between the analog loop filter and the digital reconstruction filter. Any mismatch leads to a degradation in SQNR, ultimately limiting the achievable resolution improvement by the exponential phase. We investigate this effect of coefficient mismatch on LE ADC performance using both frequency-domain analysis and simulation results and use both to derive an empirically found expression for the nonideal performance of LE I-DS ADCs.

First-order LE I-DS ADCs with coefficient mismatch: Figure 5.14 illustrates the SQNR of a first-order linear-exponential I-DS ADC as a function of the number of cycles in the exponential phase M_{EP} for different numbers of cycles in the linear phase M_{LP}. A 4-bit quantizer is employed. The frequency-domain approach from [274] was used to predict the SQNR under coefficient mismatch, for which no well-matching closed analytical expression has been found yet. The calculated values are plotted as solid lines, while simulation results are included using x-symbols for verification and are in excellent agreement with the analysis. The simulation compares the ideally achievable performances (gray lines) with the case where coefficient mismatch is introduced ($\sigma = 0.2\%$, averaged over 200 Monte Carlo runs). In the simulations, all loop-filter coefficients and k_{E} are varied in the analog part of the I-DS ADC.

For M_{EP}=0 in Fig. 5.14, we observe the expected performance increase of the linear phase of ΔSQNR=6 dB per doubling of M_{LP} for the first-order I-DS ADC,

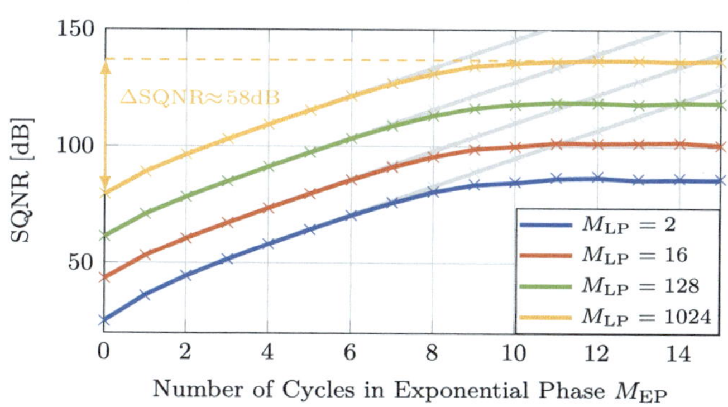

Fig. 5.14 SQNR of the first-order linear-exponential I-DS ADC over the number of cycles in the exponential phase M_{EP} for different number of cycles in the linear phase M_{LP}. A 4-bit quantizer and a −6 dBFS input are used. Solid lines: Calculated values using the frequency-domain approach. x-marks: Simulation results. The SQNR is calculated without mismatch (grey) and under a given normally distributed coefficient mismatch of $\sigma = 0.2\%$, averaged over 200 Monte Carlo runs. Licensed under CC BY 4.0. Reprinted, with permission, from Kaesser et al. [274].

cf. (3.51). Then, we see that the achievable SQNR improvement by the exponential phase does not depend on the initially achieved SQNR_{LP} and its number of cycles M_{LP}. On the contrary, for all chosen M_{LP}, the slope of the SQNR runs parallel. Then, we can observe that for a given coefficient mismatch, the achievable SQNR improvement is the same regardless of the performance from which the exponential phase starts. For the chosen example $\sigma = 0.2\%$, a maximum improvement of $\Delta SQNR_{\text{EP}} \approx 58$ dB over the initial linear phase performance can be achieved.

To illustrate this further, the maximum achievable SQNR boost ($\Delta SQNR$) and the minimum required number of exponential cycles ($M_{\text{EP,sat}}$) for this boost are plotted against a given coefficient mismatch (σ) in Fig. 5.15. $M_{\text{EP,sat}}$ is the number of exponential phase cycles required to reach the finally achievable SQNR value under the given mismatch within a 3 dB margin. The results demonstrate that—as mismatch increases—the achievable SQNR boost and the practical duration of the exponential phase diminish. This again highlights the critical role of coefficient matching in such two-phase operating incremental ADCs.

Empirical derivation of mismatch-dependent performance of the LE I-DS ADC: In Fig. 5.15, it can be observed that the maximum SQNR boost of the exponential phase drops by approximately 20 dB when the coefficient mismatch σ_{coeff} increases tenfold. While Fig. 5.15 shows the achievable SQNR boost over coefficient mismatch for the first-order architecture, it can be shown that higher-order modulators have the same linear dependency between the achievable SQNR boost due to the exponential phase and the coefficient mismatch. The only observable difference is that the dependency shows a slightly reduced $\Delta SQNR$ for increasing I-DSM order. Together with the prediction of the ideal performance of LE I-DS ADCs in (5.27)—which aligns with simulations for unscaled loop-filter architectures, i.e. when the NTF during

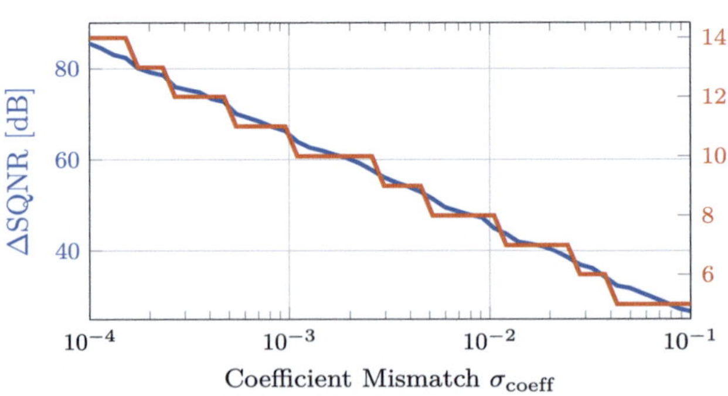

Fig. 5.15 Maximum achievable SQNR boost $\Delta SQNR$ and the minimum number of cycles required for this in the exponential phase $M_{EP,sat}$ of the first-order linear-exponential I-DS ADC over the coefficient mismatch σ. $M_{EP,sat}$ is defined as the minimum number of cycles required in the exponential phase to be closer than 3 dB to the final settled SQNR value under mismatch. The SQNR values are calculated using the frequency-domain approach. $\Delta SQNR$ is the SQNR boost achieved by the exponential phase at $M_{\text{EP,sat}}$. A 4-bit quantizer is used.

the linear phase is $(1 - z^{-1})^L$—the observations from Fig. 5.15 enable the empirical derivation of an equation to predict the performance of the LE I-DS ADC under coefficient mismatch.

$$IBN_{\mathrm{LE|mismatch}} = \frac{\Delta^2}{12} \cdot \left(\frac{L!}{M_{\mathrm{LP}}^L}\right)^2 \cdot \left(\left(\frac{k_{\mathrm{E}}}{(1+k_{\mathrm{E}})^{M_{\mathrm{EP}}+1}}\right)^2 + (L \cdot \sigma_{\mathrm{coeff}})^2\right). \quad (5.30)$$

Using this equation to predict the maximum achievable SQNRs of the simulations in Fig. 5.14, we can obtain SQNR≈ 79 dB, 97 dB, 115 dB and 133 dB. These values match very well to maximally achievable performances in Fig. 5.14 for the 0.2% standard deviation of the coefficient mismatch for the four different M_{LP}.

LE I-DS ADCs with coefficient mismatch for variable k_{E}: We have discussed and shown in Fig. 5.12 that the achievable boost in SQNR from the exponential phase depends on both its length M_{EP} and the exponential feedback gain k_{E}. While Fig. 5.14 showed that for a given coefficient mismatch, there is a maximum useful M_{EP} before the SQNR saturates, this also applies to the feedback factor k_{E}, as illustrated in Fig. 5.16. There, the ideal calculated (dashed, using (5.26)) and simulated (solid) SQNR values of the first-order linear-exponential I-DS ADC are shown as a function of M_{EP} for different values of the exponential-phase coefficient k_{E}. As before, a 4-bit quantizer and an oversampling of M=M_{LP}+M_{EP}=256, and an input amplitude of -6 dBFS are used. The coefficient variation has been set to an exemplary $\sigma = 0.2\%$. We can see that a higher k_{E} leads to an initially faster boost in SQNR for a given duration of the exponential phase. But given a certain mismatch, the finally achievable SQNR by the exponential phase remains the same. Thus, we can extend the previous statement: given a certain coefficient mismatch, there is a combination of a maximally useful gain k_{E} and duration M_{EP} of the exponential phase to achieve

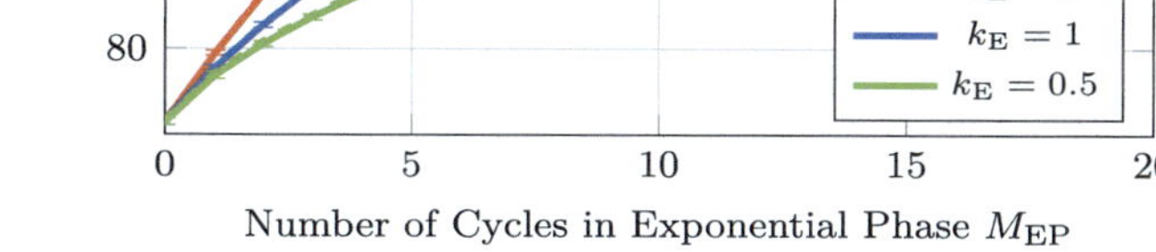

Fig. 5.16 Calculated (dashed, using (5.26)) and simulated (solid, average and 1-sigma error bars) SQNR values of the first-order linear-exponential I-DS ADC over M_{EP} for different values of k_{E}. 4-bit quantizer, $M = 256$, coefficient variation $\sigma = 0.2\%$ and an input amplitude of −6 dBFS. ©2023 IEEE. Reprinted, with permission, from Kaesser et al. [273].

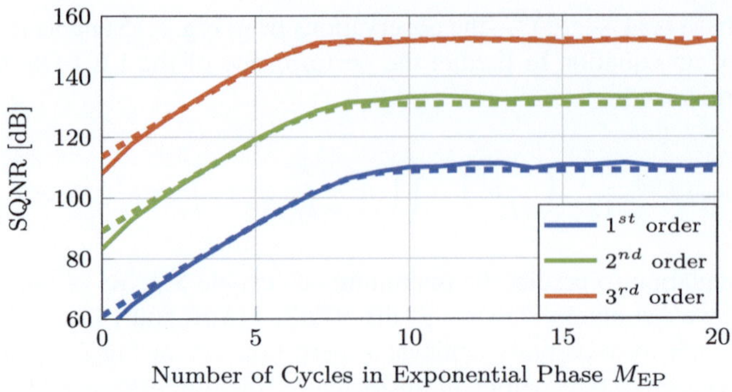

Fig. 5.17 Calculated (dashed, using (5.30)) and simulated (solid, average, 100 Monte Carlo runs) SQNR values of the linear-exponential I-DSs ADC over M_{EP} for modulator orders. $\sigma = 0.2\%$, 5-bit quantizer, M_{LP}=50, k_{E}=1 and an input amplitude of -10 dBFS.

the maximum boost $\Delta SQNR$ by the LE operation. Using (5.30), we can predict the maximally achievable SQNR under 0.2% coefficient variation to be ≈ 121 dB, which matches with the simulations in Fig. 5.16 very well.

LE I-DS ADCs with coefficient mismatch for higher-order I-DSs: As we have only considered the impact of coefficient mismatch on the first-order LE I-DS ADC so far, but we have discussed higher-order LE operation in Fig. 5.13, we finally explore the impact of mismatch on higher-order LE I-DS ADCs in Fig. 5.17. The calculated (dashed, using (5.30)) and simulated (solid) SQNR values are plotted over the exponential-phase duration M_{EP} for different modulator orders. The simulations are performed for an exemplary coefficient variation of $\sigma = 0.2\%$, a 5-bit quantizer, M_{LP}=50, and $k_{\text{E}} = 1$. An input amplitude of -10 dBFS is used to keep the higher-order modulators stable. Similar to the first-order LE I-DS ADC, where we achieved more or less linear-phase SQNR with more or less M_{LP}, also in the case of higher-order I-DS ADCs, the SQNR boost achieved by the exponential phase is similar for each I-DSM order. This leads again to a similar conclusion as before: the higher the linear-phase SQNR, the higher the finally achievable SQNR, whereas the SQNR boost by the exponential phase is limited by the coefficient mismatch. The calculations again match nicely with the simulations.

5.4 Dynamically Reconfigured Architectures using Non-Uniform Sample Weights

While Sects. 5.1, 5.2 and 5.3 introduced architectures with extended resolution by either multi-stage or multi-step operation, only one of them realized this by dynamic reconfiguration of the I-DSM: the linear-exponential ADC. Therein, the loop filter was reconfigured to realize an exponential decrease of the remaining quantization error. In this section, we will highlight further dynamically reconfigured I-DS ADC concepts, which rely on the non-uniform sample weighting using higher-order CoI reconstruction filters.

5.4.1 Non-uniform Reconstruction Filter Weights and Impact on Signal and Nonidealities

To start, we briefly review the non-uniform sample weights of higher-order CoI reconstruction filters which were discussed in detail in Sect. 3.5.2.1. Figure 5.18 reproduces Fig. 3.21 for better readability of this section, where the sample weights of CoI filters from first to fourth order for an oversampling of M=150 are shown. For higher-order CoI filters, the samples at the end of each Nyquist conversion are weighted much lower than those at the beginning.

This fact and its implications have been variously recognized in the state of the art. For example, [275] stated that while conventional DWA rotationally selects DAC unit elements to average the effects of mismatch, I-DS ADCs "with order higher than one have strictly decreasing digital filter coefficients" and "therefore, mismatch noise from the earlier samples appears in the output with higher gains than the later ones". This motivated [275] to propose an adapted DWA algorithm for I-DSM. When [191] derived the calculation of input referred noise, he stated that "every conversion has a weighting associated with each sample, and for higher-order modulators, earlier samples have a higher weighting than later samples because the digital filter has unequal weighting coefficients for higher-order modulators." While this was recognized as a distinct difference from free-running DSM, it was [276] who proposed using this fact to reduce the impact of nonidealities in later cycles of each conversion step in an I-DS ADC by reconfiguring a single-bit I-DSM into a multi-bit I-DSM, where

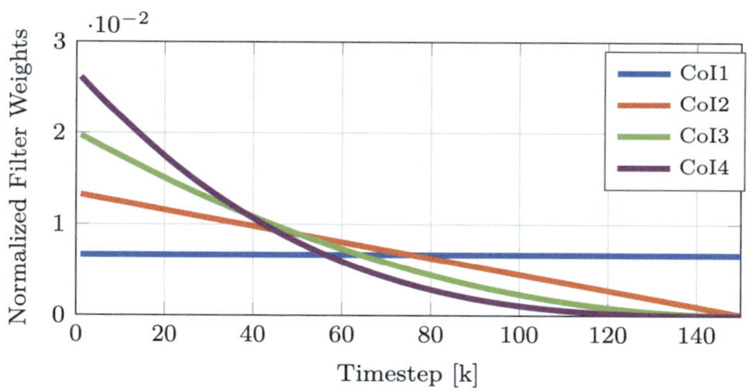

Fig. 5.18 Normalized filter weights of first- to fourth-order CoI filters vs. cycle instant $k = 1 \ldots M$ for an exemplary $M = 150$. Reproduced from Fig. 3.21 for better readability.

the initial single-bit cycles would operate with high linearity, while the late multi-bit cycles would reduce the remaining quantization error, but the lower weighting of those cycles would reduce the impact of multi-bit DAC nonlinearity.

A similar idea was proposed in [277], where circuit noise was dynamically increased toward the end of each conversion cycle with the goal of reducing loop-filter power consumption dynamically. In follow-up work, [217, 278] proposed to use multi-bit quantization only in later cycles due to the relaxed DAC linearity requirements, or to dynamically reduce opamp dynamics like GBW and SR to further save power.

The general idea of these works is to dynamically allow more nonidealities at the end of each Nyquist conversion in higher-order I-DSM, since they are suppressed by the lower sample weights of the higher-order CoI filter. Three different reconfiguration techniques are presented in the following: namely the dynamically increased thermal noise by integrator slicing [277], the recuperation phase [278], and the variable bit-width quantization [217, 276].

5.4.2 *Dynamically Increased Thermal Noise by Integrator Slicing*

Understanding the impact of thermal noise is essential for designing energy-efficient I-DS ADCs. The first integrator stage typically consumes the most power because it must meet the most stringent noise requirements [25, 186, 277]. Previous research [192, 217] has analyzed how the non-uniform weighting of reconstruction-filter samples affects the IBN at the output of a DT I-DS ADC. In such systems, the input-referred noise power of the first integrator stage, denoted as $\overline{v_{\mathrm{Cs}}^2}$, can be represented as a combination of thermal noise and $1/f$ noise from the switched-capacitor integrator.

$$\overline{v_{\mathrm{Cs}}^2} = \overline{v_{\mathrm{Cs,th}}^2} + \overline{v_{\mathrm{OTA},1/f}^2}. \tag{5.31}$$

Using a frequency-domain approach, this noise can be translated to the ADC output to calculate the output-referred noise power $\overline{D_{\mathrm{noise}^2}}$:

$$\begin{aligned} \overline{D_{\mathrm{noise}^2}} &= \frac{\overline{v_{\mathrm{Cs,th}^2}}}{f_s} \int_{-0.5}^{0.5} \left| \sum_{k=1}^{M} h_{\mathrm{rec,norm}}[k] \cdot e^{-j2\pi f k} \right|^2 df + \overline{v_{\mathrm{OTA},1/f}^2} \\ &\approx \frac{\hat{\beta}_{\mathrm{npf}}}{M} \cdot \frac{2k_B T}{C_s} \left(1 + \frac{1/6}{1 + 2R_{\mathrm{sw}} g_m} \right) + \overline{v_{\mathrm{OTA},1/f}^2} \end{aligned} \tag{5.32}$$

where C_s is the sampling capacitor, R_{sw} represents the sampling switch resistance, and g_m is the transconductance of the operational transconductance amplifier (OTA). White noise is increased by the NPF from Sect. 4.3.1 and reduced by oversampling, which results in the factor $\hat{\beta}_{\text{npf}}/M$. $1/f$ noise remains unaffected by the digital filter and is also not spread out by oversampling.

The time-domain analysis from [192] similarly shows that white noise, with a variance of $\sigma^2_{\text{noise,in}}$, can be mapped to the ADC output using normalized reconstruction filter weights:

$$\sigma^2_{\text{noise,out}} = \sum_{k=1}^{M} \sigma^2_{\text{noise,in}} \cdot h^2_{\text{rec,norm}}[k]. \tag{5.33}$$

This understanding inspired a power-saving strategy that varies the noise characteristics dynamically [217]. The input noise variance is dynamically increased during each Nyquist conversion cycle. This reduces power consumption while increasing the output noise of the I-DS ADC. However, the power reduction is linear with respect to the activation time of the higher noise, regardless of when this occurs. In contrast, the impact on the output SNR depends on the point in the Nyquist conversion cycle at which the noise is increased, with a smaller impact toward the end.

Building on this idea, the concept of *integrator slicing* was implemented [217, 277]. After the first integrator of an I-DSM is designed to offer low-enough noise to achieve full performance, it can be divided into multiple smaller slices, each selectively enabled over the course of a Nyquist conversion cycle. When parts of the integrator are deactivated, power consumption is reduced, but more noise is introduced. The concept of the slicing idea is visualized in Fig. 5.19 for a third-order DT CIFF I-DSM. An exemplary simulation is shown in Fig. 5.20. When integrator 1 is designed to fulfill the noise requirements, this sets the reference power consumption

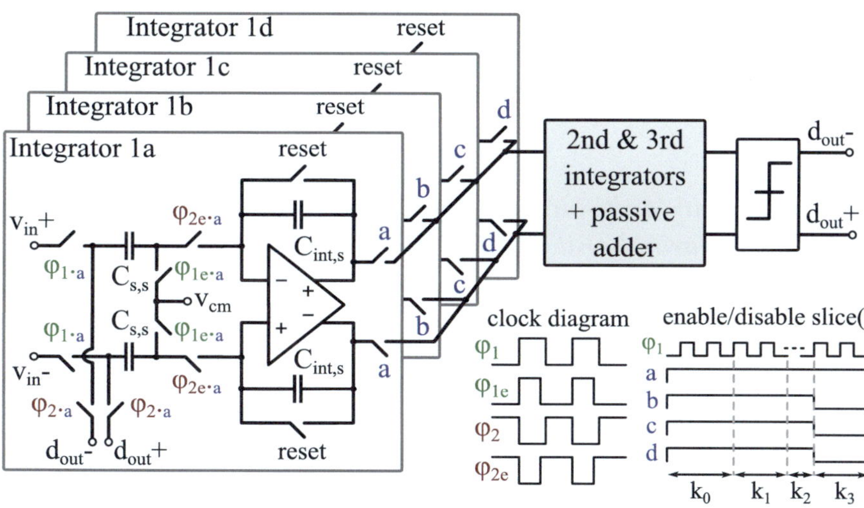

Fig. 5.19 Third-order DT I-DSM employing integrator slicing to dynamically allow more thermal noise during each Nyquist conversion. ©2025 IEEE. Modified, with permission, from Ismail et al. [199].

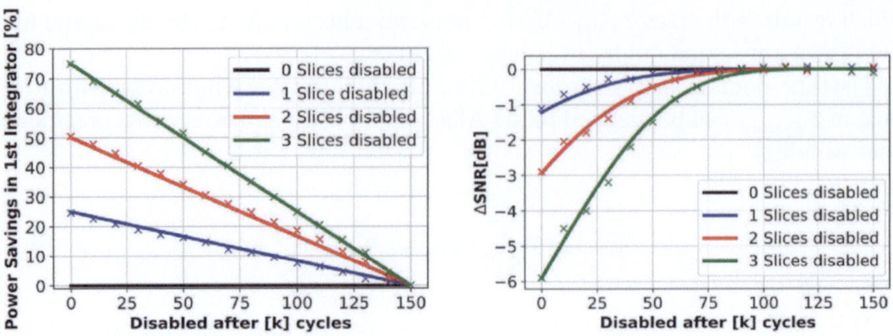

(a) (b)

Fig. 5.20 Power savings and SNR drop due to slicing of the first integrator in four pieces. (a) Calculated (cf. [217, Eq. 14]) (solid) and measured (crossed) power savings by disabling 0 to 3 slices during operation. (b) Calculated (cf. [217, Eq. 16]) (solid) and measured (crossed) SNR drop by disabling 0 3 slices during operation. ©2019 IEEE. Reprinted, with permission, from Vogelmann et al. [217].

and the reference SNR of the I-DSM. When half of the first integrator, i.e. 2 of 4 slices, are deactivated all the time (left side of Fig. 5.20), 50% of the power is saved, but $\Delta SNR = -3dB$ performance loss is seen. When the same 2 slices are deactivated for the final half of the Nyquist conversion cycle, i.e. after 75 of 150 clock cycles, still 25% of the power is saved, but almost no resulting drop ΔSNR remains.

Vogelmann et al. [217] derived an optimal switching sequence for the 4 slices, where 4 slices are operated for the initial cycles, then successively one, two, and three slices are deactivated. The prototype I-DSM, implemented in 180-nm CMOS technology, demonstrated that using an optimum switching sequence for the dynamic increase of noise and decrease of power leads to a loss of only $\Delta SNR = -0.5dB$ while the power consumption of the overall DSM was reduced by 33%.

5.4.3 *Dynamic Multi-bit DAC Nonlinearity Suppression using Variable Bit Width I-DS ADCs*

Wang et al. [276] proposed a two-step operation for I-DS ADCs. In a first phase, the I-DSM operates with single-bit quantization and features an intrinsically linear single-bit feedback DAC, which is configured into a multi-bit I-DSM in a second phase. This reduces the remaining quantization error while the influence of the uncalibrated and thus nonlinear multi-bit feedback DAC is reduced as the weights of the higher-order CoI-reconstruction filter decrease. The idea is illustrated in Fig. 5.21.

The same idea of a variable bit width (VBW) operation was phrased in [278] in combination with the recuperation phase technique, cf. Sect. 5.4.4, and it was elaborated further in [217]. While the latter work aimed to reduce loop-filter power and

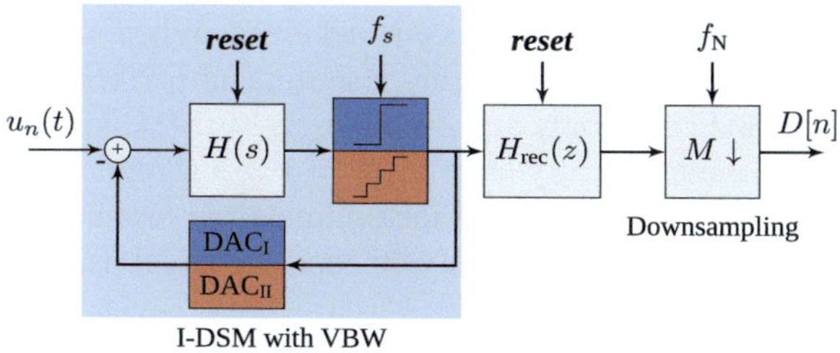

Fig. 5.21 Block diagram of a CT I-DSM with a variable-bit-width quantizer. ©2025 IEEE. Modified, with permission, from Ismail et al. [199].

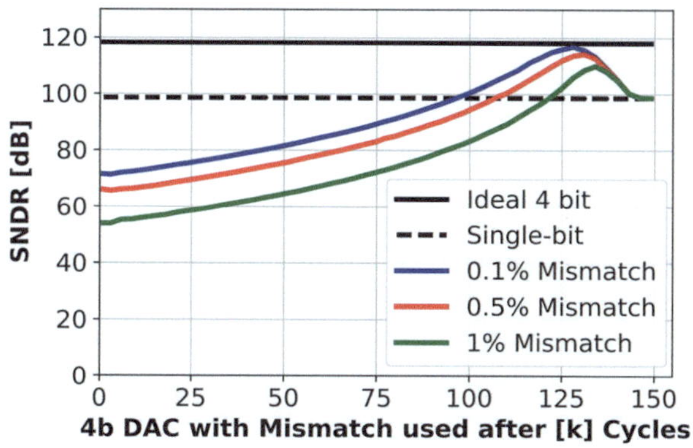

Fig. 5.22 Simulation of a third-order I-DSM combined with CoI3 reconstruction filter operating with VBW from single- to 4-bit internal quantization and DAC. $M = 150$ and multi-bit quantization is activated after k cycles. DAC unit-element mismatch with Gaussian distribution and $\sigma = 0.1/0.5/1\%$ standard deviation. ©2019 IEEE. Reprinted, with permission, from Vogelmann et al. [217].

allow more noise in later cycles, cf. Sect. 5.4.2, it also elaborated on the dynamic increase of other nonidealities, e.g. amplifier dynamics (like SR or GBW) as well as DAC nonlinearity. Figure 5.22 shows an exemplary simulation from [217], where a third-order I-DSM is combined with a CoI3 filter and simulated for $M = 150$ clock cycles. After $k \in [0, 150]$ clock cycles, the quantizer is reconfigured from single- to 4-bit operation. As the first clock cycles are operated in single-bit mode, the loop filter has been scaled for a free-running OOB gain of 1.5 using [144]. Consequently, the theoretical SQNR improvement is that of the quantizer resolution, i.e. 3 bits, resulting in about $\Delta SQNR \approx 18dB$. This is illustrated in Fig. 5.22 with the two horizontal lines indicating the ideal performance with single- or 4-bit internal quantization. The simulation results are then plotted for 3 different choices of the standard deviation of the multi-bit DAC. If multi-bit operation and thus the nonlinear multi-bit DAC are active from the beginning on ($k = 0$), the DAC nonlinearity determines the nonideal SNDR; e.g. for $\sigma_{DAC} = 0.1\%$, this results in about 12-bit DAC linearity and consequently in $SNDR \approx 72dB$. On the other end, if multi-bit operation is never active

$k = M = 150$, the single-bit performance is achieved. When an optimal activation point for the multi-bit operation is chosen, a significant performance improvement is obtained. For a quite commonly achievable $\sigma_{\mathrm{DAC}} = 0.1\%$, ideal multi-bit SQNR is realized. Even for unrealistically large $\sigma_{\mathrm{DAC}} = 1\%$, a significant SQNR improvement is observed. It should be noted that the optimal time-instant for reconfiguration depends on the I-DSM, DAC nonlinearity, M, and the target performance improvement and is best found by simulation.

The idea of dynamically configuring intrinsically linear single-bit (or 1.5-bit) operation to nonlinear multi-bit operation was implemented in prototypes in [25, 258]. Mokhtar et al. [25] implemented the original idea by means of a third-order DT I-DSM combined with a CoI3 reconstruction filter working with $M = 120$. The I-DSM was scaled for single-bit stability, and only the quantizer and DAC were reconfigured at an optimal instant to a 4-bit I-DSM toward the end of the conversion. While in pure single-bit operation the I-DS ADC would have achieved SQNR$= 92dB$, with dynamic reconfiguration it achieved spurious-free dynamic range (SFDR)$> 105dB$ and an SNR$> 98dB$. Staying in single-bit operation, this would have demanded more than double the oversampling used.

Mokhtar et al. [258] employed the reconfiguration idea within a 3–0 SMASH I-DSM. In a first phase, the third-order modulator and the zeroth-order second stage operated with intrinsically linear 1.5-bit quantizers, while only in a second phase, the second stage was reconfigured to a 4-bit quantizer. Interestingly, both the 1.5-bit and the 1.5-bit plus 4-bit quantizer realization were combined in a single 5-bit SAR ADC. Although no explicit linearization of the multi-bit DAC was used, the prototype achieved an $SFDR$ of almost $100dB$.

It should be mentioned at this point that the remaining drawback of this reconfiguration technique is that its single-bit operation in the first phase requires the loop filter to be scaled for single-bit stability with its disadvantages concerning SQNR and MSA, and only benefits from the increased intrinsic resolution of the internal quantizer.

5.4.4 Recuperation-Phase Method

The *recuperation phase* was introduced in [278]. The idea is to turn off the input signal for the last M_{RP} samples at the end of each Nyquist conversion. In fact, [16, Sect. 2.1.3] already discussed turning off the input signal for a few final conversion cycles to refine the output by ensuring that the quantization noise is available in analog form. [278] noted that the absence of the input signal reduces the internal swings of the modulator, resulting in increased MSA and reduced residual quantization error. This is illustrated in Fig. 5.23, where an exemplary single-bit third-order CIFF I-DS ADC with $M = 150$ is simulated with (red) and without (blue) RP. The RP is activated after 125 samples for $M_{\mathrm{RP}} = 25$ samples. It can be seen that in this example the MSA increases by almost 2 dB and the peak SQNR increases by more than 3 dB when RP is activated.

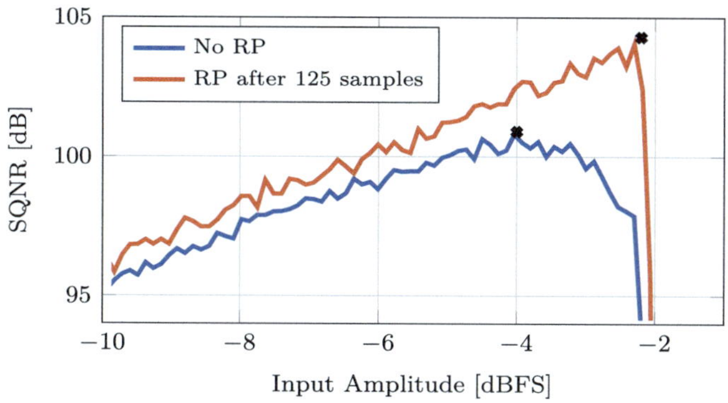

Fig. 5.23 SQNR sweep of the exemplary single-bit third-order CIFF I-DS ADC with M=150. ©2025 IEEE. Reprinted, with permission, from Ismail et al. [199].

In addition to improving MSA and SQNR, RP also has a positive impact on other dynamically reconfigured I-DSM architectures, such as the VBW, the ER, or the linear-exponential I-DS. ADC [199], which is described in Sects. 5.4.4.4, 5.4.4.5 and 5.4.4.6. A disadvantage of RP was also reported in [278]. Obviously, the longer the input signal is deactivated, the more signal power is lost in the output of the I-DS ADC. Due to the lower weights at the end of a conversion, this attenuation is not linear with the number of RP cycles M_{RP} and is analyzed in Sect. 5.4.4.1. The extent to which RP improves I-DS ADC performance depends on the I-DSM architecture used, the order, and the loop-filter scaling, and is usually found by simulation [278].

5.4.4.1 Attenuation Analysis

The recuperation phase attenuates the input signal power depending on the length of the recuperation phase M_{RP} and the order of the CoI filter, which determines the weights toward the end of a conversion. The low-frequency attenuation factor α can be calculated using

$$\alpha = \sum_{i=1}^{M_{on}} h_{\mathrm{rec,norm}}[i] \tag{5.34}$$

where $h_{\mathrm{rec,norm}}$ is the normalized filter weight of the reconstruction filter and M_{on} is the number of samples in which the input signal is turned on ($M_{\mathrm{on}} = M - M_{\mathrm{RP}}$). For CoI filters of different orders L it is also possible to derive an approximated equation [278]:

$$\alpha = 1 - \frac{\prod_{i=1}^{L}(M_{\mathrm{RP}} - i)}{\prod_{i=1}^{L}(M - i)}. \tag{5.35}$$

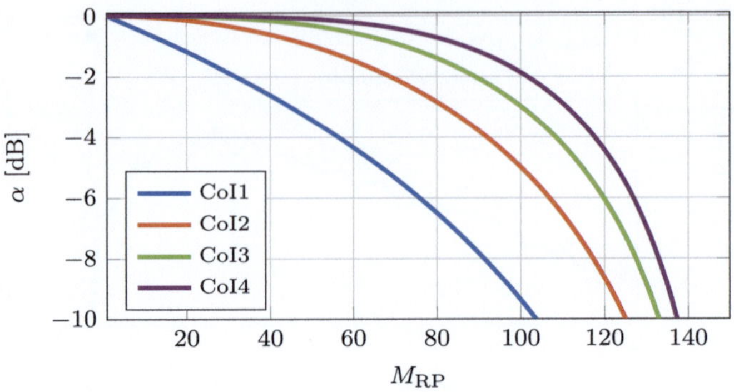

Fig. 5.24 Signal attenuation α over the length of the recuperation phase M_{RP} as a function of the order of the CoI filters.

The attenuation factor α is plotted in Fig. 5.24 for different RP lengths M_{RP} and different orders of the CoI reconstruction filter. The effect of both the RP length M_{RP} and the order of the CoI filter due to the different sample weighting is obvious: the higher the order of the CoI filter, the less the later samples are weighted, and thus the attenuation is less pronounced for a given RP length. Therefore, the combination of RP with lower-order I-DS ADCs must be carefully chosen!

5.4.4.2 Quantizer Gain and Recuperation Phase

Having shown the performance gain of RP in Fig. 5.23 and the signal attenuation penalty of using RP, the question remains as to where exactly the performance improvement comes from. The loopfilter itself remains unchanged when using RP. So, only the effective quantizer gain is left to influence the NTF. As we saw in Sect. 3.3.4, the effective quantizer gain in an incrementally operating DSM depends on the time it is acquired and on the operating state. We found that omitting the first few cycles after reset allows the calculation of q to accurately match simulation and calculation. Not surprisingly, the effective quantizer gain changes dynamically when the input signal is deactivated during RP. Figure 5.25 shows a simulation example where the same exemplary single-bit third-order CIFF I-DS ADC with $M = 150$ from Fig. 5.23 is simulated again with different input signal amplitudes. The effective quantizer gain is then calculated once with RP enabled and once without. For small amplitudes, $q \approx 1.9$ in both cases. For amplitudes approaching the MSA, when no RP is used (blue), the effective quantizer gain starts to drop slowly as the quantizer is more and more overloaded [5], and at about MSA≈-3 dBFS, q drops sharply before the I-DSM becomes unstable due to overload. This MSA corresponds to the simulated SQNR performance shown in Fig. 5.23. If RP is activated for the last 25 samples, the quantizer gain in Fig. 5.25 is calculated using only the samples during the activation of RP [200]. Now the quantizer gain looks very different at large input

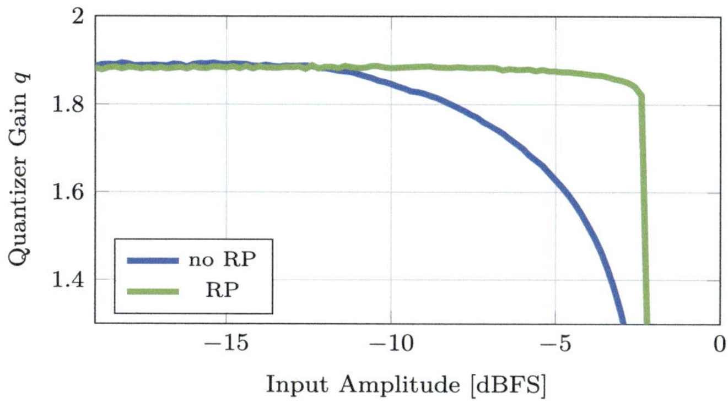

Fig. 5.25 Calculated effective quantizer gain q over the input amplitude for the exemplary single-bit third-order CIFF I-DS ADC without (blue) and with RP (green). In case RP is used, the effective quantizer gain is calculated only from samples during the activated RP.

amplitudes (green): it remains approximately constant at $q \approx 1.9$ until it suddenly drops at an input amplitude of about -2 dBFS, which corresponds to the MSA where the single-bit I-DSM becomes unstable even without an input signal. Again, this is in perfect agreement with the course of the SQNR plot in Fig. 5.23 and explains the main reason for the performance increase when RP is used.

The reason the quantizer gain remains constant up to the MSA when using RP is further illustrated with Fig. 5.26. There, for the same exemplary single-bit third-order I-DS ADC with $M = 150$, the maximum quantizer input observed at all time steps $k \in [0..150]$ over a long simulation interval is plotted. Without employing RP and using a low input amplitude of -20 dBFS, the quantizer input is bound between $\pm V_{\mathrm{FS}}$ during the entire Nyquist conversion cycle (grey). For higher input signals, such as

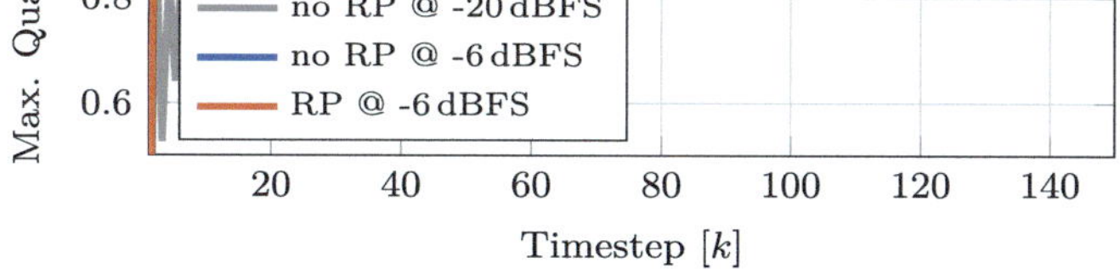

Fig. 5.26 Maximum quantizer input swing at timestep k of 1000 conversion cycles of a sinusoidal input signal at 0.3 f_{s} with different maximum amplitudes. For an amplitude close to MSA, the swing becomes bound during the enabled RP while it exceeds FS if not enabled. ©2025 IEEE. Reprinted, with permission, from Ismail et al. [199].

the exemplarily chosen -6 dBFS, the quantizer input starts exceeding the full-scale range (blue), which is the reason the effective quantizer gain drops in Fig. 5.25. If RP is activated for the last 25 samples, it can be seen that the effective quantizer gain is again limited to the full-scale range within the RP cycles (red) due to the absence of any input signal. This results in a constant quantizer gain during the last operating cycles over all amplitudes, as shown in Fig. 5.25.

5.4.4.3 Overall STF and Recuperation Phase

Another effect of the RP operation on I-DS ADC is its influence on the overall STF. The overall STF of a DT or CT I-DS ADC, including RP, can be calculated using

$$STF_{\mathrm{dt,rp}}(f) = STF_{\mathrm{dt,inc}}(f) \circledast \left(\frac{\sin\left(M_{\mathrm{on}}\pi\, ^{f}/_{f_s}\right)}{\sin\left(\pi\, ^{f}/_{f_s}\right)} \cdot e^{-j\pi f/f_s (M_{\mathrm{on}}-1)} \right) \tag{5.36}$$

$$STF_{\mathrm{ct,rp}}(f) = STF_{\mathrm{ct,inc}}(f) * \left(\frac{\sin\left(T_{\mathrm{on}}\pi f\right)}{T_{\mathrm{on}}\pi f} \cdot e^{-j\pi f T_{\mathrm{on}}} \right) \tag{5.37}$$

with $T_{\mathrm{on}} = {}^{M_{on}}/_{f_s}$ being the on-time of the input signal. For the exemplary single-bit third-order CIFF I-DS ADC with a CoI3 reconstruction filter and $M = 150$, the overall STFs are plotted for different RP lengths for the case of the CT loop filter in Fig. 5.27. In the low-frequency region, we can observe the derived attenuation α from (5.35), which was derived in the time domain in [278] and equals $\alpha = STF_{\mathrm{dt/ct,rp}}[0]$. A more pronounced RP also leads to out-of-band rippling of the overall STF.

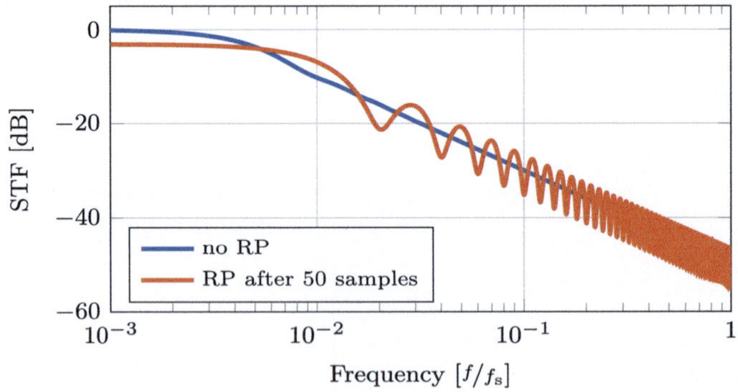

Fig. 5.27 Overall STFs of the exemplary single-bit third-order CIFF I-DS ADC with M=150 and different lengths of the recuperation phase. The notches and DC attenuation increase with the length of applied RP. ©2021 Wagner. Modified, with permission, from Wagner [27].

5.4.4.4 Recuperation Phase in Extended-Counting I-DS ADCs

It was already noted in [16] that disabling the input signal at the end of a Nyquist conversion cycle might be beneficial to generate an accurate residual quantization error at the (last) integrator output of the I-DSM. In Sect. 5.4.4.2, we have shown that a core benefit of the recuperation phase is the limitation of loop-filter dynamics and the avoidance of quantizer overload, especially at high input signals. Since in EC/ER I-DS ADCs it is the dynamics of the last operating cycle that are passed through an ISG toward the $ADC_{EC/ER}$, cf. Fig. 5.6. It therefore seems beneficial to employ RP prior to EC/ER operation.

This is illustrated in Fig. 5.28, where a third-order single-bit I-DS ADC scaled for an OBG of 1.5 with M = 150 is simulated and the maximum swing at the output of the last integrator is plotted over all sampling points $k \in [1, 150]$ observed in 500 conversion cycles for a sinusoidal input signal at 0.3 f_s with an amplitude of −3 dBFS close to MSA. Similar to Fig. 5.26, in which the quantizer input was observed during the simulation, it can be seen in Fig. 5.28 that the swing decreases when RP is activated. The consequence is twofold: not only does RP prevent the I-DSM from reaching overload, cf. Fig. 5.23, but the reduction of the last integrator output, i.e. the residue, allows a larger ISG before requantization by the $ADC_{EC/ER}$, thereby enabling higher resolution.

In Fig. 5.29, the simulated SQNR of the exemplary third-order single-bit I-DS ADC with extended-range operation is plotted. An 8-bit ADC_{ER} is employed, and the RP is applied after 125 of the $M = 150$ cycles in every Nyquist period. If we apply an ISG=2, the exemplary ER I-DS ADC achieves an SQNR≈141 dB. If the ISG is increased to ISG=3 (without activating RP), the performance degrades heavily even for small input signals (red), which is attributed to the overload of the ADC_{ER}. With

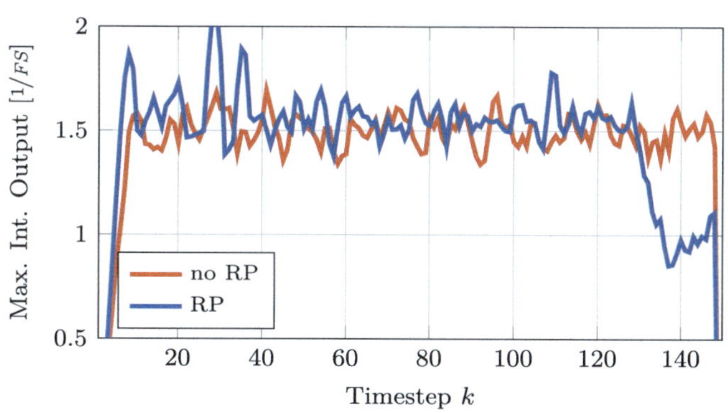

Fig. 5.28 Maximum swing at the last integrator output at timestep k of 500 conversion cycles of a sinusoidal input signal at 0.3 f_s at an amplitude of −3 dB (close to MSA). The swing becomes bounded during the enabled RP while it exceeds FS if not enabled. A third-order I-DS ADC with single-bit quantizer and coefficients scaled for an OBG of 1.5 is used. ©2025 IEEE. Reprinted, with permission, from Ismail et al. [199].

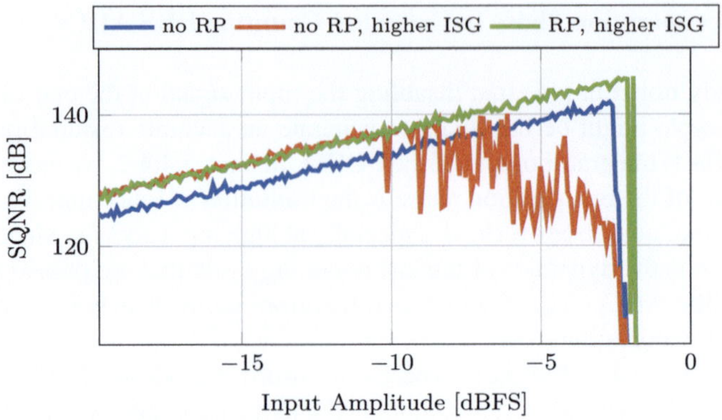

Fig. 5.29 Simulated SQNR of a third-order single-bit I-DS ADC with extended counting at $M = 150$. An 8-bit ADC is used for the EC. ISG=2 or ISG=3. RP is applied after 125 samples. ©2025 IEEE. Reprinted, with permission, from Ismail et al. [199].

RP enabled, the ISG can be increased to 3 without driving the ADC_{ER} into overload. This improves the IBN by about 3.5 dB, while the RP furthermore slightly improves the MSA. In consequence, the maximum SQNR is increased by approximately 5 dB in this simulation example.

The extent to which the swing is diminished and how much improvement RP in combination with EC/ER operation allows naturally depends on the specific architecture chosen, with, e.g. giving more advantage to higher-order or single-bit DSM, and needs evaluation by a designer.

5.4.4.5 Influence of RP on Linear-Exponential I-DS ADCs

In Sect. 5.3, we reviewed the operation of the LE I-DS ADC. Reconfiguring the I-DSM to include exponential feedback around the internal quantizer exponentially decreases the remaining quantization error [246, 267]. The block diagram in Fig. 5.10a shows the first-order LE I-DS ADC implementation from [246]. In simulations Figs. 5.12 and 5.13, we have seen that as the exponential feedback coefficient k_E increases, the suppression of quantization noise becomes more efficient, but at the same time, the quantizer input swing becomes significantly larger with increasing k_E, leading to instability.

This is further illustrated by simulation in Fig. 5.30, where the first-order LE I-DS ADC with a 3-bit internal quantizer is used. The OSR is $M = M_{LP} + M_{EP} = 100$, and the exponential phase is activated for only $M_{EP} = 10$ final clock cycles. Then, Fig. 5.30 shows the achievable MSA over an increasing exponential feedback coefficient k_E. We observe that while large k_E leads to high improvements of the SQNR, cf. Fig. 5.13, its excessive increase leads to stability problems and degrades the MSA.

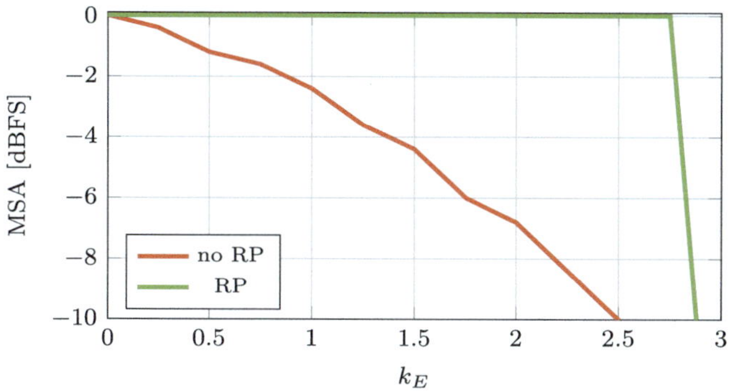

Fig. 5.30 Achievable MSA of the first-order LE I-DS ADC from Fig. 5.10a with an 8-level quantizer for increasing exponential feedback factor k_{E} with and without RP at $M = 100$. RP and exponential operation are activated for the last 10 samples. Corrected from [199].

This observation motivates to combine the LE operation with RP, as the exponential operation is supposed to suppress the quantization error but has a negative effect on the MSA. Thus, taking the input signal away during the exponential phase promises an advantage. Simulation results are plotted in Fig. 5.30 (green), where we observe that the achievable MSA over increasing exponential feedback factor k_{E} is significantly larger when RP is applied together with the exponential phase. Furthermore, it can be observed that instability induced by the exponential phase becomes largely independent from the input amplitude when RP is applied. This can be explained as follows: the first-order I-DSM operates stable within the linear phase until the input reaches full scale, while no signal content is present in the exponential phase due to the RP. Therefore, stability only depends on the amount of the fed back quantization error and therefore on the factor k_{E}. In the simulation example, the modulator becomes unstable for k_{E} values larger than around 2.5.

The benefit of the combination of the LE operation with a RP is further illustrated in Fig. 5.31, where the corresponding SQNR is simulated for the same first-order LE I-DS ADC, once with $k_{\mathrm{E}} = 1$ and without applying RP, once with RP applied and k_{E} increased to 2.5. The maximum achievable SQNR with combined RP and LE operation is boosted by more than 40 dB over the boost by LE operation alone, while the MSA is that of the ideal first-order I-DSM.

While we showed in Fig. 5.16 that the final SQNR of LE I-DS ADCs is entirely determined by the coefficient mismatch, the additional boost from increasing k_{E} appears practically ineffective. However, since the SQNR improvement can result from either the exponential phase duration (M_{EP}) or the exponential feedback factor (k_{E}), the maximum achievable SQNR (for a given mismatch) can be reached with a shorter M_{EP} if a higher k_{E} is enabled through LE operation in combination with RP. A shorter M_{EP} offers the benefit of a reduced NPF, as demonstrated in Sect. 5.3. Finally, note that the loss of output signal power due to the use of RP, as discussed in Sect. 5.4.4.1, is negligible when RP is combined with LE operation, since the exponential phase duration M_{EP} is very short.

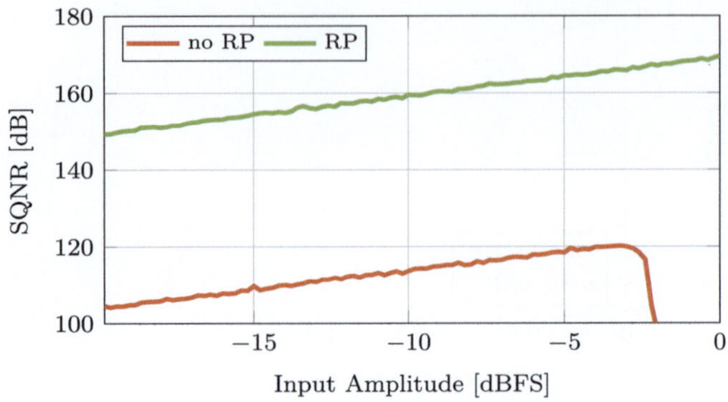

Fig. 5.31 Simulated SQNR of the first-order linear-exponential I-DS ADC with M=100 and a 8-level quantizer. RP and LE are turned on for the last 10 samples. Red: no RP, $k_E = 1$. Green: RP, $k_E = 2.5$. Corrected from [199].

5.4.4.6 Influence of RP on VBW I-DS ADCs

In Sect. 5.4.3, we have reviewed the idea of variable bit width (VBW), where a single-bit I-DSM is used in a first phase achieving intrinsically high linearity due to the single-bit outermost feedback DAC. Reconfiguring it to a multi-bit I-DSM in the final phase reduces the quantization error and offers the additional benefit that the inherent nonlinearity of the multi-bit feedback DAC is effectively suppressed, due to the lower filter weights of higher-order reconstruction filters toward the end of the conversion cycle. As the nonlinearity of the DAC causes major harmonic distortion of the input signal, it seems natural that taking the input signal away while the multi-bit DAC is employed should give an advantage. In consequence, we analyze the combination of the VBW with the RP operation (Fig. 5.32). For illustration, we simulate four different I-DSM with $M = 150$:

- an I-DSM with unaggressive single-bit scaling (OBG = 1.5) and permanent single-bit operation $M_{SB} = 150$,
- an I-DSM with unaggressive single-bit scaling (OBG = 1.5) and permanent multi-bit operation $M_{MB} = 150$,
- an I-DSM with unaggressive single-bit scaling (OBG = 1.5) and VBW operation, $M_{SB} = 125$, $M_{MB} = 25$,
- an I-DSM with aggressive multi-bit scaling (OBG = 3) and permanent multi-bit operation $M_{MB} = 150$.

All four I-DSM configurations are operated with RP (solid lines) and without RP (dashed lines) during the last 25 cycles. Simulations (colored lines) are compared to calculations (gray lines, (3.52)) using different values for the effective quantizer gain q: $q = 1.85$ for single-bit operation and $q = 1$ for multi-bit operation.

While the permanent single-bit operation achieves the lowest SQNR, the use of RP still leads the slightly improved MSA and SQNR (blue dashed versus solid), as it

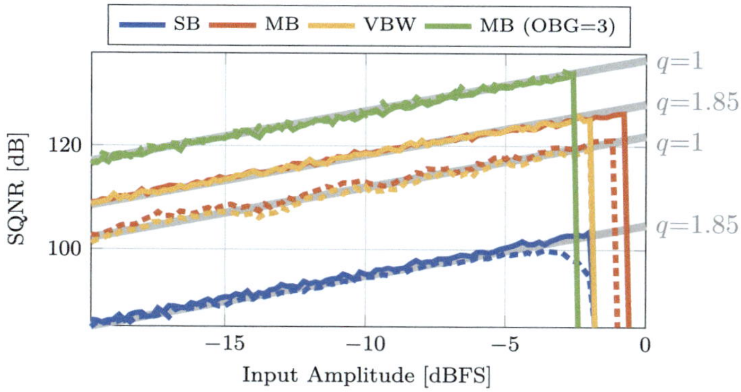

Fig. 5.32 Simulated SQNR for the exemplary third-order CIFF I-DS ADC with different quantizer options and for M = 150. The coefficients are scaled for an OBG of 1.5 (except the green line using OBG=3). The dynamically enabled multi-bit quantizer and the recuperation phase are activated after 125 samples. Solid: with RP. Dashed: without RP. Gray: calculations using (3.52). ©2025 IEEE. Modified, with permission, from Ismail et al. [199].

was already discussed in Sect. 5.4.4 and shown in Fig. 5.23). Going over to a permanent multi-bit or the VBW operation with single-bit scaling (dashed red and dashed yellow), we see that both achieve the same performance with a negligible advantage of the permanent multi-bit operation regarding MSA. This is easily explained as the MSA in the case of VBW is defined by the initial single-bit phase.

Combining RP with VBW (solid yellow) or permanent multi-bit operation (solid red) can further increase the SQNR, which can be explained by the changed effective quantizer gain of the I-DSM: If the I-DSM is scaled with OBG=1.5, but a multi-bit quantizer is used, while at the same time RP is activated to reduce the swing at the quantizer input, this will make the multi-bit quantizer work as a single-bit quantizer with only one LSB toggling. This changes the effective quantizer gain from multi-bit (q=1) to single-bit ($q\approx$1.85). This explains the small step of the achievable SQNR by using RP with VBW. The same improvement of SQNR for the single-bit scaled but permanently multi-bit operating I-DSM is of no practical relevance, since the DAC nonlinearity is not suppressed by this operation, and if the DAC nonlinearity wouldn't be an issue, a multi-bit scaling of the I-DSM would be chosen. To demonstrate the latter, the aggressively scaled I-DSM is simulated using multi-bit quantization. The more aggressive scaling gives the expected additional improvement in SQNR, and this is independent of using RP or not. With the more aggressive scaling, the quantizer input will even swing a few LSB without an input signal, resulting in a permanent effective quantizer gain q=1.

In conclusion, when using VBW to combine the advantages of intrinsically linear single-bit operation with the lower quantization noise of multi-bit operation, the use of RP is advantageous and gives an SQNR improvement equivalent to the ratio of the effective quantizer gain in single-bit versus multi-bit operation. For this, the RP phase must be sufficiently long, which is best found by simulation.

5.5 Comparison of MASH, EC/ER, and Linear-Exponential Incremental ADCs

In the previous sections, we discussed some advanced architectures for I-DS ADCs. We have reviewed their potential to improve resolution beyond that of an ordinary single-loop I-DSM, but we have also highlighted some of their drawbacks, i.e. their sensitivity to nonidealities, especially with respect to analog/digital filter mismatch. In this section, a comparison is made using an example design specification. The SMASH architecture is omitted as we have seen its much greater sensitivity to coefficient mismatch in the two outermost feedback DACs, resulting in unshaped quantization noise leakage from the first stage to the overall output, cf. Sect. 5.1.3. Without this mismatch, the SMASH can outperform the others. Including the DAC mismatch it is by far worse. With high design effort a much better mismatch between the two feedback DACs can be achieved than the mismatch between the analog and digital filters [258], but even with excellent matching in the outermost two DACs of about 0.1%, SoA implementations barely achieve SQNR$\leq$90 dB.

System example: With the intention of a fair comparison, all architectures are based on the same second-order I-DS ADC structure: a second-order unscaled CIFF modulator (OBG=4) equipped with a 10-level quantizer. Note that the 10-level QTZ is used, as a 3-bit (8-level) quantizer will not work stable with the LE I-DS ADC, while a 4-bit (16-level) quantizer would increase the ideal SQNRs even more. The oversampling ratio is set to M=50, enabling a maximum achievable SQNR$\approx$82 dB at an MSA$\approx$-0.4 dBFS for the second-order I-DS in combination with a CoI reconstruction filter. The enhanced architectures are setup as follows:

- MASH 2–1: A first-order second stage is used with a 10-level quantizer and ISG=6. The QE MASH architecture is chosen, cf. Sect. 5.1.1, requiring an additional DAC for QE extraction from the first stage. A third-order CoI reconstruction filter is applied, resulting in a NPF of 1.76. To improve stability of the first stage, a recuperation phase of $M_{RP} = 5$ is used.
- ER 2–1: Similar to the MASH structure, a first-order second stage I-DS ADC with a 10-level quantizer and ISG=6 are used for the extended range operation. As the first-order I-DS ADC operates on a sampled value of the last output value of the second integrator of the first stage, the second stage operates with M_2=M=50 after the first stage has finished. Due to this pipelined operation, the latency is doubled. Also, a sample-and-hold circuit is necessary to store the residue during the second-stage operation. A second-order CoI reconstruction filter is employed after the first stage, resulting in an NPF of 1.35. Operation and stability are again improved through a recuperation phase with $M_{RP} = 5$.
- LE I-DS ADC: An exponential phase is activated for the final 10 samples of the conversion cycle of the second-order I-DSM (M_{LP}=40, M_{EP}=10). The total oversampling is still M=50. The exponential feedback coefficient k_{E} is set to 1. The

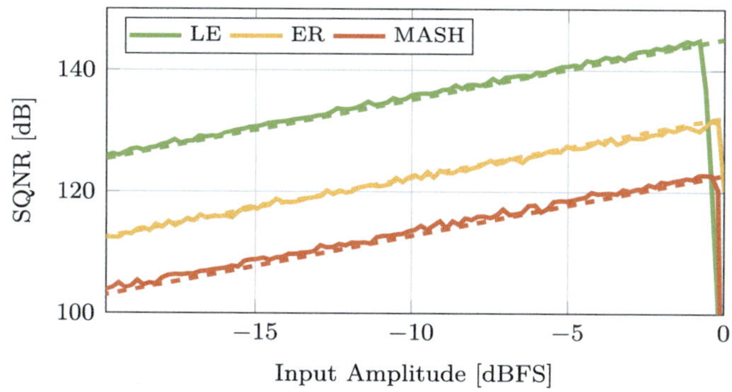

Fig. 5.33 Simulated (solid) and calculated (dashed) SQNR of different multi-stage/step architectures: a MASH 2–1, an ER 2–1 and a second-order LE I-DS ADC are used. Furthermore, the oversampling is set to M=50, the inter-stage gain to ISG=6, the first stage is scaled for OBG=4 and a 10-level quantizer in the first and second stage stage are used. For the calculations the equations previously derived in this chapter were used. Dashed green: (5.27). Dashed yellow: (5.16). Dashed red: (5.5).

length of the recuperation phase matches the exponential phase ($M_{RP} = M_{\text{EP}} = 10$) to improve stability. A matched reconstruction filter is used, leading to an NPF of 1.6.

The resulting SQNR performance of each architecture is shown in Fig. 5.33. It can be seen that the MSA of the LE ADC is slightly smaller compared to the MASH and ER ADCs, but the maximum achievable ideal SQNR is by far the largest when using the LE operation.

Influence of nonideal coefficient variation: To assess the robustness of the three exemplarily designed architectures to coefficient mismatch, a Monte Carlo simulation is performed by introducing a standard deviation σ to all scaling coefficients. For each σ value, 25 simulations are performed, and the average SQNR as well as 1-sigma error bars are shown in Fig. 5.34.

Firstly, the ideal performance of the LE architecture clearly surpasses that of the ER and MASH I-DS ADCs for very low mismatch values, as shown in Fig. 5.33. However, as mismatch increases, it is notable that the LE architecture exhibits greater sensitivity compared to the ER I-DS ADCs. In general, this performance degradation arises from increasingly dominant noise leakage from the second-order first stage into the overall output. The reduced performance of the LE I-DS ADC can be explained by (5.30), which shows that its IBN is limited by $(L \cdot \delta)^2$, as opposed to the δ^2 dependence in the ER I-DS ADC, as shown in (5.17). Interestingly, the MASH architecture demonstrates superior tolerance to larger mismatches compared to both the ER and LE approaches. While this might seem counterintuitive at first, the advantage arises from the use of a third-order CoI reconstruction filter in the 2–1 MASH, in contrast to the second-order CoI filter used in the ER architecture and the matched filter (functionally similar to a second-order CoI) employed in the LE I-DS

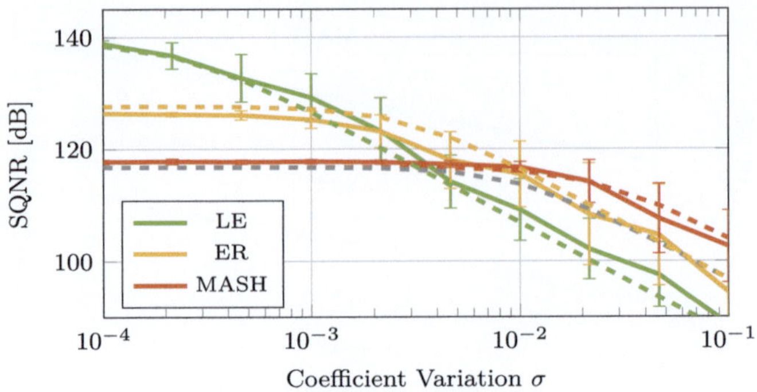

Fig. 5.34 Simulated (Solid) and Calculated (Dashed) SQNR of Different Multi-stage/step Architectures Over the Coefficient Variation σ: a MASH 2–1, an ER 2–1 and a second-order LE I-DS ADC are used. For each σ 25 Monte Carlo runs are simulated. The oversampling is set to M=50, ISG=6. The first stage is scaled for OBG=4 and a 10-level quantizer in the first and second stage are used. Dashed green: (5.30). Dashed yellow: (5.17). Dashed red: (5.7).

ADC. As shown in the derivation of (5.7), the quantization noise leaked from the first stage of the MASH architecture is filtered by the third-order CoI3 reconstruction filter, resulting in greater attenuation compared to the second-order filtering in the LE and ER designs. This observation is further supported by the gray and red dashed lines in Fig. 5.34, which depict the calculated SQNR degradation of the MASH architecture using (5.7). The red line corresponds to the use of a CoI3 filter, while the gray line represents a scenario with only a CoI2 filter applied to the noise leakage from the first stage. Notably, these calculations align well with the simulation results.

Conclusion: We observe that for the given example, the LE I-DS ADC shows superior performance boost compared to the MASH and ER architectures. In a practical implementation, the additional exponential feedback loop needs implementation within the first stage, and practical issues like loop-delay, adder implementation, etc. need to be regarded. Especially for CT implementations, the sensitivity to mismatch becomes very critical, and all architectures suffer similar performance drops. While the MASH shows a slightly better noise leakage suppression, this comes at the disadvantage of a higher noise penalty factor.

5.6 Further Incremental ADC Architectural Variants

5.6.1 Incremental Zoom ADC

Another advanced ADC architecture, often based on incremental ADCs, is the zoom ADC, which was first proposed in [219]. The basic idea is that a coarse ADC determines a coarse range where the input signal is, and then selects an appropriate

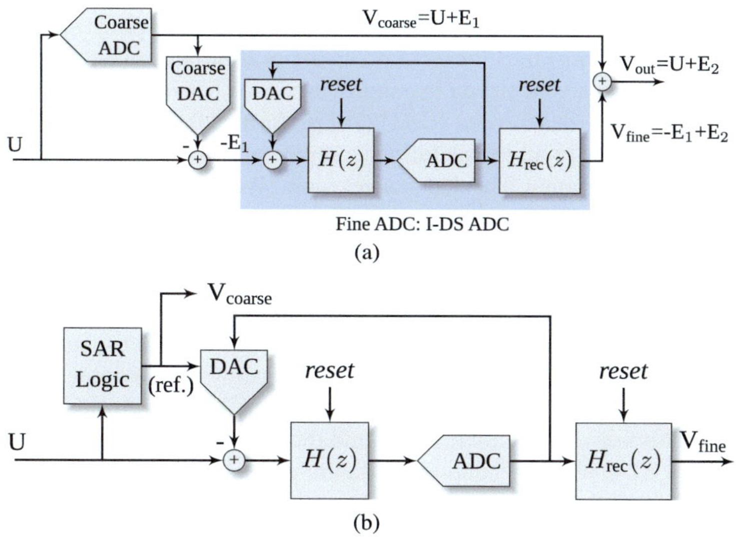

Fig. 5.35 (a) Conventional two-step ADC and (b) incremental zoom ADC.

quantization range for a fine ADC: thereby, the coarse ADC helps the fine ADC to zoom its input conversion range around the current input signal, thus the name was derived. From the day of its publication, there has been a dispute among researchers if this is simply a conventional two-step ADC, where a coarse ADC makes a coarse quantization, and then with the help of a DAC, it creates a residue for a fine ADC to quantize with higher resolution. Thus, it is shortly discussed how a zoom ADC differs from a conventional two-step ADC with the help of Fig. 5.35. In a conventional two-step ADC, cf. Fig. 5.35a, the coarse ADC together with an explicit DAC generate a residue ($-E_1$ in Fig. 5.35a), which is passed to the fine ADC to perform quantization of the residue. A known disadvantage of such two-step ADCs is that the DAC has to provide the linearity and noise requirements equivalent to the overall resolution.

The (original) zoom ADC as in Fig. 5.35b works differently: a coarse ADC performs a coarse quantization, but its digital output is used to select reference voltages for the feedback DAC of an I-DS based ADC. Thus, there is no actual analog residue formed, and the whole conversion is performed by the I-DS ADC. As indicated in Fig. 5.35b, the zoom ADC merges the feedback DAC of the DSM based ADC with the DAC of a residue generating coarse stage. The coarse ADC is only employed to *zoom* the reference voltage range of the I-DS ADC to a closer region around the input signal. The claimed advantages of the zoom ADC were the relaxed resolution requirements of the fine ADC since it works with a small reference voltage range at any time, which leads to a relatively shorter conversion time. Furthermore, the integrator swing can be lowered which allows the use of more energy-efficient amplifiers. On the downside, even though there is no need for an explicit coarse DAC realizing an accurate residue, the zoom ADC requires selectable reference voltages, whose accuracy (noise and linearity) must be as good as the overall performance of the ADC. In the SoA, this was, e.g. achieved by employing DEM [219] or DWA [279, 280].

The original *incremental* zoom ADC [219] performed the two steps sequentially, and a too fast changing input signal during incremental ADC operation could take the input out of the zoom range. This led to a very low Nyquist frequency of 25S/s. This was improved in later designs leading to the dynamic zoom ADCs, where the incremental ADC was replaced by a free-running ADC and the coarse ADC operated more often to keep track between input signal and reference voltage selection [279, 280]. While these early designs were implemented as DT ADCs, [189] presented the first CT zoom ADC with all the advantages of the CT domain, like anti-aliasing filtering, resistive input, etc., but no incremental ADC was used anymore. Follow-up work also realized the zoom ADC with different sub-ADCs; e.g. [281] realized the fine ADC as incremental noise-shaping SAR ADC, and the coarse ADC was implemented as a digital-slope ADC with a second-order digital prediction filter. Unfortunately, the authors missed that the coarse prediction brings memory into the system, which is not periodically reset. Thus, the claim that the overall ADC operates in incremental mode, is more than questionable, which should have been seen as channel-leakage if the incremental ADC, e.g. multiplexes between two different input signals. Also, the zoom ADC has been wrongly described as an ADC generating a residue signal, e.g. in [282, Fig. 1], which would make it a standard two-step ADC.

This brief summary shows that when even the zoom ADC was originally designed for incremental operation with an I-DS main ADC, this has been blurred into many other variants and can no longer be claimed to be a distinct incrementally operating ADC.

5.6.2 *Incremental ADC Operation Without Periodic Reset of the DSM*

As demonstrated in [283], it is possible to multiplex a free-running DSM without resetting its loop or decimation filter. The study further indicates that if the DT STF is designed to be unity, crosstalk suppression is primarily limited by analog circuit imperfections. Taking a complementary approach, [284] shows that periodic resets of the loop and reconstruction filter in I-DS ADCs can be eliminated, while still preserving incremental operation. As we have reviewed in Sect. 4.3.5, the impact of a nonideal reset on the incremental ADC performance is dominated by the first-sample weight of the digital reconstruction filter. While [244] analyzed the additional inband noise due to a nonideal reset in an I-DS ADC, the complete absence of the reset leads to an additional IBN given by [284]

$$IBN_{noReset} = \alpha^2 \cdot \frac{\Delta^2}{12} \tag{5.38}$$

where α is the first sample weight of the employed reconstruction filter. Consequently, if a reconstruction filter with a small first sample weight is used, the impact of a nonideal or even omitted reset becomes negligible.

This can be achieved with using, e.g. an at least second-order sinc filter, which is demonstrated using system-level simulations in [284]. The applicability of this approach in time-domain multiplexed scenarios with multiple input channels was validated. It also leads to single-shot operation of DSM without reset, if the right reconstruction filter is employed. In [284], the free-running configuration was shown to maintain clear separation between channels and exhibited no cross-talk or harmonic distortion, further demonstrating the practical viability of reset-free I-DS ADCs in multi-channel systems.

Appendix A
Performance Measures and Definitions

This book concludes with an appendix that provides an overview of core ADC performance metrics, as they are used throughout this book [4, 285]. Figure A.1 shows an exemplary spectrum and a SNDR plot that will be used to explain those metrics in the following. For the spectrum, a single or two-tone sinusoidal input signal is applied to a converter and the output spectrum is measured. We read output signal power, noise floor, integrated inband noise, the power of harmonic distortion, and other tones. For the SNR plot, the input signal of a single sinusoidal input is successively increased and the SQNR, SNR and/or SNDR are plotted. The SQNR is always better than the SNR as circuit noise will influence or even dominate the performance. If the SNDR is significantly worse than the SNR, we can infer that the converter is non-linearity limited. If the input signal cannot be increased up to FS, we read that the converter tends to overloading and needs a certain back-off for operation. The maximum achievable SNR and SNDR are used to calculate the effective resolution and thus the ENOB of a converter.

Signal Usually, the test signal is a single sinusoidal with frequency f_{sig} well below the Nyquist frequency for oversampling converters. This includes several harmonics into the inband range. The signal powers are usually expressed in dBFS (dB full scale). Two-tone signals are used to assess linearity performance at the band edge. More realistic test signals are sometimes used for ADC dedicated for certain applications, like receiver ADCs.

Inband noise (IBN) describes the total output noise in the band of interest and therefore the minimum resolvable signal power. It directly reflects the influence on the architecture and nonideal behavior and other properties of the ADC. It includes quantization noise, circuit noise, and often also all other disturbers and distortion are included as well.

M. Ortmanns et al., *Incremental Delta-Sigma ADCs*,
https://doi.org/10.1007/978-3-032-09468-1

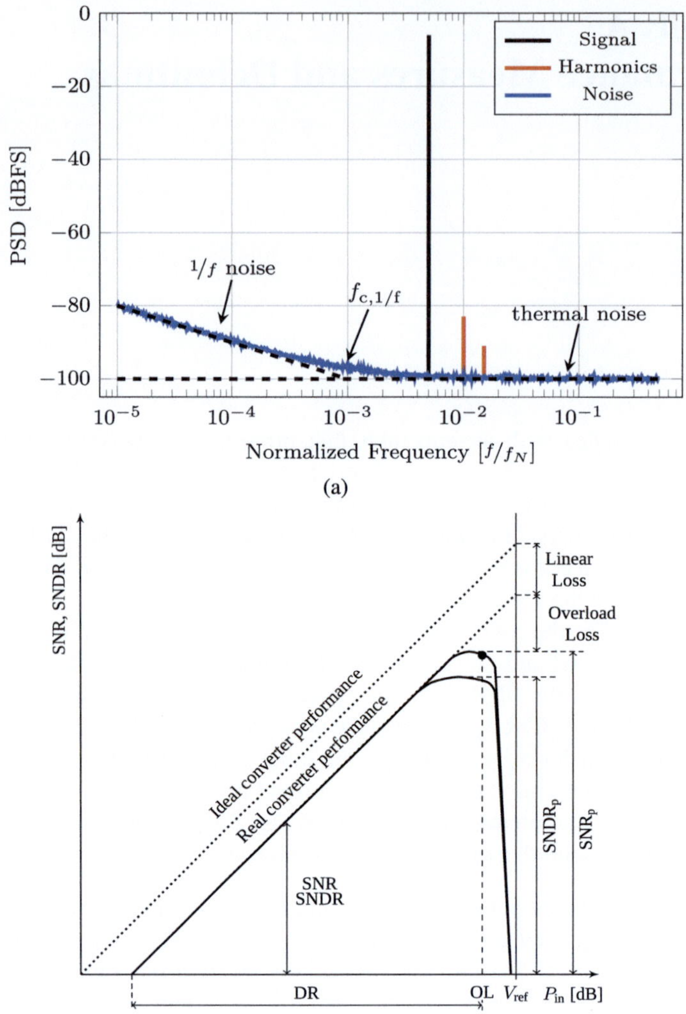

Fig. A.1 (a) Illustrative spectrum of an ADC output signal showing single-sinusoidal input signal, harmonic distortion, and noise. (b) Illustrative SNR versus input amplitude of a DS ADC.

SNR is the ratio of the signal power to the in-band noise power, specified for a certain input amplitude. The power contained in the bins of harmonic components (only if visible) is then excluded. The maximum achievable SNR, the peak SNR_{p}, is often reported. It is described by

$$SNR = 10 \cdot \log_{10}\left(\frac{P_{\text{sig,Out}}}{P_{\text{noise}}}\right) \tag{A.1}$$

Signal-to-quantization-noise ratio (SQNR) is equally defined as the SNR, but only quantization noise is considered.

Signal-to-interference ratio including noise and distortion (SINAD) or SNDR is the ratio of the signal power to all noise and—in contrast to the SNR—all distortion components.

SFDR is defined as the ratio of the input signal amplitude to the amplitude of the largest harmonic or spurious spectral component.

Effective number of bits (ENOB) is one of the most important measures, as it reflects how many useful bits the ADC can provide. For a sinusoidal input signal, it is defined based on (2.15) as

$$ENOB[bit] = \frac{SNDR[dB] - 1.76\,\mathrm{dB}}{6.02\,\mathrm{dB}} \tag{A.2}$$

Total harmonic distortion (THD) is a measure of the total power contained in the harmonics relative to the fundamental tone.

Maximum stable amplitude (MSA) is defined as the maximum input amplitude of a single sinusoidal signal, for which the ADC is stable, i.e. operates correctly and is not overloaded. For most ADC the MSA is equal to the full-scale range, but a DSM becomes unstable for a full-scale input signal and a certain back-off has to be taken into account. The definition of the MSA is not always equally done, but as stated in [4] most publications define stable operation, until the SNR falls 3 to 6 dB below SNR_p for higher amplitudes, which equals a loss of 0.5 to 1 bit from the maximum SNDR.

DR is defined as the ratio of the MSA and the smallest detectable input sinusoidal signal.

Figure of merit (FOM is used as a measure to compare different ADC designs. Basically, there are two major FoMs definitions in the state of the art, the Walden and the Schreier FoM. While for low-to-medium resolutions, the Walden FoM is still used, the Schreier FoM is almost exclusively used nowadays as it is well suited to compare thermal noise-limited designs. It equally treats $\times$ 2 in noise (i.e. DR), a $\times$ 2 in bandwidth and a $\times$ 2 in power with 3 dB each.

$$FOM_\mathrm{S}\,[\mathrm{dB}] = SNDR\,[\mathrm{dB}] + 10 \cdot \log_{10}\left(\frac{f_\mathrm{b}\,[\mathrm{Hz}]}{P\,[\mathrm{W}]}\right). \tag{A.3}$$

Integral nonlinearity (INL) describes the maximum deviation in step width (width between transitions) from the ideal value of one LSB. Whereas not used for conventional free-running DSM, the measure is useful for incremental implementations. This is because a free-running DSM has only an averaging dynamic operation, whereas an incremental DSM feature true sample-sample conversion.

Differential nonlinearity (DNL) describes the maximum deviation of the step midpoints from the ideal step midpoints (or the maximum deviation of the transition points from ideal). Similar as the INL, also the DNL is not used on free-running DSM, but is employed for incremental ADC.

Appendix B
Example Designs

Within this chapter, an example of a generally applicable design process should be given that can be followed if a basic I-DS ADC is required. It is shown for both the DT and the CT domain. While automated design tools exist (c.f. Sect. 2.4), a manual process is shown in a step-by-step guide to give more insight. Thereby, the MATLAB code to derive the exemplary sets of coefficients, to calculate the transfer functions and to estimate the performance will be given.

The examples in the following sections are based on third-order incremental Delta-Sigma modulators (I-DSMs) with single- and multi-bit (4-bit) quantization. The I-DSMs are combined with CoI3 reconstruction filters unless otherwise specified and $M=50$ for the 4-bit and $M=150$ for single-bit operations. The loop-filters have been implemented as cascade-of-integrators with feedforward compensation (CIFF) architecture, and the block diagram of the used DT I-DSM in CIFF architecture is shown in Fig. B.1. It includes a direct feed-in low-distortion path. The loop-filter scaling was obtained using the DS toolbox with OBG=1.5 for the single- and OBG=3 for the 4-bit DSM. For all of the following code examples the MATLAB Schreier toolbox is required.

B.1 Third-Order DT I-DS ADCs

At first the design process will be shown for the DT I-DS ADC.

M. Ortmanns et al., *Incremental Delta-Sigma ADCs*,
https://doi.org/10.1007/978-3-032-09468-1

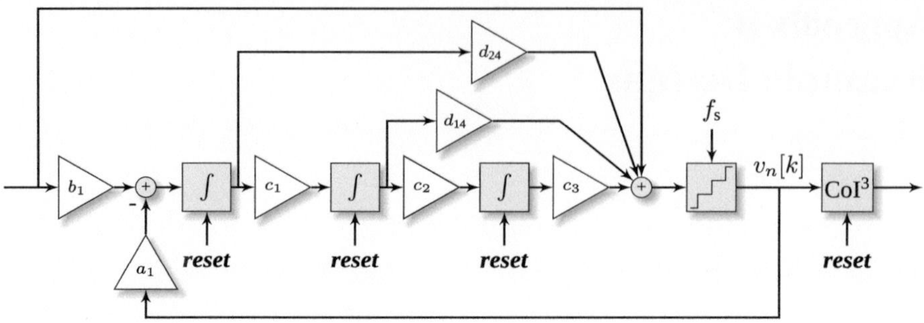

Fig. B.1 Third-order CIFF I-DSM with reconstruction filter.

B.1.1 Modulator Coefficients

The scaling coefficients for the I-DS ADC can be obtained using the MATLAB Schreier toolbox as it was done throughout the book. The coefficients of the example are shown in Table B.1. In (a) and (b) the coefficients are scaled to achieve the same NTF and STF. However, in (b) the coefficients are also scaled for to achieve unity swing at the integrators.

The coefficients can be obtained using the following MATLAB code (exemplary shown for the multi-bit case):

```
%% Derive DT coefficients
% Parameters
order = 3;
archDT = 'CIFF';
% Multi-bit coefficients
OSR = 50;
qLevels = 16;
OBG = 3;
% Get NTF
NTF = synthesizeNTF(order, OSR, 0, OBG);
% Get unscaled coefficients
[a,g,b,c] = realizeNTF(NTF,archDT);
ABCD_dt = stuffABCD(a,g,b,c,archDT);
% Add low distortion path
ABCD_dt(end,order+1) = 1;
% Get scaled coefficients
ABCDs_dt = scaleABCD(ABCD_dt, qLevels);
```

It is not possible to simulate I-DS ADCs in the Schreier toolbox so far. However, to get a first estimate about the performance (especially the MSA) of the DSM it might be useful to look at the free-running amplitude sweep. In Sect. 4.1.1.1 we showed

Table B.1 Coefficients obtained by Schreier's toolbox for a DT DSM (single-bit OBG=1.5, 4-bit OBG=3). (a) Coefficients not scaled for internal swings. (b) Coefficients are scaled to unity swings.

	b_1	a_1	c_1	c_2	c_3	d_{14}	d_{24}
OBG=1.5	1	1	1	1	0.04	0.80	0.29
OBG=3	1	1	1	1	0.41	1.98	1.50

(a)

	b_1	a_1	c_1	c_2	c_3	d_{14}	d_{24}
OBG=1.5	0.36	0.36	0.26	0.17	2.69	2.20	3.06
OBG=3	0.31	0.31	0.92	0.53	2.77	6.49	5.38

(b)

that the MSA of the I-DS ADC can differ from the free-running one. However, as we use a CIFF architecture with the low-distortion path the MSA should be the same. To calculate and plot the free-running amplitude sweep the following code can be used:

```
%% Amplitude Sweep Freerunning
[snr,amp] = simulateSNR(NTF,OSR,[],0,qLevels);
figure;
plot(amp,snr);
grid on;
xlabel('Input Level (dBFS)');
ylabel('SQNR (dB)');
title('Amplitude Sweep')
[peak_snr,peak_amp] = peakSNR(snr,amp);
msg = sprintf('peak SQNR = %4.1fdB  ...
    \n@ amp=%4.1fdB  ',peak_snr,peak_amp);
text(peak_amp-10,peak_snr,msg, ...
    'hor','right', 'vertical','middle');
msg = sprintf('OSR=%d ',OSR);
text(0,5,msg,'hor','right');
```

B.1.2 Transfer-Function Calculation

The transfer functions of an I-DS are useful to estimate its performance regarding quantization noise (NTF), thermal noise (NPF, STF), and many more. Therefore, the following code shows how to calculate the transfer functions for the exemplary DT I-DS ADC:

```
%% Calculate DT TFs
points = 10000;

% TF COI3
TF_REC = tf(1,[1 -1],1, 'variable', 'z^-1')^order;
weights = impz(TF_REC.numerator{1},TF_REC.denominator{1},OSR);
TF_REC = TF_REC/sum(weights);
% ABCD matrices
A = ABCD_dt(1:order,1:order);
Bin = ABCD_dt(1:order,order+1);
Bdac = ABCD_dt(1:order,order+2);
C = ABCD_dt(end,1:order);
Din = ABCD_dt(end,order+1);

% Quantizer Gain
qGain = 1;
% Calculate frequencies
freqs = linspace(-0.5, 0.5, points);
% Generate polynomials
[b1, a1] = ss2tf(A, Bin, C, Din, 1);
[b2, a2] = ss2tf(A, Bdac, C, 0 , 1);
% DSM Freerunning TFs
MOD_FF_FREE = qGain*tf(b1, a1, 1);
MOD_FB_FREE = tf(b2, a2, 1);
MOD_NTF_FREE = 1/(1-qGain*MOD_FB_FREE);
MOD_STF_FREE = MOD_FF_FREE*MOD_NTF_FREE;
% Incremental TFs
% STF
STF = calcTfIncDt(MOD_STF_FREE, TF_REC, OSR, freqs);
% NTF
NTF = calcTfIncDt(MOD_NTF_FREE, TF_REC, OSR, freqs);

% Plot TFs
figure;
hold on;
grid on;
plot(freqs,20*log10(abs(STF)));
plot(freqs,20*log10(abs(NTF)));
set(gca,'XScale','log');
xlabel('Frequency [1/fs]');
ylabel('Magnitude [dB]');
title('Transfer Functions')
```

It has to be mentioned here that the variable *qGain* describes the quantizer gain of the I-DS ADC. In multi-bit modulators it is usually unity but in case a single-bit quantizer is used it has to be derived by simulation. Furthermore, in this code the function *calcTfIncDt* is used. The code of this function is as following:

```
function[transFunc]=calcTfIncDt(TF_MOD, TF_REC, OSR, freqs)
h_REC = impz(TF_REC.Numerator{1},TF_REC.Denominator{1},OSR);
h_MOD = impz(TF_MOD.Numerator{1},TF_MOD.Denominator{1},OSR);
h_TF = conv(h_MOD, h_REC);
h_TF = h_TF(1:OSR);
% DT with freqs from -0.5 to 0.5
tmp = zeros(size(freqs));
tmp(1:OSR) = h_TF';
transFunc = fftshift(fft(tmp));
end
```

B.1.3 Performance Estimation

Having derived the transfer functions it is now possible to estimate the performance of the I-DS ADC as described in Sect. 3.4.2 using the NTF to predict the performance or the approximate equation as it was derived in Sect. 3.4.3.

```
%% Predict IBN
quantStep = 2/(qLevels-1);
deltf = abs(freqs(3)-freqs(2));
% Frequency domain estimation
IBN_est__ntf = (sum(abs(NTF).^2)*(quantStep^2)/12*deltf);
IBN_est__ntf = 10*log10(IBN_est__ntf)
% Time domain estimation
OBG_factor = 2^order;
for i=1:order
    OBG_factor = OBG_factor+(-1)^i*ABCD_dt(end,i)*2^(order-i);
end
IBN_est_approx = (quantStep^2)/12/OBG_factor*...
    (factorial(order)/(OSR^order*qGain*ABCD_dt(end,order)))^2;
IBN_est_approx = 10*log10(IBN_est_approx)
```

B.2 Third-Order CT I-DS ADCs

In the CT case, the coefficients can also be obtained using the Schreier toolbox. The CT coefficients are the result of an IIT so the NTF of the CT DSM matched the DT one.

Table B.2 Coefficients obtained by Schreier's toolbox for a CT DSM (single-bit OBG=1.5, multi-bit OBG=3).

	b_1	a_1	c_1	c_2	c_3	d_{14}	d_{24}
OBG=1.5	1	1	1	1	0.04	0.67	0.24
OBG=3	1	1	1	1	0.41	1.37	1.09

The coefficients shown in Table B.2 can be obtained using the following MATLAB code (exemplary done for the multi-bit coefficients):

```
%% Derive DT coefficients
% Parameters
order = 3;
archDT = 'CIFF';
% Multi-bit coefficients
OSR = 50;
qLevels = 16;
OBG = 3;
% Get NTF
NTF = synthesizeNTF(order, OSR, 0, OBG);
% Get unscaled DT coefficients
[a,g,b,c] = realizeNTF(NTF,archDT);
ABCD_dt = stuffABCD(a,g,b,c,archDT);
% Get unscaled CT coefficients
ABCD_ct = realizeNTF_ct(NTF,archCT);
% Add low distortion path
ABCD_ct(end,order+1) = 1;
```

The Schreier toolbox doesn't offer a function to scale the coefficients for unity swings at the integrators. However, ifNRZ DACs are used the same scaling ratio as in the DT case can also be applied to the CT case. If another feedback DAC form is deployed the scaled coefficients have to be derived by simulation.

The performance estimation of the CT DSM can be done the same way it was shown for the DT one. The transfer function calculations however must be done differently and will therefore be shown next.

B.2.1 Transfer-Function Calculation

The NTF of the CT I-DS ADC can be calculated in the same way as in the DT case. Since the IIT is used to calculate the CT coefficients the NTF is anyways the same as in the DT case. However, the calculation of the STF changes and can be done using the following MATLAB code:

```
%% Calculate CT TFs
points = 100000;
% TF COI3
TF_REC = tf(1,[1 -1],1, 'variable', 'z^-1')^order;
weights = impz(TF_REC.numerator{1},TF_REC.denominator{1},OSR);
TF_REC = TF_REC/sum(weights);
% ABCD matrices
% DT equivalent
A_dt = ABCD_dt(1:order,1:order);
Bdac_dt = ABCD_dt(1:order,order+2);
C_dt = ABCD_dt(end,1:order);
% CT
A_ct = ABCD_ct(1:order,1:order);
Bin_ct = ABCD_ct(1:order,order+2);
C_ct = ABCD_ct(end,1:order);
Din_ct = ABCD_ct(end,order+1);
% Quantizer Gain
qGain = 1;
% Calculate frequencies
% for perfect accuracy frequency range should be
% from -infinity to infinity
freqs = linspace(-20, 20, points);
% Generate polynomials
[b1, a1] = ss2tf(A_ct, Bin_ct, C_ct, Din_ct, 1);
[b2, a2] = ss2tf(A_dt, Bdac_dt, C_dt, 0,1);

% STF (frequency method)
% define s and z
s = 1i*2*pi*freqs;
s_vec = ((ones(length(b1),1)*s).^ ...
    (((length(b1)-1):-1:0)'*ones(1,length(s))));
z = exp(s);
z_vec = ((ones(length(b1),1)*z).^ ...
    ((((length(b1)-1):-1:0))'*ones(1,length(z))));
% Calculate free-running transfer functions
H1 = (b1*s_vec)./(a1*s_vec);
H2 = (b2*z_vec)./(a2*z_vec);
f_ntf_free = 1./(1-qGain*H2);
f_stf_free = qGain*H1.*f_ntf_free;
f_stf_free = conj(f_stf_free');
f_stf_free(isnan(f_stf_free)) = 0;
h_REC = impz(TF_REC.Numerator{1},TF_REC.Denominator{1},OSR);
f_rec = squeeze(freqresp(tf(h_REC', 1, 1), freqs, 'Hz'));
% Incremental STF via convolution with rect
f_rect = conj(OSR*sinc(freqs*(OSR)).*exp(-1i*pi*freqs*OSR));
```

```
STF = abs(freqs(2)-freqs(1))* ...
    conv(f_stf_free.*f_rec, f_rect, 'same');

% NTF (same as in DT)
H2 = tf(b2, a2, 1, 'variable', 'z^-1');
MOD_NTF_FREE = 1/(1-qGain*H2);
freqs_dt = freqs(find(abs(freqs) < 0.5));
NTF = calcTfIncDt(MOD_NTF_FREE, TF_REC, OSR, freqs_dt);

% Plot TFs
figure;
hold on;
grid on;
plot(freqs,20*log10(abs(STF)));
plot(freqs_dt,20*log10(abs(NTF)));
xlim([0.0001 2]);
set(gca,'XScale','log');
xlabel('Frequency [1/fs]');
ylabel('Magnitude [dB]');
title('Transfer Functions')
```

References

1. H. Inose, Y. Yasuda, and J. Murakami, “A telemetering system by code modulation - Δ- Σ modulation,” *IRE Transactions on Space Electronics and Telemetry*, vol. SET-8, no. 3, pp. 204–209, 1962.
2. S. R. Norsworthy, R. Schreier, and G. C. Temes, *Delta-Sigma Data Converters: Theory, Design, and Simulation*. IEEE Press, 1997.
3. R. Schreier and G. C. Temes, *Understanding Delta-Sigma data converters*. Piscataway, NJ: IEEE Press, 2005, vol. 74.
4. M. Ortmanns and F. Gerfers, *enContinuous-Time Sigma-Delta A/D Conversion: Fundamentals, Performance Limits and Robust Implementations*. Springer Science & Business Media, 2006.
5. S. Pavan, R. Schreier, and G. C. Temes, *Understanding Delta-Sigma Data Converters.*, 2nd ed. Wiley-IEEE Press, 2017.
6. R. Koch and B. Heise, “A 120khz sigma/delta a/d converter,” in *1986 IEEE International Solid-State Circuits Conference. Digest of Technical Papers*, vol. XXIX, 1986, pp. 138–139.
7. C. Zhang, L. Breems, Q. Liu, G. Radulov, M. Bolatkale, S. Bajoria, R. Rutten, and A. van Roermund, “A 6GS/s 0.5GHz BW continuous-time 2-1-1 MASH Δ- Σ modulator with phase-boosted current-mode ELD compensation in 40nm CMOS,” in *ESSCIRC 2021 - IEEE 47th European Solid State Circuits Conference (ESSCIRC)*, 2021, pp. 491–494.
8. M. Ortmanns, “Wideband and low-power delta-sigma adcs: State of the art, trends and implementation examples,” in *ESSCIRC 2021 - IEEE 47th European Solid State Circuits Conference (ESSCIRC)*, 2021, pp. 28–35.
9. R. Plassche and R. v. d. Grift, “A five digit a/d converter,” in *1977 IEEE International Solid-State Circuits Conference. Digest of Technical Papers*, vol. XX, 1977, pp. 102–103.
10. R. van de Plassche and R. E. J. van Der Grift, “A five-digit analog-digital converter,” *IEEE Journal of Solid-State Circuits*, vol. 12, no. 6, pp. 656–662, 1977.
11. R. van de Plassche, “A sigma-delta modulator as an a/d converter,” *IEEE Transactions on Circuits and Systems*, vol. 25, no. 7, pp. 510–514, 1978.
12. J. Robert, G. C. Temes, V. Valencic, R. Dessoulavy, and P. Deval, “A 16-bit low-voltage CMOS A/D converter,” *IEEE Journal of Solid-State Circuits*, vol. 22, no. 2, pp. 157–163, 1987.

M. Ortmanns et al., *Incremental Delta-Sigma ADCs*,
https://doi.org/10.1007/978-3-032-09468-1

13. J. Robert and P. Deval, "A second-order high-resolution incremental A/D converter with offset and charge injection compensation," *IEEE Journal of Solid-State Circuits*, vol. 23, no. 3, pp. 736–741, 1988.
14. Z. Tan, C.-H. Chen, Y. Chae, and G. C. Temes, "Incremental delta-sigma adcs: A tutorial review," *IEEE Transactions on Circuits and Systems I: Regular Papers*, vol. 67, no. 12, pp. 4161–4173, 2020.
15. G. C. Temes and C.-H. Chen, *Incremental Data Converters for Sensor Interfaces*. Hoboken, NJ: John Wiley & Sons, Inc., 2024. [Online]. Available: https://doi.org/10.1002/9781394178452
16. J. Márkus, "Higher-order incremental delta-sigmaanalog-to-digital converters," phdthesis, Budapest University of Technology and Economics, 2005.
17. B. Murrmann, "ADC Performance Survey 1997-2020," [Online]. Available: https://web.stanford.edu/~murmann/adcsurvey.html
18. B. E. Jonsson. (2012, Aug.) A/d-converter performance evolution. [Online; accessed 9-April-2025]. [Online]. Available: https://converterpassion.wordpress.com/2012/08/26/adc-performance-evolution-thermal-figure-of-merit-fom/
19. Y. Chae, C. M. Lopez, K. A. Makinwa, M. Ortmanns, and W. Sansen, "A glimpse of the history of analog ics: A tale of amplifiers, data converters, and sensor interfaces," *IEEE Solid-State Circuits Magazine*, vol. 15, no. 3, pp. 43–52, 2023.
20. R. Walden, "Analog-to-digital converter technology comparison," in *Proceedings of 1994 IEEE GaAs IC Symposium*, 1994, pp. 217–219.
21. S. Rabii and B. Wooley, "A 1.8-v digital-audio sigma-delta modulator in 0.8-/spl mu/m cmos," *IEEE Journal of Solid-State Circuits*, vol. 32, no. 6, pp. 783–796, 1997.
22. T. Sundstrom, B. Murmann, and C. Svensson, "Power dissipation bounds for high-speed nyquist analog-to-digital converters," *IEEE Transactions on Circuits and Systems I: Regular Papers*, vol. 56, no. 3, pp. 509–518, 2009.
23. C. C. Lee and M. P. Flynn, "A 14b 23ms/s 48mw resetting $\delta\sigma$ adc with 87db sfdr 11.7b enob & 0.5mm2 area," in *2008 IEEE Symposium on VLSI Circuits*, 2008, pp. 182–183.
24. M. A. Mokhtar, P. Vogelmann, J. Wagner, and M. Ortmanns, "Incremental sturdy-MASH sigma-delta modulator with reduced sensitivity to DAC mismatch," in *Proc. IEEE Int. Symp. Circuits and Systems (ISCAS)*, 2019, pp. 1–5.
25. M. A. Mokhtar, O. Ismail, P. Vogelmann, J. G. Kauffman, and M. Ortmanns, "A 40 ks/scalibration-free incremental $\Delta\Sigma$ adc achieving 104 db dr and 105.7 db sfdr," in *ESSCIRC 2023- IEEE 49th European Solid State Circuits Conference (ESSCIRC)*, 2023, pp. 401–404.
26. J. Steensgaard, R. Reay, R. Perry, D. Thomas, G. Tu, and G. Reitsma, "A 24b 2ms/s sar adc with 0.03ppm inl and 106.3db dr in 180nm cmos," in *2022 IEEE International Solid-State Circuits Conference (ISSCC)*, vol. 65, 2022, pp. 168–170.
27. J. Wagner, "On the optimization of sigma-delta adcs," Ph.D. dissertation, University of Ulm, 2021. [Online]. Available: https://oparu.uni-ulm.de/xmlui/123456789/38160
28. A. V. V. Oppenheim, R. W. V. Schafer, and J. R. V. Buck, Eds., *EnglischDiscrete-time signal processing*, 2nd ed., ser. Prentice Hall signal processing series. Upper Saddle River, NJ: Prentice-Hall Internat., 1999.
29. B. Widrow, "A study of rough amplitude quantization by means of nyquist sampling theory," *IRE Transactions on Circuit Theory*, vol. 3, no. 4, pp. 266–276, 1956.
30. W. R. Bennet, "Spectra of quantized signals," *Bell System Technical Journal*, vol. 27, pp. 446–472, 1948.
31. A. Sripad and D. Snyder, "A necessary and sufficient condition for quantization errors to be uniform and white," *IEEE Transactions on Acoustics, Speech, and Signal Processing*, vol. 25, no. 5, pp. 442–448, 1977.
32. R. M. Gray and D. L. Neuhoff, "Quantization," *IEEE Transactions on Information Theory*, vol. 44, no. 6, pp. 2325–2383, 1998.
33. E. F. Stikvoort, "Some remarks on the stability and performance of the noise shaper or sigma-delta modulator," *IEEE Transactions on Communications*, vol. 36, no. 10, pp. 1157–1162, 1988.

34. S. Ardalan and J. Paulos, "An analysis of nonlinear behavior in delta - sigma modulators," *IEEE Transactions on Circuits and Systems*, vol. 34, no. 6, pp. 593–603, 1987.
35. L. Jie, X. Tang, J. Liu, L. Shen, S. Li, N. Sun, and M. P. Flynn, "An overview of noise-shaping sar adc: From fundamentals to the frontier," *IEEE Open Journal of the Solid-State Circuits Society*, vol. 1, pp. 149–161, 2021.
36. S. Lindfors and K. Halonen, "Two-step quantization in multibit Delta-Sigma modulators," *IEEE Transactions on Circuits and Systems II: Analog and Digital Signal Processing*, vol. 48, no. 2, pp. 171–176, 2001, conference Name: IEEE Transactions on Circuits and Systems II: Analog and Digital Signal Processing.
37. L. Samid, M. Ortmanns, Y. Manoli, and F. Gerfers, "A new kind of low-power multibit third order continuous-time lowpass sigma delta modulator," in *2002 IEEE International Symposium on Circuits and Systems. Proceedings (Cat. No.02CH37353)*, vol. 3, 2002, pp. III–III.
38. M. Jang, X. Tang, Y. Lim, J. G. Kauffman, N. Sun, M. Ortmanns, and Y. Chae, "Design techniques for energy-efficient analog-to-digital converters," *IEEE Open Journal of the Solid-State Circuits Society*, vol. 3, pp. 145–161, 2023.
39. S. Jantzi, R. Schreier, and M. Snelgrove, "Bandpass sigma-delta analog-to-digital conversion," *IEEE Transactions on Circuits and Systems*, vol. 38, no. 11, pp. 1406–1409, 1991.
40. Y.-H. Chang, C.-Y. Wu, and T.-C. Yu, "Chopper-stabilized sigma-delta modulator," in *1993 IEEE International Symposium on Circuits and Systems (ISCAS)*, 1993, pp. 1286–1289 vol.2.
41. Y. Jung, S. Oh, J. Koo, S. Park, J.-H. Suh, D. Cho, S. Ha, and M. Je, "A 187db foms 46fj/conv. 2nd-order highpass Δ- Σ capacitance-to-digital converter," in *2023 IEEE Symposium on VLSI Technology and Circuits (VLSI Technology and Circuits)*, 2023, pp. 1–2.
42. R. Schreier and M. Snelgrove, "Bandpass sigma-delta modulation," *Electronics Letters*, vol. 25, no. 23, pp. 1560–1561, 1989.
43. E. Hogenauer, "An economical class of digital filters for decimation and interpolation," *and Signal Processing IEEE Transactions on Acoustics, Speech*, vol. 29, no. 2, pp. 155–162, 1981.
44. G. Mitteregger, C. Ebner, S. Mechnig, T. Blon, C. Holuigue, E. Romani, A. Melodia, and V. Melini, "A 14b 20 mw 640 mhz cmos ct Δ - Σ adc with 20mhz signal bandwidth and 12b enob," in *Solid-State Circuits Conference, 2006. ISSCC 2006. Digest of Technical Papers. IEEE International*, 2006, pp. 131–140.
45. A. Baluni and S. Pavan, "Analysis and design of a 20-mhz bandwidth continuous-time delta-sigma modulator with time-interleaved virtual-ground-switched fir feedback," *IEEE Journal of Solid-State Circuits*, vol. 56, no. 3, pp. 729–738, 2021.
46. T. Frey and M. Bossert, *Signal- und Systemtheorie*, 2nd ed. Wiesbaden: Vieweg + Teubner, 2008.
47. J. Candy, "A use of double integration in sigma delta modulation," *IEEE Transactions on Communications*, vol. 33, no. 3, pp. 249–258, 1985.
48. N. Tan and S. Eriksson, "A fully differential switched-current delta-sigma modulator using a single 3.3-v power-supply voltage," in *1994 IEEE International Symposium on Circuits and Systems (ISCAS)*, vol. 5, 1994, pp. 485–488 vol.5.
49. J. A. Cherry and W. M. Snelgrove, *Continuous-time delta-sigma modulators for high-speed A/D conversion: theory, practice and fundamental performance limits*. Springer Science & Business Media, 1999, vol. 521.
50. J. M. de la Rosa, "Sigma-delta modulators: Tutorial overview, design guide, and state-of-the-art survey," *IEEE Transactions on Circuits and Systems I: Regular Papers*, vol. 58, no. 1, pp. 1–21, 2011.
51. R. Schreier and B. Zhang, "Delta-sigma modulators employing continuous-time circuitry," *IEEE Transactions on Circuits and Systems I: Fundamental Theory and Applications*, vol. 43, no. 4, pp. 324–332, 1996.
52. J. A. Cherry and W. M. Snelgrove, "Excess loop delay in continuous-time delta-sigma modulators," *IEEE Transactions on Circuits and Systems II: Analog and Digital Signal Processing*, vol. 46, no. 4, pp. 376–389, 1999.

53. H. Aboushady and M.-. Louerat, "Systematic approach for discrete-time to continuous-time transformation of $\sigma\delta$ modulators," in *Proc. 2002 IEEE Int Symp. Circuits and Systems*, vol. 4, 2002, p. IV.
54. S. Loeda, H. M. Reekie, and B. Mulgrew, "On the design of high-performance wide-band continuous-time sigma-delta converters using numerical optimization," *IEEE Transactions on Circuits and Systems I: Regular Papers*, vol. 53, no. 4, pp. 802–810, 2006.
55. T. Brückner, M. Kiebler, M. Lorenz, C. Zorn, W. Mathis, and M. Ortmanns, "Calculating transfer functions of CT sigma-delta modulators with arbitrary DAC waveforms," in *Proc. IEEE Int. Symp. Circuits and Systems (ISCAS2013)*, 2013, pp. 1264–1267.
56. R. Ritter, J. G. Kauffman, J. Becker, and M. Ortmanns, "A 10 MHz bandwidth, 70 dB SNDR continuous time delta-sigma modulator with digitally improved reconfigurable blocker rejection," *IEEE Journal of Solid-State Circuits*, vol. 51, no. 3, pp. 660–670, 2016.
57. G. R. Ritchie, "Higher order interpolative analog to digital converters," Ph.D. dissertation, Pennsylvania Univ., Philadelphia, 1977.
58. R. Ritter and M. Ortmanns, "Continuous-Time Delta-Sigma ADCs With Improved Interferer Rejection," *IEEE Journal on Emerging and Selected Topics in Circuits and Systems*, vol. 5, no. 4, pp. 500–513, Dec. 2015, conference Name: IEEE Journal on Emerging and Selected Topics in Circuits and Systems.
59. C. Pan, H. San, and T. Shibata, "A 2nd-order delta sigma ad modulator using dynamic amplifier and dynamic sar quantizer," in *2016 International Symposium on Intelligent Signal Processing and Communication Systems (ISPACS)*, 2016, pp. 1–5.
60. A. Gharbiya and D. Johns, "On the implementation of input-feedforward delta-sigma modulators," *IEEE Transactions on Circuits and Systems II: Express Briefs*, vol. 53, no. 6, pp. 453–457, 2006.
61. A. Marques, V. Peluso, M. S. Steyaert, and W. M. Sansen, "Optimal parameters for $\delta\sigma$ modulator topologies," *IEEE Transactions on Circuits and Systems II: Analog and Digital Signal Processing*, vol. 45, no. 9, pp. 1232–1241, 1998.
62. P. F. Ferguson, A. Ganesan, and R. W. Adams, "One bit higher order sigma-delta a/d converters," in *IEEE International Symposium on Circuits and Systems*, 1990, pp. 890–893 vol.2.
63. S. Loeda, J. Harrison, F. Pourchet, and A. Adams, "A 10/20/30/40 mhz feed-forward fir dac continuous-time Δ- Σ adc with robust blocker performance for radio receivers," in *2015 Symposium on VLSI Circuits (VLSI Circuits)*, 2015, pp. C262–C263.
64. J. Silva, M. UK, J. Steensgaard, and G. Temes, "Wideband low-distortion delta-sigma adc topology," *Electronics Letters*, vol. 37, pp. 737 – 738, 07 2001.
65. C. H. Chen, T. He, Y. Zhang, and G. C. Temes, "Incremental analog-to-digital converters for high-resolution energy-efficient sensor interfaces," *IEEE Journal on Emerging and Selected Topics in Circuits and Systems*, vol. 5, no. 4, pp. 612–623, 2015.
66. K. Philips, P. Nuijten, R. Roovers, A. van Roermund, F. Chavero, M. Pallares, and A. Torralba, "A continuous-time $\Sigma\Delta$ adc with increased immunity to interferers," *Solid-State Circuits, IEEE Journal of*, vol. 39, no. 12, pp. 2170–2178, 2004.
67. J. Kauffman, P. Witte, J. Becker, and M. Ortmanns, "An 8.5 mw continuous-time $\Delta\Sigma$ modulator with 25 mhz bandwidth using digital background dac linearization to achieve 63.5 db sndr and 81 db sfdr," *Solid-State Circuits, IEEE Journal of*, vol. 46, no. 12, pp. 2869–2881, 2011.
68. J. Wagner, S. Reich, R. Ritter, J. Anders, and M. Ortmanns, "Finite GBW in single OpAmp CT $\Sigma\Delta$ modulators," in *2016 IEEE International Conference on Electronics, Circuits and Systems (ICECS)*, Dec. 2016, pp. 468–471.
69. M. A. Mokhtar, M. Merkle, J. G. Kauffman, and M. Ortmanns, "A comparative study of noise behavior in single-opamp resonators in delta-sigma modulators," in *2021 28th IEEE International Conference on Electronics, Circuits, and Systems (ICECS)*, 2021, pp. 1–4.
70. K. Matsukawa, K. Obata, Y. Mitani, and S. Dosho, "A 10 mhz bw 50 fj/conv. continuous time $\Delta\Sigma$ modulator with high-order single opamp integrator using optimization-based design method," in *2012 Symposium on VLSI Circuits (VLSIC)*, 2012, pp. 160–161.

71. S. Zeller, C. Muenker, and R. Weigel, “A 0.039mm2 inverter-based 1.82mw 68.6db-sndr 10mhz-bw ct-$\Sigma\Delta$-adc in 65nm cmos,” in *2013 Proceedings of the ESSCIRC (ESSCIRC)*, 2013, pp. 319–322.
72. L. Breems, M. Bolatkale, H. Brekelmans, S. Bajoria, J. Niehof, R. Rutten, B. Oude-Essink, F. Fritschij, J. Singh, and G. Lassche, “A 2.2ghz continuous-time $\Delta\Sigma$ adc with -102dbc thd and 25mhz bw,” in *2016 IEEE International Solid-State Circuits Conference (ISSCC)*, 2016, pp. 272–273.
73. K. Xing, W. Wang, Y. Zhu, C. H. Chan, and R. P. Martins, “A 10.4mw 50mhz-bw 80db-dr single-opamp third-order ctsdm with sab-eld-merged integrator and 3-stage opamp,” in *2020 IEEE Symposium on VLSI Circuits*, 2020, pp. 1–2.
74. R. W. Adams, “EnglishDesign and Implementation of an Audio 18-Bit Analog-to-Digital Converter Using Oversampling Techniques,” *EnglishJournal of the Audio Engineering Society*, vol. 34, no. 3, pp. 153–166, 1986, publisher: Audio Engineering Society.
75. Y. Geerts, M. Steyaert, and W. M. C. Sansen, *enDesign of Multi-Bit Delta-Sigma A/D Converters*. Springer Science & Business Media, 2002, google-Books-ID: 9NSFL_8o6sAC.
76. J. G. Kenney and L. R. Carley, “Clans: a high-level synthesis tool for high resolution data converters,” in *Proc. [1988] IEEE Int. Conf. Computer-Aided Design (ICCAD-89) Digest of Technical Papers*, 1988, pp. 496–499.
77. F. Medeiro, B. Pérez-Verdú, and A. Rodríguez-Vázquez, *Top-Down Design of High-Performance Sigma-Delta Modulators*, ser. The Springer International Series in Engineering and Computer Science. Springer US, 1999.
78. F. Gerfers, M. Ortmanns, and Y. Manoli, “A 1.5-v 12-bit power-efficient continuous-time third-order $\Sigma\Delta$ modulator,” *IEEE Journal of Solid-State Circuits*, vol. 38, no. 8, pp. 1343–1352, 2003.
79. J. Bastos, A. Marques, M. Steyaert, and W. Sansen, “A 12-bit intrinsic accuracy high-speed CMOS DAC,” *IEEE Journal of Solid-State Circuits*, vol. 33, no. 12, pp. 1959–1969, 1998, conference Name: IEEE Journal of Solid-State Circuits.
80. A. Sanyal and N. Sun, “Dynamic element matching techniques for static and dynamic errors in continuous-time multi-bit $\delta\sigma$ modulators,” *IEEE Journal on Emerging and Selected Topics in Circuits and Systems*, vol. 5, no. 4, pp. 598–611, 2015.
81. M. Runge, J. Edler, T. Kaiser, and F. Gerfers, “A 18ms/s 76db sndr 93db sfdr ct Δ- Σ modulator with input voltage tracking 2nd-order gmvc filter and shared fir dac in 22nm fdsoi cmos,” in *2021 IEEE Custom Integrated Circuits Conference (CICC)*, 2021, pp. 1–2.
82. Y. Manoli, “A self calibration method for fast, high resolution a/d and d/a converters,” in *ESSCIRC ’88: Fourteenth European Solid-State Circuits Conference*, 1988, pp. 110–113.
83. T. Cataltepe, A. Kramer, L. Larson, G. Temes, and R. Walden, “Digitally corrected multi-bit sigma delta data converters,” in *IEEE International Symposium on Circuits and Systems,*, 1989, pp. 647–650 vol.1.
84. D. Groeneveld, H. Schouwenaars, H. Termeer, and C. Bastiaansen, “A self-calibration technique for monolithic high-resolution d/a converters,” *IEEE Journal of Solid-State Circuits*, vol. 24, no. 6, pp. 1517–1522, 1989.
85. L. Carley, “A noise-shaping coder topology for 15+ bit converters,” *IEEE Journal of Solid-State Circuits*, vol. 24, no. 2, pp. 267–273, 1989, conference Name: IEEE Journal of Solid-State Circuits.
86. R. Baird and T. Fiez, “Linearity enhancement of multibit Delta-Sigma A/D and D/A converters using data weighted averaging,” *IEEE Transactions on Circuits and Systems II: Analog and Digital Signal Processing*, vol. 42, no. 12, pp. 753–762, 1995, conference Name: IEEE Transactions on Circuits and Systems II: Analog and Digital Signal Processing.
87. P. Kiss, U.-K. Moon, J. Steensgaard, J. Stonick, and G. Temes, “Multibit Delta-Sigma ADC with mixed-mode DAC error correction,” in *ISCAS 2001. The 2001 IEEE International Symposium on Circuits and Systems (Cat. No.01CH37196)*, vol. 1, 2001, pp. 280–283 vol. 1.
88. P. Witte and M. Ortmanns, “A background dac error estimation in sigma-delta adcs using a pseudo random noise based correlation technique,” in *2009 IEEE International Symposium on Circuits and Systems*, 2009, pp. 1549–1552.

89. J. G. Kauffman, P. Witte, M. Lehmann, J. Becker, Y. Manoli, and M. Ortmanns, "A 72 db dr, ct Δ- Σ modulator using digitally estimated, auxiliary dac linearization achieving 88 fj/conv-step in a 25 mhz bw," *IEEE Journal of Solid-State Circuits*, vol. 49, no. 2, pp. 392–404, 2014.
90. M. De Bock, X. Xing, L. Weyten, G. Gielen, and P. Rombouts, "Calibration of dac mismatch errors in $\sigma\delta$ adcs based on a sine-wave measurement," *IEEE Transactions on Circuits and Systems II: Express Briefs*, vol. 60, no. 9, pp. 567–571, 2013.
91. M. De Bock, A. Babaie-Fishani, and P. Rombouts, "Improved offline calibration for dac mismatch in low osr ΣΔ adcs with distributed feedback," in *2014 IEEE International Symposium on Circuits and Systems (ISCAS)*, 2014, pp. 938–941.
92. S. Ho, C.-L. Lo, J. Ru, and J. Zhao, "A 23 mw, 73 db dynamic range, 80 mhz bw continuous-time delta-sigma modulator in 20 nm cmos," *IEEE Journal of Solid-State Circuits*, vol. 50, no. 4, pp. 908–919, 2015.
93. S.-H. Wu, T.-K. Kao, Z.-M. Lee, P. Chen, and J.-Y. Tsai, "A 160mhz-bw 72db-dr 40mw continuous-time Δ- Σ modulator in 16nm cmos with analog isi-reduction technique," in *2016 IEEE International Solid-State Circuits Conference (ISSCC)*, 2016, pp. 280–281.
94. H. Zhu, W. Yang, G. Engel, and Y.-B. Kim, "A two-parameter calibration technique tracking temperature variations for current source mismatch," *IEEE Transactions on Circuits and Systems II: Express Briefs*, vol. 64, no. 4, pp. 387–391, 2017.
95. M. Runge and F. Gerfers, "A digital compensation method canceling static and non-linear time-variant feedback dac errors in ΣΔ analog-to-digital converters," in *2017 IEEE International Symposium on Circuits and Systems (ISCAS)*, 2017, pp. 1–4.
96. M. Runge and F. Gerfers, "Correlation based time-variant dac error estimation in continuous-time ΣΔ adcs with pseudo random noise," in *2018 IEEE International Symposium on Circuits and Systems (ISCAS)*, 2018, pp. 1–5.
97. D. Kong and I. Galton, "Adaptive cancellation of static and dynamic mismatch error in continuous-time dacs," *IEEE Transactions on Circuits and Systems I: Regular Papers*, vol. 65, no. 2, pp. 421–433, 2018.
98. T. He, M. Ashburn, S. Ho, Y. Zhang, and G. Temes, "A 50mhz-bw continuous-time ΔΣ adc with dynamic error correction achieving 79.8db sndr and 95.2db sfdr," in *2018 IEEE International Solid - State Circuits Conference - (ISSCC)*, 2018, pp. 230–232.
99. S. Pavan and R. Theertham, "Improved offline calibration of dac mismatch errors in delta–sigma data converters," *IEEE Transactions on Circuits and Systems II: Express Briefs*, vol. 66, no. 10, pp. 1618–1622, 2019.
100. P. Payandehnia, T. He, Y. Wang, and G. C. Temes, "Digital correction of dac nonlinearity in multi-bit feedback a/d converters: Invited tutorial," in *2020 IEEE Custom Integrated Circuits Conference (CICC)*, 2020, pp. 1–8.
101. D. Basak, S. Kalani, Y. Zhang, and K.-P. Pun, "An automatic on-chip calibration technique for static and dynamic dac error correction in high-speed continuous-time delta-sigma modulators," *IEEE Access*, vol. 7, pp. 172 097–172 109, 2019.
102. A. Abdelaal, J. Kauffman, J. Becker, M. D. Longa, F. Conzatti, and M. Ortmanns, "A calibration-free 80mhz ct dsm using dual quantization and isi shuffler achieving 106.2db sfdr," in *Proceedings of the IEEE Custom Integrated Circuits Conference (CICC)*, Boston, USA, 2025.
103. L. Dorrer, F. Kuttner, P. Greco, P. Torta, and T. Hartig, "A 3-mw 74-db snr 2-mhz continuous-time delta-sigma adc with a tracking adc quantizer in 0.13-/spl mu/m cmos," *IEEE Journal of Solid-State Circuits*, vol. 40, no. 12, pp. 2416–2427, 2005.
104. M. Z. Straayer and M. H. Perrott, "A 12-bit, 10-mhz bandwidth, continuous-time $\sigma\delta$ adc with a 5-bit, 950-ms/s vco-based quantizer," *IEEE Journal of Solid-State Circuits*, vol. 43, no. 4, pp. 805–814, 2008.
105. T. Hayashi, Y. Inabe, K. Uchimura, and T. Kimura, "A multistage delta-sigma modulator without double integration loop," in *1986 IEEE International Solid-State Circuits Conference. Digest of Technical Papers*, vol. XXIX, 1986, pp. 182–183.

106. Y. Matsuya, K. Uchimura, A. Iwata, T. Kobayashi, M. Ishikawa, and T. Yoshitome, "A 16-bit oversampling a-to-d conversion technology using triple-integration noise shaping," *IEEE Journal of Solid-State Circuits*, vol. 22, no. 6, pp. 921–929, 1987.
107. L. Longo and M. Copeland, "A 13 bit isdn-band oversampled adc using two-stage third order noise shaping," in *Proceedings of the IEEE 1988 Custom Integrated Circuits Conference*, 1988, pp. 21.2/1–21.2/4.
108. G. Yin and W. Sansen, "A high-frequency and high-resolution fourth-order $\Sigma\Delta$ a/d converter in bicmos technology," in *ESSCIRC '93: Nineteenth European Solid-State Circuits Conference*, vol. 1, 1993, pp. 1–4.
109. T. Leslie and B. Singh, "An improved sigma-delta modulator architecture," in *IEEE International Symposium on Circuits and Systems*, 1990, pp. 372–375 vol.1.
110. B. Brandt and B. Wooley, "A 50-mhz multibit sigma-delta modulator for 12-b 2-mhz a/d conversion," *IEEE Journal of Solid-State Circuits*, vol. 26, no. 12, pp. 1746–1756, 1991.
111. A. Marques, V. Peluso, M. Steyaert, and W. Sansen, "Optimal parameters for cascade /spl delta//spl sigma/ modulators," in *1997 IEEE International Symposium on Circuits and Systems (ISCAS)*, vol. 1. IEEE, 1997, pp. 61–64 vol.1.
112. N. Maghari, S. Kwon, G. Temes, and U. Moon, "Sturdy mash - modulator," *Electronics Letters*, vol. 42, pp. 1269 – 1270, 02 2006.
113. J. Zhang, S.-W. Sin, Y. Liu, F. Ye, G. Wang, M. Ortmanns, and L. Qi, "On the synthesis of continuous-time sturdy mash delta-sigma modulators," *IEEE Transactions on Circuits and Systems II: Express Briefs*, vol. 70, no. 2, pp. 356–360, 2023.
114. L. Qi, A. Jain, D. Jiang, S.-W. Sin, R. P. Martins, and M. Ortmanns, "A 76.6-db-sndr 50-mhz-bw 29.2-mw multi-bit ct sturdy mash with dac non-linearity tolerance," *IEEE Journal of Solid-State Circuits*, vol. 55, no. 2, pp. 344–355, 2020.
115. O. Oliaei, "Design of continuous-time sigma-delta modulators with arbitrary feedback waveform," *IEEE Transactions on Circuits and Systems II: Analog and Digital Signal Processing*, vol. 50, no. 8, pp. 437–444, 2003.
116. B. Putter, "/spl sigma//spl delta/ adc with finite impulse response feedback dac," in *2004 IEEE International Solid-State Circuits Conference (IEEE Cat. No.04CH37519)*, 2004, pp. 76–77 Vol.1.
117. S. Pavan, "Finite-impulse-response (fir) feedback in continuous-time delta-sigma converters," in *2018 IEEE Custom Integrated Circuits Conference (CICC)*, 2018, pp. 1–8.
118. Z. Cao, T. Song, and S. Yan, "A 14 mw 2.5 ms/s 14 bit $\sigma\delta$ modulator using split-path pseudo-differential amplifiers," *IEEE Journal of Solid-State Circuits*, vol. 42, no. 10, pp. 2169–2179, 2007.
119. Y. Zhang, C. Chen, T. He, and G. C. Temes, "A 35μw 96.8db sndr 1 khz bw multi-step incremental adc using multi-slope extended counting with a single integrator," in *2016 IEEE Symposium on VLSI Circuits (VLSI-Circuits)*, 2016, pp. 1–2.
120. A. Sukumaran and S. Pavan, "Low Power Design Techniques for Single-Bit Audio Continuous-Time Delta Sigma ADCs Using FIR Feedback," *IEEE Journal of Solid-State Circuits*, vol. 49, no. 11, pp. 2515–2525, 2014, conference Name: IEEE Journal of Solid-State Circuits.
121. J. Gealow, M. Ashburn, C.-H. Lou, S. Ho, P. Riehl, A. Shabra, J. Silva, and Q. Yu, "A 2.8 mw $\Delta\Sigma$ adc with 83 db dr and 1.92 mhz bw using fir outer feedback and tia-based integrator," in *2011 Symposium on VLSI Circuits - Digest of Technical Papers*, 01 2011, pp. 42–43.
122. Y. Zhang, C.-H. Chen, T. He, and G. C. Temes, "A continuous-time delta-sigma modulator for biomedical ultrasound beamformer using digital eld compensation and fir feedback," *IEEE Transactions on Circuits and Systems I: Regular Papers*, vol. 62, no. 7, pp. 1689–1698, 2015.
123. S. Pavan, "Continuous-time delta-sigma modulator design using the method of moments," *IEEE Transactions on Circuits and Systems I: Regular Papers*, vol. 61, no. 6, pp. 1629–1637, 2014.
124. A. Abdelaal, M. Pietzko, M. A. Mokhtar, J. G. Kauffman, and M. Ortmanns, "FIR Filter with Symmetric Non-Equal Coefficients for CT Delta-Sigma Modulators," in *2021 IEEE International Symposium on Circuits and Systems (ISCAS)*, 2021, pp. 1–5, iSSN: 2158-1525.

125. L. Risbo, "Stability predictions for high-order $\sigma\delta$ modulators based on quasilinear modeling," in *Proc. IEEE Int. Symp. Circuits and Systems - ISCAS '94*, vol. 5, 1994, pp. 361–364 vol.5.
126. B. Agrawal and K. Shenoi, "Design methodology for $\Sigma\Delta$m," *IEEE Transactions on Communications*, vol. 31, no. 3, pp. 360–370, 1983.
127. K.-H. Chao, S. Nadeem, W. Lee, and C. Sodini, "A higher order topology for interpolative modulators for oversampling a/d converters," *IEEE Transactions on Circuits and Systems*, vol. 37, no. 3, pp. 309–318, 1990.
128. R. Schreier, "An empirical study of high-order single-bit delta-sigma modulators," *IEEE Transactions on Circuits and Systems II: Analog and Digital Signal Processing*, vol. 40, no. 8, pp. 461–466, 1993.
129. W. L. Lee, "A novel higher order interpolative modulator topology for high resolution oversampling a/d converters," *Master's thesis, Massachusetts Institute of Technology*, 1987.
130. J. G. Kenney and L. R. Carley, "Design of multibit noise-shaping data converters," *Analog Integrated Circuits and Signal Processing*, vol. 3, no. 3, p. 259, 1993. [Online]. Available: http://dx.doi.org/10.1007/BF01239365
131. S. Hein and A. Zakhor, "On the stability of sigma delta modulators," *IEEE Transactions on Signal Processing*, vol. 41, no. 7, pp. 2322–2348, 1993.
132. R. Baird and T. Fiez, "Stability Analysis of High-Order Delta-Sigma Modulation for ADC's," *IEEE Trans. Circuits Syst.*, vol. 41, no. 1, pp. 59–62, 1994.
133. L. Risbo, "$\Sigma\Delta$ modulators - stability analysis and optimization," phdthesis, Technical University of Denmark, 1994.
134. J. Anders, J.-K. Bremer, and W. Mathis, "Mixed-logic dynamical system modeling of $\Sigma\Delta$ - modulators and its application to stability analysis," in *2009 IEEE International Symposium on Circuits and Systems*. IEEE, 2009, pp. 3122–3125.
135. C. Zorn, S. Stegemann, and W. Mathis, "An approach to stability analysis of transfer systems with switched feedback," in *ICSES 2010 International Conference on Signals and Electronic Circuits*, 2010, pp. 181–184.
136. C. Zorn, S. Stegemann, T. Brücknery, M. Ortmanns, and W. Mathis, "A novel optimization method for ct sigma-delta-modulators using a switched system model," in *2011 IEEE International Symposium of Circuits and Systems (ISCAS)*, 2011, pp. 1848–1851.
137. C. Zorn, T. Brückner, M. Ortmanns, and W. Mathis, "Performance tuning of multi-bit continuous time $\Sigma\Delta$-modulators using a switched system model," in *2012 19th IEEE International Conference on Electronics, Circuits, and Systems (ICECS 2012)*, 2012, pp. 420–423.
138. P. Rombouts, M. De Bock, J. De Maeyer, and L. Weyten, "A describing function study of saturated quantization and its application to the stability analysis of multi-bit sigma delta modulators," *IEEE Transactions on Circuits and Systems I: Regular Papers*, vol. 60, no. 7, pp. 1740–1752, 2013.
139. J. Lota, M. Al-Janabi, and I. Kale, "Nonlinear model-based approach for accurate stability prediction of one-bit higher-order delta–sigma modulators," *IEEE Transactions on Instrumentation and Measurement*, vol. 62, no. 4, pp. 686–692, 2013.
140. J. Lota, M. Al-Janabi, and I. Kale, "Nonlinear stability prediction of multibit delta–sigma modulators for sinusoidal inputs," *IEEE Transactions on Instrumentation and Measurement*, vol. 63, no. 1, pp. 18–26, 2014.
141. N. He, F. Kuhlmann, and A. Buzo, "Double-loop sigma-delta modulation with dc input," *IEEE Transactions on Communications*, vol. 38, no. 4, pp. 487–495, 1990.
142. J. Wagner, M. Ortmanns, and J. M. de la Rosa, "Man or machine — design automation of delta-sigma modulators," in *Proc. IEEE Int. Symp. Circuits and Systems (ISCAS)*, 2018, pp. 1–5.
143. V. Peluso, A. Marques, M. Steyaert, and W. Sansen, "Optimal parameters for single loop /spl delta//spl sigma/ modulators," in *1997 IEEE International Symposium on Circuits and Systems (ISCAS)*, vol. 1. IEEE, 1997, pp. 57–60 vol.1.
144. R. Schreier, "The Delta-Sigma Toolbox," [Online]. Available: http://www.mathworks.com/matlabcentral/fileexchange/19

145. M. Keller, A. Buhmann, J. Sauerbrey, M. Ortmanns, and Y. Manoli, "A comparative study on excess-loop-delay compensation techniques for continuous-time sigma-delta modulators," *Circuits and Systems I: Regular Papers, IEEE Transactions on*, vol. 55, no. 11, pp. 3480–3487, 2008.
146. M. Ortmanns, F. Gerfers, and Y. Manoli, "Compensation of the influence of finite gain-bandwidth on continuous-time sigma-delta modulators," in *Proc. 10th IEEE Int. Conf. Electronics, Circuits and Systems the 2003*, vol. 3, 2003, pp. 950–953 Vol.3.
147. T. Brückner, C. Zorn, J. Anders, J. Becker, W. Mathis, and M. Ortmanns, "A GPU-accelerated web-based synthesis tool for CT sigma-delta modulators," *IEEE Transactions on Circuits and Systems I: Regular Papers*, vol. 61, no. 5, pp. 1429–1441, 2014.
148. H. Tang and A. Doboli, "High-level synthesis of delta-sigma modulator topologies optimized for complexity, sensitivity, and power consumption," *IEEE Transactions on Computer-Aided Design of Integrated Circuits and Systems*, vol. 25, pp. 597–607, 2006.
149. J. Anders, W. Mathis, and M. Ortmanns, "A new optimization approach for the automatic design of Σ Δ-modulators," in *Proc. IEEE Int. Symp. Circuits and Systems*, 2008, pp. 1432–1435.
150. K. Francken and G. G. E. Gielen, "A high-level simulation and synthesis environment for $\delta\sigma$ modulators," *IEEE Transactions on Computer-Aided Design of Integrated Circuits and Systems*, vol. 22, no. 8, pp. 1049–1061, 2003.
151. J. Ruiz-Amaya, J. M. de la Rosa, F. V. Fernandez, F. Medeiro, R. del Rio, B. Perez-Verdu, and A. Rodriguez-Vazquez, "High-level synthesis of switched-capacitor, switched-current and continuous-time $\sigma\delta$ modulators using simulink-based time-domain behavioral models," *IEEE Transactions on Circuits and Systems I: Regular Papers*, vol. 52, no. 9, pp. 1795–1810, 2005.
152. G. Liñán-Cembrano, N. Loureno, N. Horta, and J. M. d. l. Rosa, "Design automation of analog and mixed-signal circuits using neural networks – a tutorial brief," *IEEE Transactions on Circuits and Systems II: Express Briefs*, pp. 1–1, 2023.
153. P. Díaz-Lobo, G. Liñán-Cembrano, and J. M. d. l. Rosa, "On the use of artificial neural networks for the automated high-level design of $\sigma\delta$ modulators," *IEEE Transactions on Circuits and Systems I: Regular Papers*, pp. 1–0, 2023.
154. P. Manrique-Merchán, G. Liñán-Cembrano, and J. de La Rosa, "Aisad: A user-friedly ai-assisted matlab tool for the high-level design of $\Sigma\Delta$ modulators," in *2024 39th Conference on Design of Circuits and Integrated Systems (DCIS)*, 2024, pp. 1–6.
155. G. Venturini, "python-deltasigma," [Online]. Available: http://www.python-deltasigma.io
156. B. Cortés-Delgadillo, P. A. Rodríguez-Navas, L. I. Guerrero-Linares, and J. M. de la Rosa, "Embedding matlab optimizers in simsides for the high-level design of $\sigma\delta$ modulators," *IEEE Transactions on Circuits and Systems II: Express Briefs*, vol. 65, pp. 547–551, 2018.
157. J. Wagner, P. Vogelmann, and M. Ortmanns, "An automated design environment for ct incremental sigma-delta adcs," in *Proc. Circuits and Systems (ICECS) 2020 IEEE Int. Conf. Electronics*, 2020.
158. P. Kaesser, J. Wagner, O. Ismail, and M. Ortmanns, "Inner transfer functions in incremental $\Delta\Sigma$ adcs," *IEEE Transactions on Circuits and Systems II: Express Briefs*, pp. 1–1, 2024.
159. G. Liñán-Cembrano and J. M. de la Rosa, "Optimized design of $\sigma\delta$ modulators using deep-learning and simulated annealing," *Electronics Letters*, vol. 61, no. 1, p. e70256, 2025. [Online]. Available: https://ietresearch.onlinelibrary.wiley.com/doi/abs/10.1049/ell2.70256
160. F. Medeiro, B. Perez-Verdu, A. Rodriguez-Vazquez, and J. Huertas, "A vertically integrated tool for automated design of $\Sigma\Delta$ modulators," *Solid-State Circuits, IEEE Journal of*, vol. 30, no. 7, pp. 762 – 772, 1995.
161. K. Francken, P. Vancorenland, and G. Gielen, "DAISY: a simulation-based high-level synthesis tool for Delta-Sigma modulators," in *IEEE/ACM International Conference on Computer Aided Design. ICCAD - 2000. IEEE/ACM Digest of Technical Papers (Cat. No.00CH37140)*, 2000, pp. 188–192, iSSN: 1092-3152.
162. A. Buhmann, M. Keller, M. Maurer, M. Ortmanns, and Y. Manoli, "Disco - a toolbox for the discrete-time simulation of continuous-time sigma-delta modulators using matlab^{TM}," in *Proc. 50th Midwest Symp. Circuits and Systems*, 2007, pp. 1082–1085.

163. T.-H. Kuo, K.-D. Chen, and J.-R. Chen, "Automatic coefficients design for high-order sigma-delta modulators," *Circuits and Systems II: Analog and Digital Signal Processing, IEEE Transactions on*, vol. 46, no. 1, pp. 6 –15, 1999.
164. H.-M. Wang and T.-H. Kuo, "An automatic coefficient design methodology for high-order bandpass sigma-delta modulator with single-stage structure," *Circuits and Systems II: Express Briefs, IEEE Transactions on*, vol. 53, no. 7, pp. 580 – 584, 2006.
165. I. Williams, L.A. and B. Wooley, "MIDAS-a functional simulator for mixed digital and analog sampled data systems," in *Circuits and Systems, 1992. ISCAS '92. Proceedings., 1992 IEEE International Symposium on*, vol. 5, 1992, pp. 2148 –2151 vol.5.
166. V. Liberali, V. Dias, M. Ciapponi, and F. Maloberti, "TOSCA: a simulator for switched-capacitor noise-shaping A/D converters," *Computer-Aided Design of Integrated Circuits and Systems, IEEE Transactions on*, vol. 12, no. 9, pp. 1376 –1386, sep 1993.
167. G. Brauns, R. Bishop, M. Steer, J. Paulos, and S. Ardalan, "Table-based modeling of delta-sigma modulators using ZSIM," *Computer-Aided Design of Integrated Circuits and Systems, IEEE Transactions on*, vol. 9, no. 2, pp. 142 –150, 1990.
168. C.-M. Hsieh and H.-W. Chiu, "Sigma delta modulator design automation," in *Communications, Circuits and Systems, 2007. ICCCAS 2007. International Conference on*, July 2007, pp. 1034–1038.
169. P. Benabes, M. Keramat, and R. Kielbasa, "A methodology for designing continuous-time sigma-delta modulators," in *European Design and Test Conference, 1997. ED TC 97. Proceedings*, Mar 1997, pp. 46–50.
170. P. Benabes, P. Aldebert, and R. Kielbasa, "A matlab based tool for bandpass continuous-time sigma-delta modulators design," in *Circuits and Systems, 1998. ISCAS '98. Proceedings of the 1998 IEEE International Symposium on*, vol. 6, 1998, pp. 274–277 vol.6.
171. O. Yetik, O. Saglamdemir, S. Talay, and G. Dundar, "A coefficient optimization and architecture selection tool for $\Sigma\Delta$ modulators in MATLAB," in *Design, Automation Test in Europe Conference Exhibition, 2007. DATE '07*, April 2007, pp. 1–6.
172. G. Altinok, M. Al-Janabi, and I. Kale, "A GUI driven $\Sigma - \Delta$ modulator design, evaluation and measurement tool with a view to practical implementation," in *Instrumentation and Measurement Technology Conference, 2009. I2MTC '09. IEEE*, 2009, pp. 1022–1025.
173. J. Robert and V. Valencic, "Offset and charge injection compensation in an incremental analog to digital converter," in *ESSCIRC '85: 11th European Solid-State Circuits Conference*, 1985, pp. 45–48.
174. J. Robert and P. Deval, "A second-order high-resolution incremental A/D converter with offst and charge injection compensation," in *Proc. ESSCIRC '87: 13th European Solid-State Circuits Conf*, 1987, pp. 109–112.
175. S. Tao and A. Rusu, "A power-efficient continuous-time incremental sigma-delta adc for neural recording systems," *IEEE Transactions on Circuits and Systems I: Regular Papers*, vol. 62, no. 6, pp. 1489–1498, 2015.
176. J. Markus, J. Silva, and G. C. Temes, "Theory and applications of incremental $\Delta\Sigma$ converters," *IEEE Transactions on Circuits and Systems I: Regular Papers*, vol. 51, no. 4, pp. 678–690, 2004.
177. P. Vogelmann, J. Wagner, and M. Ortmanns, "A 14b, Twofold Time-Interleaved Incremental $\Delta\Sigma$ ADC Using Hardware Sharing," *IEEE Transactions on Circuits and Systems I: Regular Papers*, vol. 67, no. 11, pp. 3681–3692, 2020, conference Name: IEEE Transactions on Circuits and Systems I: Regular Papers.
178. J. M. Garcia, "Digitally enhanced continuous-time sigma-delta analogue-to-digital converters," Ph.D. dissertation, KTH School of Information and Communication Technology, 2012.
179. S. Tao, "Power-efficient continuous-time incremental sigma-delta analog-to-digital converters," Ph.D. dissertation, KTH School of Information and Communication Technology, 2015.
180. M. Mollazadeh, K. Murari, G. Cauwenberghs, and N. Thakor, "Micropower cmos integrated low-noise amplification, filtering, and digitization of multimodal neuropotentials," *IEEE Transactions on Biomedical Circuits and Systems*, vol. 3, no. 1, pp. 1–10, 2009.

181. Y. M. Chi and G. Cauwenberghs, "Micropower integrated bioamplifier and auto-ranging adc for wireless and implantable medical instrumentation," in *2010 Proceedings of ESSCIRC*, Sept 2010, pp. 334–337.
182. J. Uhlig, R. Schuffny, H. Neubauer, J. Hauer, and J. Haase, "A low-power continuous-time incremental 2nd-order-MASH Σ Δ-modulator for a CMOS imager," in *Proc. 16th IEEE Int. Conf. Electronics Circuits and Systems - (ICECS 2009)*, 2009, pp. 33–36.
183. J. Garcia, S. Rodriguez, and A. Rusu, "A low-power CT incremental 3rd order $\sigma\delta$ ADC for biosensor applications," *IEEE Transactions on Circuits and Systems I: Regular Papers*, vol. 60, no. 1, pp. 25–36, 2013.
184. J. Garcia, S. Rodriguez, and A. Rusu, "On continuous-time incremental adcs with extended range," *IEEE Transactions on Instrumentation and Measurement*, vol. 62, no. 1, pp. 60–70, 2013.
185. S. Tao and A. Rusu, "A comparative design study of continuous-time incremental sigma-delta adc architectures," *International Journal of Circuit Theory and Applications*, vol. 44, no. 12, pp. 2147–2163, 2016. [Online]. Available: https://onlinelibrary.wiley.com/doi/full/10.1002/cta.2218
186. M. A. Mokhtar, P. Vogelmann, M. Haas, and M. Ortmanns, "A 94.3-dB SFDR, 91.5-dB dr, and 200-kS/s CT incremental delta–sigma modulator with differentially reset FIR feedback," *IEEE Solid-State Circuits Letters*, vol. 2, no. 9, pp. 87–90, 2019.
187. M. A. Mokhtar, P. Vogelmann, M. Haas, and M. Ortmanns, "A 94.3-dB SFDR, 91.5-dB dr, and 200-kS/s CT incremental delta–sigma modulator with differentially reset FIR feedback," in *Proc. ESSCIRC 2019 - IEEE 45th European Solid State Circuits Conf. (ESSCIRC)*, 2019, pp. 87–90.
188. S. Wu, Y. Shu, A. Y. Chiou, W. Huang, Z. Chen, and H. Hsieh, "9.1 a current-sensing front-end realized by a continuous-time incremental adc with 12b sar quantizer and reset-then-open resistive dac achieving 140db dr and 8ppm inl at 4ks/s," in *2020 IEEE International Solid-State Circuits Conference - (ISSCC)*, 2020, pp. 154–156.
189. B. Gönen, S. Karmakar, R. van Veldhoven, and K. A. A. Makinwa, "A continuous-time zoom ADC for low-power audio applications," *IEEE Journal of Solid-State Circuits*, vol. 55, no. 4, pp. 1023–1031, 2020.
190. Z. Lu, M. Zhao, and Z. Tan, "A 188.6-μw continuous-time incremental delta-sigma adc with extended counting achieving 95.2-db sndr and 175.4-db fomsndr," in *2024 IEEE Custom Integrated Circuits Conference (CICC)*, 2024, pp. 1–2.
191. T. C. Caldwell and D. A. Johns, "Incremental data converters at low oversampling ratios," *IEEE Transactions on Circuits and Systems I: Regular Papers*, vol. 57, no. 7, pp. 1525–1537, 2010.
192. J. Steensgaard, Z. Zhang, W. Yu, A. Sarhegyi, L. Lucchese, D. I. Kim, and G. C. Temes, "Noise –power optimization of incremental data converters," *IEEE Transactions on Circuits and Systems I: Regular Papers*, vol. 55, no. 5, pp. 1289–1296, 2008.
193. S. Bannwarth, A. Wenzler, and W. Mathis, "On the Transfer Behaviour of Incremental Sigma Delta Converters," in *VXV International Symposium on Theoretical Engineering*, 2009, pp. 1–5.
194. J. Wagner, P. Vogelmann, and M. Ortmanns, "On the signal filtering property of CT incremental sigma-delta ADCs," *IEEE Transactions on Circuits and Systems II: Express Briefs*, p. 1, 2019.
195. C. Lyden, J. Ryan, C. A. Ugarte, J. Kornblum, and Fan Ma Yung, "A single shot /spl sigma//spl delta/ architecture for multiplexed applications," in *IEE Colloquium on Oversampling and Sigma-Delta Strategies for DSP*, Nov 1995, pp. 6/1–6/4.
196. S. Bannwarth, A. Wenzler, and W. Mathis, "On the transfer behaviour of incremental sigma delta converters," in *Proc. VXV Int. Symp. Theoretical Engineering*, 2009, pp. 1–5. [Online]. Available: https://ieeexplore.ieee.org/document/5760639?arnumber=5760639
197. Y. Yamamoto and M. Nagahara, "Digital control," *Wiley Encyclopedia of Electrical and Electronics Engineering*, vol. vol. 5, pp. 445–457, 1999.
198. M. Keller, A. Buhmann, M. Ortmanns, and Y. Manoli, "A method for the discrete-time simulation of continuous-time sigma-delta modulators," in *2007 IEEE International Symposium on Circuits and Systems*, 2007, pp. 241–244.

199. O. Ismail, P. Kaesser, and M. Ortmanns, "Reducing the impact of non-idealities on incremental $\delta\sigma$ adcs by reconfiguration: A review," in *2025 IEEE Custom Integrated Circuits Conference (CICC)*, 2025, pp. 1–8.
200. P. Kaesser, J. Wagner, O. Ismail, B. Driemeyer, and M. Ortmanns, "Quantizer gain in incremental delta-sigma adcs," in *2023 30th IEEE International Conference on Electronics, Circuits and Systems (ICECS)*, 2023, pp. 1–4.
201. J. Wagner, P. Vogelmann, and M. Ortmanns, "Performance evaluation of incremental sigma-delta adcs based on their ntf," *IEEE Transactions on Circuits and Systems II: Express Briefs*, vol. 67, pp. 2813–2817, 2020.
202. P. Kaesser and M. Ortmanns, "Performance prediction of incremental delta-sigma adcs," *IEEE Open Journal of Circuits and Systems: Regular Papers*, 2025.
203. J. Markus, P. Deval, V. Quiquempoix, J. Silva, and G. C. Temes, "Incremental delta-sigma structures for DC measurement: an overview," in *Proc. IEEE Custom Integrated Circuits Conf. 2006*, 2006, pp. 41–48.
204. M. Sporer, I.-G. Vasilaş, A. Adžemović, N. Graber, S. Reich, C. Gueli, M. Eickenscheidt, I. Diester, T. Stieglitz, and M. Ortmanns, "Neurobus - architecture for an ultra-flexible neural interface," *IEEE Transactions on Biomedical Circuits and Systems*, pp. 1–16, 2024.
205. C. Lyden, "Single shot sigma-delta analog to digital converter," U.S. Patent 5 189 419, 23, 1993.
206. C. Lyden, J. Ryan, C. A. Ugarte, J. Kornblum, and F. M. Yung, "A single shot sigma delta analog to digital converter for multiplexed applications," in *Proc. IEEE 1995 Custom Integrated Circuits Conf*, 1995, pp. 203–206.
207. S. Hein, K. Ibraham, and A. Zakhor, "New properties of sigma-delta modulators with DC inputs," *IEEE Transactions on Communications*, vol. 40, no. 8, pp. 1375–1387, 1992.
208. S. Hein and A. Zakhor, *Sigma delta modulators: nonlinear decoding algorithms and stability analysis*. Boston [u.a.]: Kluwer Academic, 1993, bibliotheks-Zentrale.
209. S. Kavusi, H. Kakavand, and A. E. Gamal, "On incremental sigma-delta modulation with optimal filtering," *IEEE Transactions on Circuits and Systems I: Regular Papers*, vol. 53, no. 5, pp. 1004–1015, 2006.
210. G. Guo, D. Wu, Y. Shen, L. Pan, and J. Xu, "An optimal filter with optional resolution used in incremental ADC for sensor application," in *Proc. Communications and Networks (CECNet) 2011 Int. Conf. Consumer Electronics*, 2011, pp. 1042–1045.
211. D. D. Kim and M. A. Brooke, "Integrator bound-based nonlinear decoder for delta-sigma modulator," in *Proc. 49th IEEE Int. Midwest Symp. Circuits and Systems*, vol. 1, 2006, pp. 457–460.
212. H. A. Wake, D. D. Kim, and M. A. Brooke, "Mixed-signal implementation of a nonlinear decoder for delta-sigma encoded stream," in *Proc. IEEE Int. Symp. Circuits and Systems*, 2006, pp. 4 pp.–.
213. L. G. McIlrath, "A robust o(n log n) algorithm for optimal decoding of first-order sigma;-delta; sequences," *IEEE Transactions on Signal Processing*, vol. 50, no. 8, pp. 1942–1950, 2002.
214. S. Maréchal, F. Krummenacher, and M. Kayal, "Optimal filtering of incremental first-order sigma-delta modulators with sweep input," in *Proc. Circuits and Systems 2010 17th IEEE Int. Conf. Electronics*, 2010, pp. 539–542.
215. S. Maréchal, F. Krummenacher, and M. Kayal, "Optimal filtering of an incremental second-order MASH11 sigma-delta modulator," in *Proc. and Systems 2011 18th IEEE Int. Conf. Electronics, Circuits*, 2011, pp. 240–243.
216. B. Wang, M. Law, S. B. Belhaouari, and A. Bermak, "Near-optimal decoding of incremental delta-sigma adc output," *IEEE Transactions on Circuits and Systems I: Regular Papers*, pp. 1–11, 2020.
217. P. Vogelmann, J. Wagner, M. Haas, and M. Ortmanns, "A dynamic power reduction technique for incremental $\Delta\Sigma$ modulators," *IEEE Journal of Solid-State Circuits*, pp. 1–13, 2019.
218. P. Kaesser, O. Ismail, J. Becker, and M. Ortmanns, "Does the maximum stable amplitude depend on reconstruction filters in incremental $\Delta\Sigma$adcs?" in *2025 IEEE International Symposium on Circuits and Systems (ISCAS)*, 2025, pp. 1–5.

219. Y. Chae, K. Souri, and K. A. A. Makinwa, "A 6.3 μW 20 bit incremental zoom-ADC with 6 ppm INL and 1 μV offset," *IEEE Journal of Solid-State Circuits*, vol. 48, no. 12, pp. 3019–3027, 2013.
220. B. Wang, M.-K. Law, and J. Schneider, "$L_2min^{2/2s}$: Efficient linear reconstruction filter for incremental delta-sigma adcs," *IEEE Transactions on Signal Processing*, vol. 71, pp. 3229–3241, 2023.
221. W. Yu, M. Aslan, and G. C. Temes, "82 dB SNDR 20-channel incremental ADC with optimal decimation filter and digital correction," in *Proc. IEEE Custom Integrated Circuits Conf. 2010*, 2010, pp. 1–4.
222. J. Garcia and A. Rusu, "An extended-range incremental CT $\sum \Delta$ ADC with optimized digital filter," in *Proc. Thirteenth Int. Symp. Quality Electronic Design (ISQED)*, 2012, pp. 179–184.
223. P. Vogelmann, "EnglischHigh-resolution incremental $\Delta\Sigma$ adcs for biomedical applications," Hochschulschrift, University of Ulm, Ulm, 2021, tag der Verteidigung: 02.07.2021.
224. J. Uhlig and R. Schüffny, "An empirical study of the stability of 4th-order incremental- Σ Δ-adcs," in *Proc. NORCHIP*, Nov. 2011, pp. 1–4.
225. A. Mohamed and J. Anders, "Stability Analysis of Incremental $\Sigma\Delta$ Modulators using Mixed-Logic Dynamical Systems and Optimal Control Theory," in *2020 IEEE International Symposium on Circuits and Systems (ISCAS)*, 2020, pp. 1–5, iSSN: 2158-1525.
226. J. Wagner, P. Vogelmann, and M. Ortmanns, "On the optimization of DT incremental sigma-delta modulators in combination with CoI reconstruction filters," in *Proc. New Generation of CAS (NGCAS)*, 2018, pp. 82–85.
227. A. Mohamed, "Design and implementation of a high speed incremental sigma delta modulator in 180nm cmos," Master thesis, University of Example, December 2017.
228. S. Pavan, T. Halder, and A. Kannan, "Continuous-time incremental delta-sigma modulators with fir feedback," *IEEE Transactions on Circuits and Systems I: Regular Papers*, vol. 68, no. 8, pp. 3222–3231, 2021.
229. P. Kaesser, O. Ismail, J. Becker, and M. Ortmanns, "Compensation of excess loop delay in continuous-time incremental $\Delta\Sigma$ adcs," in *2025 IEEE International Symposium on Circuits and Systems (ISCAS)*, 2025, pp. 1–5.
230. S. Pavan, "Continuous-Time Delta-Sigma Modulator Design Using the Method of Moments," *IEEE Transactions on Circuits and Systems I: Regular Papers*, vol. 61, no. 6, pp. 1629–1637, 2014, conference Name: IEEE Transactions on Circuits and Systems I: Regular Papers.
231. M. A. Mokhtar, P. Vogelmann, A. Abdelaal, J. G. Kauffman, and M. Ortmanns, "FIR DACs in CT Incremental Delta-Sigma Modulators," in *2020 IEEE International Symposium on Circuits and Systems (ISCAS)*, 2020, pp. 1–5, iSSN: 2158-1525.
232. A. Mohamed, A. Sakr, and J. Anders, "Fir feedback in continuous- time incremental sigma-delta adcs," in *2019 17th IEEE International New Circuits and Systems Conference (NEWCAS)*, 2019, pp. 1–4.
233. M. Pietzko, J. Wagner, A. Abdelaal, J. G. Kauffman, and M. Ortmanns, "Influence of excess loop delay on the stf of continuous-time delta-sigma modulators," in *2021 IEEE International Symposium on Circuits and Systems (ISCAS)*, 2021, pp. 1–5.
234. J. Reiss, "Understanding sigma-delta modulation: The solved and unsolved issues," *AES: Journal of the Audio Engineering Society*, vol. 56, 01 2008.
235. J. Markus, J. Silva, and G. Temes, "Theory and applications of incremental Delta-Sigma converters," *IEEE Transactions on Circuits and Systems I: Regular Papers*, vol. 51, no. 4, pp. 678–690, 2004.
236. M. Ortmanns, F. Gerfers, and Y. Manoli, "Compensation of finite gain-bandwidth induced errors in continuous-time sigma-delta modulators," *IEEE Transactions on Circuits and Systems I: Regular Papers*, vol. 51, no. 6, pp. 1088–1099, 2004.
237. S. Pavan and R. S. Rajan, "Interreciprocity in linear periodically time-varying networks with sampled outputs," *IEEE Transactions on Circuits and Systems II: Express Briefs*, vol. 61, no. 9, pp. 686–690, 2014.
238. H. Tao, L. Toth, and J. Khoury, "Analysis of timing jitter in bandpass sigma-delta modulators," *IEEE Transactions on Circuits and Systems II: Analog and Digital Signal Processing*, vol. 46, no. 8, pp. 991–1001, 1999.

239. V. Peluso, M. Steyaert, and W. Sansen, *enDesign of Low-Voltage Low-Power CMOS Delta-Sigma A/D Converters*. Springer Science & Business Media, 1999, google-Books-ID: HdSP21JvP88C.
240. C. Enz and G. Temes, "Circuit techniques for reducing the effects of op-amp imperfections: autozeroing, correlated double sampling, and chopper stabilization," *Proceedings of the IEEE*, vol. 84, no. 11, pp. 1584–1614, 1996.
241. V. Quiquempoix, P. Deval, A. Barreto, G. Bellini, J. Markus, J. Silva, and G. C. Temes, "A low-power 22-bit incremental ADC," *IEEE Journal of Solid-State Circuits*, vol. 41, no. 7, pp. 1562–1571, 2006.
242. P. Kaesser, O. Ismail, D.-P. Wiens, and M. Ortmanns, "Offset cancellation in incremental $\Delta\Sigma$ adcs," in *2024 IEEE International Symposium on Circuits and Systems (ISCAS)*, 2024, pp. 1–5.
243. V. Quiquempoix and P. Deval, "Fractal sequencing schemes for offset cancellation in sampled data acquisition systems," Patent US Patent 6,909,388, 2005.
244. O. Ismail, P. Kaesser, J. G. Kauffman, and M. Ortmanns, "Non-ideal reset in incremental delta-sigma adcs," in *2024 22nd IEEE Interregional NEWCAS Conference (NEWCAS)*, 2024, pp. 118–122.
245. Y. Liu, E. Bonizzoni, and F. Maloberti, "High-order multi-bit incremental converter with smart-DEM algorithm," in *Proc. IEEE Int. Symp. Circuits and Systems (ISCAS2013)*, 2013, pp. 157–160.
246. B. Wang, S. Sin, S. U, F. Maloberti, and R. P. Martins, "A 550-μw 20-kHz bw 100.8-dB SNDR linear-exponential multi-bit incremental $\Sigma\Delta$ ADC with 256 Clock cycles in 65-nm CMOS," *IEEE Journal of Solid-State Circuits*, pp. 1–12, 2019.
247. Y. Liu, E. Bonizzoni, A. D'Amato, and F. Maloberti, "A 105-db sndr, 10 ksps multi-level second-order incremental converter with smart-dem consuming 280 μW and 3.3-v supply," in *2013 Proceedings of the ESSCIRC (ESSCIRC)*, 2013, pp. 371–374.
248. O. Ismail, P. Kaesser, J. G. Kauffman, and M. Ortmanns, "Dac element mismatch shaping algorithms in incremental delta-sigma adcs," in *2024 IEEE International Symposium on Circuits and Systems (ISCAS)*, 2024, pp. 1–5.
249. Y. Wang, S. Dey, T. He, L. Shi, J. Zheng, M. Kareppagoudr, Y. Zhang, K. Sobue, K. Hamashita, K. Tomioka, and G. Temes, "A hybrid continuous time incremental and sar two-step adc with 90.5db dr over 1mhz bw," in *2021 IEEE Asian Solid-State Circuits Conference (A-SSCC)*, 2021, pp. 1–3.
250. R. Schreier, "Bandpass delta-sigma data converters," in *Conference Record of The Twenty-Ninth Asilomar Conference on Signals, Systems and Computers*, vol. 1, 1995, pp. 94–97 vol.1.
251. C. H. Chen, Y. Zhang, T. He, P. Y. Chiang, and G. C. Temes, "A micro-power two-step incremental analog-to-digital converter," *IEEE Journal of Solid-State Circuits*, vol. 50, no. 8, pp. 1796–1808, 2015.
252. R. Gao, M. Guo, S.-W. Sin, L. Qi, B. Wang, G. Wang, and R. P. Martins, "Weightings in incremental adcs: A tutorial review," in *2023 IEEE Custom Integrated Circuits Conference (CICC)*, 2023, pp. 1–8.
253. K. Seo, I. Jang, K. Noh, and S. Ryu, "An incremental zoom sturdy mash adc," in *2017 IEEE 60th International Midwest Symposium on Circuits and Systems (MWSCAS)*, 2017, pp. 1013–1016.
254. J.-S. Huang, S.-C. Kuo, and C.-H. Chen, "A multistep multistage fifth-order incremental delta sigma analog-to-digital converter for sensor interfaces," *IEEE Journal of Solid-State Circuits*, vol. 58, no. 10, pp. 2733–2744, 2023.
255. C. H. Chen, Y. Zhang, T. He, and G. C. Temes, "Two-step multi-stage incremental ADC," *Electronics Letters*, vol. 51, no. 24, pp. 1975–1977, 2015.
256. M. Aziz, P. Kaesser, S. Ibrahim, and M. Ortmanns, "A comparative study of incremental $\Delta\Sigma$ analog-to-digital converter architectures with extended order and resolution," *Electronics*, vol. 14, no. 2, 2025. [Online]. Available: https://www.mdpi.com/2079-9292/14/2/372

257. Y. Wang, L. Shi, T. He, Y. Zhang, C.-H. Chen, and G. C. Temes, "Robust continuous-time mash delta sigma modulator," in *2018 IEEE 61st International Midwest Symposium on Circuits and Systems (MWSCAS)*, 2018, pp. 310–313.
258. M. A. Mokhtar, A. Abdelaal, M. Sporer, J. Becker, J. G. Kauffman, and M. Ortmanns, "A 0.9-V DAC-Calibration-Free Continuous-Time Incremental Delta-Sigma Modulator Achieving 97-dB SFDR at 2 MS/s in 28-nm CMOS," *IEEE Journal of Solid-State Circuits*, pp. 1–1, 2022.
259. D.-Y. Yoon, S. Ho, and H.-S. Lee, "A continuous-time sturdy-mash $\delta\sigma$ modulator in 28 nm cmos," *IEEE Journal of Solid-State Circuits*, vol. 50, no. 12, pp. 2880–2890, 2015.
260. C. Jansson, "A high-resolution, compact, and low-power adc suitable for array implementation in standard cmos," *IEEE Transactions on Circuits and Systems I: Fundamental Theory and Applications*, vol. 42, no. 11, pp. 904–912, 1995.
261. I. Seitzer, G. Pretzl, and N. Hamdy, *Electronic Analog-to-Digital Converters: Principles, Circuits, Devices, Testing*, ser. A Wiley-Interscience publication. Wiley, 1983. [Online]. Available: https://books.google.de/books?id=vFpTAAAAMAAJ
262. R. Harjani and T. A. Lee, "Frc: a method for extending the resolution of nyquist rate converters using oversampling," *IEEE Transactions on Circuits and Systems II: Analog and Digital Signal Processing*, vol. 45, no. 4, pp. 482–494, 1998.
263. A. Agah, K. Vleugels, P. B. Griffin, M. Ronaghi, J. D. Plummer, and B. A. Wooley, "A high-resolution low-power oversampling ADC with extended-range for bio-sensor arrays," in *Proc. IEEE Symp. VLSI Circuits*, 2007, pp. 244–245.
264. A. Agah, K. Vleugels, P. B. Griffin, M. Ronaghi, J. D. Plummer, and B. A. Wooley, "A high-resolution low-power incremental ADC with extended range for biosensor arrays," *IEEE Journal of Solid-State Circuits*, vol. 45, no. 6, pp. 1099–1110, 2010.
265. P. Rombouts, W. D. Wilde, and L. Weyten, "A 13.5-b 1.2-v micropower extended counting a/d converter," *IEEE Journal of Solid-State Circuits*, vol. 36, no. 2, pp. 176–183, 2001.
266. J. D. Maeyer, P. Rombouts, and L. Weyten, "A double-sampling extended-counting adc," *IEEE Journal of Solid-State Circuits*, vol. 39, no. 3, pp. 411–418, 2004.
267. T. Katayama, S. Miyashita, K. Sobue, and K. Hamashita, "A 1.25ms/s two-step incremental adc with 100db dr and 110db sfdr," in *2018 IEEE Symposium on VLSI Circuits*. IEEE, 2018, pp. 205–206.
268. J. Kim, W. Jung, S. Lim, Y. Park, W. Choi, Y. Kim, C. Kang, J. Shin, K. Choo, W. Lee, J. Heo, B. Kim, S. Kim, M. Kwon, K. Yoo, J. Seo, S. Ham, C. Choi, and G. Han, "A 14b extended counting ADC implemented in a 24mpixel aps-c CMOS image sensor," in *Proc. IEEE Int. Solid-State Circuits Conf*, 2012, pp. 390–392. [Online]. Available: https://ieeexplore.ieee.org/document/6177060/
269. Y. Oike and A. El Gamal, "CMOS image sensor with per-column $\Sigma\Delta$ ADC and programmable compressed sensing," *IEEE Journal of Solid-State Circuits*, vol. 48, no. 1, pp. 318–328, 2013.
270. Y. Zhang, C. H. Chen, T. He, and G. C. Temes, "A 16 b multi-step incremental analog-to-digital converter with single-opamp multi-slope extended counting," *IEEE Journal of Solid-State Circuits*, vol. 52, no. 4, pp. 1066–1076, 2017.
271. Y. Zhang, C.-H. Chen, T. He, K. Sobue, K. Hamashita, and G. C. Temes, "A two-capacitor sar-assisted multi-step incremental adc with a single amplifier achieving 96.6 db sndr over 1.2 khz bw," in *2017 IEEE Custom Integrated Circuits Conference (CICC)*. IEEE, 2017, pp. 1–4.
272. B. Wang, S. Sin, U. Seng-Pan, F. Malobertr, and R. P. Martins, "A 550–μ w 20kHz bw 100.8db SNDR linear-exponential multi-bit incremental converter with 256-cycles in 65Nm CMOS," in *Proc. IEEE Symp. VLSI Circuits*, 2018, pp. 207–208.
273. P. Kaesser, O. Ismail, C. Rudorf, J. Wagner, and M. Ortmanns, "Linear-exponential i-ds adcs: Analysis, limitations and higher order," in *2023 IEEE International Symposium on Circuits and Systems (ISCAS)*, 2023, pp. 1–5.
274. P. Kaesser, O. Ismail, J. Wagner, R. F. H. Fischer, and M. Ortmanns, "Frequency-domain analysis of reconfigured incremental $\Delta\Sigma$ adcs on the example of the exponential phase," *IEEE Transactions on Circuits and Systems I: Regular Papers*, vol. 70, no. 11, pp. 4346–4356, 2023.

275. A. Mehrabi, M. Ranjbar, and O. Oliaei, "Multibit incremental data converters with reduced sensitivity to mismatch," in *Proc. IEEE Int. Symp. Circuits and Systems*, 2008, pp. 1436–1439.
276. Biao Wang, Sai-Weng Sin, Seng-Pan U, and R. P. Martins, "A high resolution multi-bit incremental converter insensitive to DAC mismatch error," in *2016 12th Conference on Ph.D. Research in Microelectronics and Electronics (PRIME)*. Lisbon, Portugal: IEEE, 2016, pp. 1–4. [Online]. Available: http://ieeexplore.ieee.org/document/7519459/
277. P. Vogelmann, M. Haas, and M. Ortmanns, "A 1.1mW 200kS/s incremental $\Delta\Sigma$ ADC with a DR of 91.5dB using integrator slicing for dynamic power reduction," in *Proc. IEEE Int. Solid - State Circuits Conf. - (ISSCC)*, 2018, pp. 236–238.
278. P. Vogelmann, J. Wagner, A. Hänsler, and M. Ortmanns, "Improved SQNR and MSA in incremental $\Delta\Sigma$ modulators by using a recuperation phase," *IEEE Transactions on Circuits and Systems II: Express Briefs*, vol. 65, no. 5, pp. 557–561, 2018.
279. B. Gönen, F. Sebastiano, R. Quan, R. van Veldhoven, and K. A. A. Makinwa, "A dynamic zoom ADC with 109-dB dr for audio applications," *IEEE Journal of Solid-State Circuits*, vol. 52, no. 6, pp. 1542–1550, 2017.
280. S. Karmakar, B. Gònen, F. Sebastiano, R. Van Veldhoven, and K. A. A. Makinwa, "A 280μw dynamic-zoom ADC with 120dB dr and 118dB SNDR in 1kHz bw," in *Proc. IEEE Int. Solid - State Circuits Conf. - (ISSCC)*, 2018, pp. 238–240.
281. Z. Wang, L. Jie, Z. Kong, M. Zhan, Y. Zhong, Y. Wang, and X. Tang, "10.6 a 150khz-bw 15-enob incremental zoom adc with skipped sampling and single buffer embedded noise-shaping sar quantizer," in *2023 IEEE International Solid-State Circuits Conference (ISSCC)*, 2023, pp. 9–11.
282. Y. Guo, J. Jin, X. Liu, and J. Zhou, "A 372 μw 10 khz-bw 109.2 db-sndr nested delta-sigma modulator using hysteresis-comparison msb-pass quantization," *IEEE Journal of Solid-State Circuits*, vol. 58, no. 9, pp. 2554–2563, 2023.
283. R. S. A. Kumar, D. Behera, and N. Krishnapura, "Reset-free memoryless delta–sigma analog-to-digital conversion," *IEEE Transactions on Circuits and Systems I: Regular Papers*, vol. 65, no. 11, pp. 3651–3661, 2018.
284. O. Ismail, P. Kaesser, M. Sporer, J. Kaufmann, and M. Ortmanns, "Incremental delta sigma adcs without periodic reset," in *2025 IEEE International Symposium on Circuits and Systems (ISCAS)*, 2025, pp. 1–5.
285. S. J. Tilden, T. E. Linnenbrink, and P. J. Green, "Overview of ieee-std-1241 "standard for terminology and test methods for analog-to-digital converters"," in *IMTC/99. Proceedings of the 16th IEEE Instrumentation and Measurement Technology Conference (Cat. No.99CH36309)*, vol. 3, 1999, pp. 1498–1503 vol.3.

Index

M. Ortmanns et al., *Incremental Delta-Sigma ADCs*,
https://doi.org/10.1007/978-3-032-09468-1

MIX
Papier aus verantwortungsvollen Quellen
Paper from responsible sources
FSC® C105338

If you have any concerns about our products,
you can contact us on
ProductSafety@springernature.com

In case Publisher is established outside the EU,
the EU authorized representative is:
Springer Nature Customer Service Center GmbH
Europaplatz 3, 69115 Heidelberg, Germany

Printed by Libri Plureos GmbH
in Hamburg, Germany